Part II

next Friday

Homework #3 due

↑ No class Friday
↑

Introductory Real Analysis

Frank Dangello
Michael Seyfried
Shippensburg University

HOUGHTON MIFFLIN COMPANY   Boston   New York

Introductory Real Analysis

Dedicated to Sue and Rosa

Sponsoring Editor: Jack Shira
Assistant Editor: Carolyn Johnson
Editorial Assistant: Christine E. Lee
Art Supervisor/Interior Design: Gary Crespo
Marketing Manager: Michael Busnach
Senior Manufacturing Coordinator: Pricilla Bailey
Cover Design: Henry Rachlin
Cover Image: © Photodisc

Printed in the U.S.A.

Library of Congress Catalog Card Number: 99-71719
ISBN: 0-395-95933-0
123456789-DC-03 02 01 00 99

Contents

Preface

To the Student

Is the prize worth the struggle? This book provides not only a complete foundation for the basic topics covered in the usual calculus sequence but also a stepping stone to higher-level mathematics. This book contains some beautiful mathematical results and opens the door for the student to see other beautiful mathematical results. This is the prize. The struggle is that the ability to see these beautiful mathematical results does not come easily. Put more succinctly, Introductory Real Analysis is a hard course.

With the possible exception of mathematical geniuses, such as Euler, Gauss, and Riemann, people are not born knowing how to do proofs. The skill to prove a mathematical result is learned just as the skill to solve problems is learned, usually with lots of practice. We suggest covering up our proof of a statement. After understanding what the statement means, attempt your own proof. If you are stuck, look at the first line of our proof and see if you can proceed. If necessary, look at another line, and so on.

The examples and exercises are essential components of this book. They will help clarify and strengthen your insight into the text material; and they will increase your mathematical maturity, which, like the ability to do proofs, is not an innate human trait. As much as possible, we have arranged the exercises in each section to correspond to the development of the text material, rather than ordering them from easy to hard. Learning to use previous results to obtain new results is part of mathematical maturity, and we freely draw on the exercises in this manner.

Throughout the book we give overviews and commentaries in order to provide insights into what we are doing and where we are going. These features take the form of either short paragraphs or more formal remarks. Our purpose for dividing some of the commentaries into remarks is to separate different concepts for the student.

In answer to our original question, those who have seen the beauty in mathematics definitely think the prize is worth the struggle.

To the Instructor

Although the only prerequisite for this book is the usual calculus sequence, another mathematics course (such as an introduction to abstract algebra) would help develop the student's sophistication. In this book we do not do analysis from scratch, nor do we construct the real numbers. Rather, we build on what the student has learned in calculus. This book is designed to be used effectively in one- or two-semester courses. The first nine chapters contain an ample amount of material for most two-semester courses, while Chapters 10 and 11 offer additional topics depending on the instructor's personal taste.

As the table of contents indicates, our approach is sequential rather than topological. Because the table of contents lists the topics covered in the book, which for the most part should be done in order, we mention some aspects of the book that are not apparent from the table of contents.

Chapter 1 is an introductory chapter whose purposes are to provide some necessary topics for proceeding in the book and to allow the students to develop some easy proofs. The only topic in Chapter 1 that most students have not seen before is the arbitrary collection of sets, which helps the students develop proofs with quantifiers. In our course, we cover Chapter 1 quickly. Because we do not fix notation until Chapter 2, the instructor can begin with Chapter 2, referring back to Chapter 1 as needed. We have meticulously provided appropriate references throughout the book.

In Section 2.4 our purpose is to show that the rational numbers are countable and that the real numbers are uncountable. Our experience is that one could get bogged down in this section since cardinality is so fascinating to students; for example, how can there be just as many real numbers in the open interval from 0 to 1 as there are on the real line?

Starting in Chapter 3, we use the terminology of neighborhoods, which provides a unifying aspect for both sequential and function limits. In our classes we usually cover about half of Section 3.8, our purpose being to understand what the limit superior and limit inferior are and what they mean in terms of sequential limits. Exercise 5 in this section is very valuable in this regard. We point out that Exercise 11 in Section 3.7 is needed only for the proof of Proposition 3.9 in Section 3.8. We use Section 3.8 in the proof of Mertens's Theorem in Chapter 7 and in Chapter 9. In Chapter 7, our Ratio and Root tests use only limits, whereas in Chapter 9 we extend these tests using limit superior and limit inferior. Of course, Chapter 9 relies heavily on Chapters 7 and 8.

Chapter 6 contains an extensive section on improper integrals with tests for convergence similar to those of infinite series. This section may be covered in depth or lightly depending on the instructor's preference. Although we do not recommend this, Chapter 7 can be covered after Chapter 3 if one is willing to allow the Integral test without formally doing improper integrals or monotone functions. Our development of the Riemann-Stieltjes integral in Chapter 10 parallels our development of the Riemann integral in Chapter 6. Many of the proofs in Chapter 10 for monotone integrators are only slight modifications of the corresponding proofs in Chapter 6. We think Chapter 6 should be covered before Chapter 10, although both chapters can be taught simultaneously. Also, Section 10.4 on functions of bounded variation can be covered without doing anything else in Chapter 10; it depends only on Chapter 4. Thus, an instructor who does not reach Chapter 10 and who has only a couple of class periods left in the semester can cover Section 10.4.

Chapter 4, of course, depends on Chapters 2 and 3. Chapter 11 gives a topological view of our previous concepts, and many of the results in Chapter 4 are now special cases of the action of a continuous function on a compact or connected set.

Accompanying the text is an Instructor's Manual, which contains complete solutions to every exercise in the text and additional exercises (also with complete solutions) suitable for student take-home problems or projects. The Instructor's Manual also contains some suggestions on the text material and on the level of difficulty of some of the exercises.

To All

A book such as this naturally contains many standard (to a mathematician) proofs that can be found almost anywhere. Some arguments are original with us in the sense that we developed them, and we have never seen them anywhere else. When we could not improve on another person's proof or development, we have used it with an appropriate reference. However, all mistakes are ours, and we would appreciate being informed of any errors detected in the book so that we can correct them.

We would like to acknowledge many people for their help in the preparation of this book. For their insightful criticisms and suggestions we thank the reviewers: Michael Berry, West Virginia Wesleyan College; David Gurney, Southeastern Louisiana University; Nathaniel F. Martin, University of Virginia; John W. Neuberger, University of North Texas; Alec Norton, University of Texas at Austin; Richard B. Thompson, University of Arizona; Guoliang Yu, University of Colorado; and Marvin Zeman, Southern Illinois University at Carbondale. We also thank our colleagues Douglas Ensley, Frederick Nordai, and William Weller of Shippensburg University for their input; our secretary Pamela McLaughlin for her assistance; and, most of all, our students for their invaluable comments when we did classroom testing of earlier versions of this book.

1

Proofs, Sets, and Functions

O*ne purpose of this chapter is to help improve the reader's ability to understand and create proofs. We attempt to do this in Section 1.1. In Section 1.2, we consider sets, including arbitrary collections of sets. Sections 1.3 and 1.4 deal with functions and mathematical induction, respectively.*

1.1 Proofs

Many of the statements we prove in this section are known to the reader. It is the technique of proof we wish to emphasize. The typical mathematical assertion that requires proof is the *conditional statement* (or the *implication*)

$$\text{if } p, \text{ then } q \quad (\text{or } p \text{ implies } q). \tag{1}$$

Here, p and q are statements each of which is either true or false but not both. In (1), p is called the *hypothesis* and q is called the *conclusion*.

Direct Proofs

The direct method of proving conditional statement (1) assumes the truth of p and deduces the truth of q. We illustrate the direct method below.

Proposition 1.1 If n is an even integer, then n^2 is an even integer.

Proof Assuming the hypothesis that n is an even integer, we have $n = 2k$ for some integer k. Then $n^2 = 4k^2 = 2(2k^2)$ is an even integer, because $2k^2$ is an integer. ∎

In Proposition 1.1, the hypothesis p is the statement "n is an even integer" and the conclusion q is the statement "n^2 is an even integer." First of all, we need to know the definitions of the terms in the proposition. In Proposition 1.1, we need to know what an even integer is. Also, in the proof, note the use of the existential quantifier "for some" (equivalent to "there exists" and symbolically denoted by ∃). That is, given an even integer n, there is only one integer k such that $n = 2k$. The universal quantifier "for all" (equivalently, "for every," "for each," "for any," and denoted symbolically by ∀) would have been incorrect in the proof above. Either at the very start of the proof or at the end of the first line of the proof, we should ask ourselves

"What is it that I must do to show that n^2 is an even integer?"

Basically, you are asking yourself "what is the next to the last statement in the proof?"—the last statement typically being "therefore q" or, in this case, "therefore n^2 is an even integer." In Proposition 1.1, we have to show that n^2

1

is two times an integer. One detail omitted in the proof above is why $2k^2$ is an integer. The reader should answer this.

Converse

The *converse* of the conditional statement "if p, then q" is the statement "if q, then p." A little thought should indicate that a conditional statement and its converse may or may not have the same truth value. As examples, the converse of the conditional statement in Proposition 1.1 is true, whereas the converse of the calculus theorem

If a function f is differentiable at a point a, then f is continuous at a

is false.

Biconditional

Given statements p and q, the *biconditional statement* "p if and only if q" (also denoted $p \Leftrightarrow q$ or p iff q) means

if p, then q and if q, then p.

To prove the biconditional statement, one must prove both conditional statements. To illustrate this, we need some terminology. Throughout the text, $\mathbb{R}$ denotes the real numbers, $\mathbb{N}$ denotes the positive integers $\{1, 2, 3, \ldots\}$, and $<$ denotes the usual ordering on $\mathbb{R}$.

> **Definition 1.1** Let A be a subset of $\mathbb{R}$. A is *unbounded above* if for each positive real number x there exists an a in A such that $x < a$. Symbolically, A is unbounded above if $\forall\, x > 0,\, \exists\, a$ in A such that $x < a$. Note that a depends on x.

Definitions are to be interpreted in the "if and only if" sense, even though it is common practice not to state them this way. For example, in Definition 1.1, the "if" is actually "if and only if."

> **Archimedean Principle** If x is a positive real number, then there exists a positive integer n such that $1/n < x$. (Symbolically, $\forall x > 0,\, \exists\, n$ in $\mathbb{N}$ such that $1/n < x$.) Note that n depends on x.

Proposition 1.2 $\mathbb{N}$ is unbounded above if and only if the Archimedean Principle holds.

Proof Assume that $\mathbb{N}$ is unbounded above. We need to show that the Archimedean Principle is true. Let $x > 0$. Since $\mathbb{N}$ is unbounded above, there exists an n in $\mathbb{N}$ such that $1/x < n$. Therefore, $1/n < x$.

Next, assume that the Archimedean Principle holds. We need to show that $\mathbb{N}$ is unbounded above. Let $x > 0$. (The reader should now ask: What do we have to do to finish the proof?) Then $1/x > 0$ (see Exercise 6). By the Archimedean Principle, there exists an n in $\mathbb{N}$ such that $1/n < 1/x$, and thus $x < n$. Therefore, $\mathbb{N}$ is unbounded above. ∎

We point out that we have not shown that $\mathbb{N}$ is unbounded above, nor have we shown that the Archimedean Principle is true. What we have shown is that "$\mathbb{N}$ is unbounded above" is a true statement if and only if the Archimedean Principle is a true statement, or that the statement "$\mathbb{N}$ is unbounded above" is logically equivalent to the Archimedean Principle.

Indirect Proofs

There are two types of indirect proofs: the contrapositive argument and the contradiction argument. The *contrapositive* of the conditional statement "if p, then q" is the statement "if not q, then not p."

A little thought should indicate that both a conditional statement and its contrapositive have the same truth value. Thus, to prove a conditional statement one can prove its contrapositive. Since we will have to negate many statements throughout the text, we state a basic rule for negation:

> Change all universal quantifiers to existential quantifiers; change all existential quantifiers to universal quantifiers; and negate the main clause.

Proposition 1.3   Let n be an integer. If n^2 is an odd integer, then n is an odd integer.

Proof   This is the contrapositive of Proposition 1.1.   ■

To prove the conditional statement "if p, then q" by contradiction, one assumes that p is true and that q is false and "hunts" for a contradiction. Once a contradiction is reached, it follows that if p is true, then q must also be true. Where do you find the contradiction? Sometimes the contradiction is clear (for example, $0 = 1$), and sometimes it is very unclear.

To illustrate proof by contradiction in the next two propositions, we assume the usual order properties on $\mathbb{R}$ and that if a is a real number, then $-a$ is the additive inverse of a (so $-a + a = 0$).

Proposition 1.4   Let a be in $\mathbb{R}$. If $a > 0$, then $-a < 0$.

Proof   Assume $a > 0$ and that the statement $-a < 0$ is false. Then $-a \geq 0$. Since $a > 0$, $-a + a > 0$ or $0 > 0$, which is a contradiction. Therefore, $-a < 0$.   ■

Proposition 1.5   Let a be in $\mathbb{R}$. If $a < \varepsilon$ for all $\varepsilon > 0$, then $a \leq 0$.

Proof   The key to this proof is the universal quantifier "for all" in the hypothesis. Assume that $a < \varepsilon$ for all $\varepsilon > 0$ and that $a > 0$. Then, by the Archimedean Principle, there exists a positive integer n such that $0 < 1/n < a$. With $\varepsilon = 1/n$ (that is, using $1/n$ as a particular value of ε), this is a contradiction. Therefore, $a \leq 0$.   ■

The Archimedean Principle will be established in Theorem 2.1. Alternatively, instead of $1/n$ above, we could use $\varepsilon = a/2$.

For the remainder of this section we need the following terminology. A *rational number* is a real number that can be expressed in the form m/n, where m and n are integers and $n \neq 0$. An *irrational number* is a real number that is not a rational number. A *prime number* (or simply a *prime*) is a positive integer greater than 1 whose only positive divisors are itself and 1. By divisors, we mean integers that divide a given number exactly (that is, with zero remainder). For example, 2, 3, 5, 7, and 11 are primes, whereas 9 is not a prime because 3 is a divisor of 9. We will need the following theorem:

> If a prime divides a product of two integers, then the prime must divide at least one of the two integers.

Theorem 1.1 $\sqrt{2}$ is an irrational number.

Proof   Suppose $\sqrt{2}$ is a rational number. Since $\sqrt{2}$ is positive, there exist positive integers m and n such that $\sqrt{2} = m/n$ and m/n is in lowest terms. (We can always reduce a fraction to its lowest terms.) Then $2n^2 = m^2$. Since 2 divides $2n^2$, 2 divides m^2. Since 2 is a prime, 2 divides m. Thus there is a positive integer k such that $m = 2k$. Then $2n^2 = 4k^2$ and $n^2 = 2k^2$. As above, 2 divides n^2 and hence 2 divides n. Thus m/n is not in lowest terms, which is a contradiction. Therefore, $\sqrt{2}$ is not a rational number.   ∎

For the next theorem we need the result that each positive integer greater than 1 is divisible by a prime.

Theorem 1.2   There are infinitely many primes. (*Note*: This result appears in Euclid's *Elements*, Book IX, Proposition 20.)

Proof   Suppose there are only finitely many distinct primes, say $p_1, p_2, \ldots, p_n$. Let $M = p_1 p_2 \cdots p_n + 1$. Then M is an integer, and so there exists a prime that divides M. Thus some p_i divides M. But p_i divides $p_1 p_2 \cdots p_n$. Therefore, p_i divides 1, which is a contradiction.   ∎

From the proofs in this section, certain things should be clear. First, one must know what the terms in the theorem mean. Second, one usually needs to know facts (axioms, propositions, theorems, etc.) to use in the proof. Third, one must know the end of the proof; and keeping the end in mind helps to prevent the line of reasoning from straying off course (McArthur).

Exercises

1. Prove that $\sqrt{6}$ is irrational.

2. Prove that $\sqrt{p}$ is irrational, where p is a prime number.

3. Let a and b be real numbers. Prove that $a^2 + b^2 = 0$ if and only if $a = 0$ and $b = 0$.

4. Let a and b be real numbers. Prove that $ab = 0$ if and only if $a = 0$ or $b = 0$.

5. Let a be a real number. Prove that if $a < 0$, then $-a > 0$.

6. Let a be a real number. Prove that if $a > 0$, then $1/a > 0$.

7. Let a be a real number. If $a^2 = a$, prove that either $a = 0$ or $a = 1$.

8. (Pigeonhole Principle) Suppose we place m pigeons in n pigeonholes, where m and n are positive integers. If $m > n$, show that at least two pigeons must be placed in the same pigeonhole. [*Hint* (from Robert Lindahl of Morehead State University): For $i = 1, 2, \ldots, n$, let x_i denote the number of pigeons that are placed in the ith pigeonhole; let x_k denote the largest of the x_i's; and let $\bar{x} = (\sum_{i=1}^{n} x_i)/n$ denote the average of the x_i's. Show that $x_k \geq \bar{x} = m/n > 1$.]

1.2 Sets

Although the first part of this section should be familiar to the reader, the technique of proof may not be. We adopt the viewpoint of "naive" set theory considering the notion of a set as already known.

Basic Results and Set Operations

A *set* is a well-defined collection of objects. By "well-defined" we mean that, given a set and an object, it is possible to tell whether the object is or is not in the set. Each object of a set is called an *element* of the set, a *point* of the set, or a *member* of the set. If A is a set and x is a point, then

$x \in A$ denotes that x is an element of A

while

$x \notin A$ denotes that x is not an element of A.

A set can be defined either by listing the elements of the set or by stating a property of its elements. For example,

$A = \{-1, 4\}$
$\quad = \{x \in \mathbb{R} : x^2 - 3x - 4 = 0\}$

where $\mathbb{R}$ denotes the set of real numbers.

The *empty set* (*void set*, *null set*), denoted by $\emptyset$, is the set with no elements. Thus

$\emptyset = \{x \in \mathbb{R} : x^2 < 0\} = \{x : x \neq x\}$

and so on.

> **Definition 1.2**   Let A and B be sets.
>
> 1. A is a *subset* of B, denoted by $A \subset B$ or $B \supset A$, if for each x in A, x is in B.
> 2. A is *equal to* B, denoted by $A = B$, if $A \subset B$ and $B \subset A$.
> 3. A is a *proper subset* of B if $A \subset B$ and $A \neq B$.

Thus, to prove that two sets are equal, one must show that each is a subset of the other. Also note that since $\emptyset$ contains no elements, it follows from the definition of a subset that $\emptyset \subset A$ for all sets A.

Definition 1.3 Let A and B be sets.

1. The *union* of A and B, denoted by $A \cup B$, is defined as
$$A \cup B = \{x : x \in A \text{ or } x \in B\}.$$
The word "or" is used in the inclusive sense, so that points that belong to both A and B also belong to the union.

2. The *intersection* of A and B, denoted by $A \cap B$, is defined as
$$A \cap B = \{x : x \in A \text{ and } x \in B\}.$$
Thus $A \cap B \subset A \cup B$.

3. A and B are *disjoint* if $A \cap B = \emptyset$.

Proposition 1.6 Let A, B, and C be sets. Then

1. $A \cup A = A$ and $A \cap A = A$;
2. $A \cup \emptyset = A$ and $A \cap \emptyset = \emptyset$;
3. $A \subset A \cup B$ and $A \cap B \subset A$;
4. $A \cup B = B \cup A$ and $A \cap B = B \cap A$ (commutative property);
5. $A \cup (B \cup C) = (A \cup B) \cup C$ and $A \cap (B \cap C) = (A \cap B) \cap C$ (associative property);
6. $A \cup (B \cap C) = (A \cup B) \cap (A \cup C)$ and $A \cap (B \cup C) = (A \cap B) \cup (A \cap C)$ (distributive property);
7. $A \subset B$ if and only if $A \cup B = B$ and $A \subset B$ if and only if $A \cap B = A$.

Proof   We prove the first equality in part 6. By the definition of equality of sets, we need to show that
$$A \cup (B \cap C) \subset (A \cup B) \cap (A \cup C) \tag{2}$$
and that
$$(A \cup B) \cap (A \cup C) \subset A \cup (B \cap C). \tag{3}$$
To show (2), let $x \in A \cup (B \cap C)$. Then either $x \in A$ or $x \in B \cap C$. If $x \in A$, then by part 3, $x \in A \cup B$ and $x \in A \cup C$. By the definition of intersection, $x \in (A \cup B) \cap (A \cup C)$. If $x \in B \cap C$, then $x \in B$ and $x \in C$. Again, by part 3, $x \in B \cup A$ and $x \in C \cup A$. By part 4, $x \in A \cup B$ and $x \in A \cup C$. Therefore, $x \in (A \cup B) \cap (A \cup C)$.

To show (3), let $x \in (A \cup B) \cap (A \cup C)$. Then $x \in A \cup B$ and $x \in A \cup C$. If $x \in A$, then by part 3, $x \in A \cup (B \cap C)$. So we may assume $x \notin A$. Then, by the definition of union, $x \in B$ and $x \in C$ and so $x \in B \cap C$. Again, by part 3, $x \in (B \cap C) \cup A = A \cup (B \cap C)$.

We now prove the first equality in part 7. We need to show that

if $A \subset B$, then $A \cup B = B$ (4)

and that

if $A \cup B = B$, then $A \subset B$.   (5)

To show (4), we assume $A \subset B$ and note that we have to show $A \cup B = B$. So we have to show that $A \cup B \subset B$ and that $B \subset A \cup B$. The latter containment follows from part 3. To show $A \cup B \subset B$, let $x \in A \cup B$. Then either $x \in A$ or $x \in B$. If $x \in A$, since $A \subset B$, $x \in B$.

To show (5), assume that $A \cup B = B$. We need to show that $A \subset B$. Let $x \in A$. Then $x \in A \cup B$ by part 3. Since $A \cup B = B$, $x \in B$.

The remaining parts of the proposition are left as exercises.   ■

Definition 1.4 Let A and B be sets. The *complement of B relative to A*, denoted by $A \setminus B$, is defined as

$$A \setminus B = \{x \in A : x \notin B\}.$$

For example, if $\mathbb{R}$ denotes the set of real numbers and $\mathbb{Q}$ denotes the set of rational numbers—that is, if

$$\mathbb{Q} = \left\{ \frac{m}{n} : m \text{ and } n \text{ are integers and } n \neq 0 \right\},$$

then $\mathbb{R} \setminus \mathbb{Q}$ is the set of irrational numbers. Also note that $\mathbb{Q} \setminus \mathbb{R} = \emptyset$.

Proposition 1.7 Let A, B, and C be sets. Then

1. $A \setminus \emptyset = A$ and $A \setminus A = \emptyset$

2. DeMorgan's Laws:

$$A \setminus (B \cap C) = (A \setminus B) \cup (A \setminus C)$$

and

$$A \setminus (B \cup C) = (A \setminus B) \cap (A \setminus C).$$

DeMorgan's Laws are generally remembered as stating that the complement of an intersection is the union of the complements and the complement of a union is the intersection of the complements.

Proof   We prove the first equality in part 2, leaving the rest as an exercise. We need to show that

$$A \setminus (B \cap C) \subset (A \setminus B) \cup (A \setminus C)$$ (6)

and that

$$(A \setminus B) \cup (A \setminus C) \subset A \setminus (B \cap C).$$ (7)

To show (6), let $x \in A \setminus (B \cap C)$. By the definition of complement, $x \in A$ and $x \notin B \cap C$. By the definition of intersection, either $x \notin B$ or $x \notin C$ (note that "and" would be incorrect here). If $x \notin B$, then $x \in A \setminus B$, and if $x \notin C$, then $x \in A \setminus C$. So, in either case, $x \in (A \setminus B) \cup (A \setminus C)$.

To show (7), let $x \in (A \setminus B) \cup (A \setminus C)$. Then either $x \in A \setminus B$ or $x \in A \setminus C$. If $x \in A \setminus B$, then $x \in A$ and $x \notin B$. Hence, $x \in A$ and $x \notin B \cap C$. Therefore, $x \in A \setminus (B \cap C)$. If $x \in A \setminus C$, then $x \in A$ and $x \notin C$. Hence, $x \in A$ and $x \notin B \cap C$. Therefore, $x \in A \setminus (B \cap C)$. ■

Arbitrary Unions and Intersections

We now generalize Definition 1.3 to an arbitrary collection of sets. For example,

$$\{A_n : n \in \mathbb{N}\} \text{ where } A_n = (0, n) \text{ for each } n \text{ in } \mathbb{N}.$$

The reader should draw these sets on the real line.

Definition 1.5   Let $\mathfrak{A}$ be a collection of sets.

1. The *union* of $\mathfrak{A}$, denoted by $\bigcup \mathfrak{A}$, is defined as

$$\bigcup \mathfrak{A} = \{x : x \in A \text{ for at least one } A \in \mathfrak{A}\}.$$

2. If $\mathfrak{A}$ is nonempty, the *intersection* of $\mathfrak{A}$, denoted by $\bigcap \mathfrak{A}$, is defined as

$$\bigcap \mathfrak{A} = \{x : x \in A \text{ for all } A \in \mathfrak{A}\}.$$

This definition extends the notions of union and intersection given previously, for if $\mathfrak{A} = \{A, B\}$, where A and B are sets, then $\bigcup \mathfrak{A} = A \cup B$ and $\bigcap \mathfrak{A} = A \cap B$. If $\mathfrak{A}$ is an empty collection of sets, then $\bigcup \mathfrak{A} = \emptyset$ and we do not define $\bigcap \mathfrak{A}$.

An equivalent formulation of Definition 1.5 can be given in terms of index sets. Let I be a set, called the *index set*. Suppose A_α is a set for each α in I. Then $\mathfrak{A} = \{A_\alpha : \alpha \in I\}$ is an *indexed collection of sets*. Notationally,

$$\bigcup \mathfrak{A} = \bigcup \{A_\alpha : \alpha \in I\} = \bigcup_{\alpha \in I} A_\alpha = \bigcup_{\alpha \in I} A_\alpha$$
$$= \{x : x \in A_\alpha \text{ for some } \alpha \in I\}$$

and

$$\bigcap \mathfrak{A} = \bigcap \{A_\alpha : \alpha \in I\} = \bigcap_{\alpha \in I} A_\alpha = \bigcap_{\alpha \in I} A_\alpha$$
$$= \{x : x \in A_\alpha \text{ for all } \alpha \in I\} \text{ (for } I \neq \emptyset).$$

If $I = \{1, 2, 3, \ldots, n\}$ for some n in $\mathbb{N}$, we write $\bigcup_{i \in I} A_i = \bigcup_{i=1}^{n} A_i$ and $\bigcap_{i \in I} A_i = \bigcap_{i=1}^{n} A_i$.

If $I = \mathbb{N}$, we write $\bigcup_{n \in \mathbb{N}} A_n = \bigcup_{n=1}^{\infty} A_n$ and $\bigcap_{n \in \mathbb{N}} A_n = \bigcap_{n=1}^{\infty} A_n$.

Example 1.1 $\bigcup_{n=1}^{\infty} (0, n) = (0, \infty)$ and $\bigcap_{n=1}^{\infty} (0, n) = (0, 1)$.

Example 1.2 $\bigcup_{n=1}^{\infty} \{n\} = \mathbb{N}$ and $\bigcap_{n=1}^{\infty} \{n\} = \emptyset$.

Example 1.3 $\bigcup_{n=1}^{\infty} (-n, n) = \mathbb{R}$ and $\bigcap_{n=1}^{\infty} (-n, n) = (-1, 1)$.

Example 1.4 $\bigcup_{n=1}^{\infty} (0, 1/n) = (0, 1)$ and $\bigcap_{n=1}^{\infty} (0, 1/n) = \emptyset$. To show the latter, suppose $x \in \bigcap_{n=1}^{\infty} (0, 1/n)$. Then $0 < x < 1/n$ for all n in $\mathbb{N}$. But, by the Archimedean Principle (which is proved in Section 2.3), since $x > 0$, there is an n_0 in $\mathbb{N}$ with $0 < 1/n_0 < x$. This contradicts $x < 1/n$ for all n in $\mathbb{N}$. Therefore, $\bigcap_{n=1}^{\infty}(0, 1/n) = \emptyset$.

Proposition 1.8   (DeMorgan's Laws) Let X be a set. Let I be a nonempty index set and let A_α be a set for each $\alpha \in I$. Then $X \setminus \bigcup_{\alpha \in I} A_\alpha = \bigcap_{\alpha \in I}(X \setminus A_\alpha)$ and $X \setminus \bigcap_{\alpha \in I} A_\alpha = \bigcup_{\alpha \in I}(X \setminus A_\alpha)$.

Proof   We prove the first equality, leaving the second as an exercise. We need to show that

$$X \setminus \bigcup_{\alpha \in I} A_\alpha \subset \bigcap_{\alpha \in I}(X \setminus A_\alpha) \tag{8}$$

and that

$$\bigcap_{\alpha \in I}(X \setminus A_\alpha) \subset X \setminus \bigcup_{\alpha \in I} A_\alpha. \tag{9}$$

To show (8), let $x \in X \setminus \bigcup_{\alpha \in I} A_\alpha$. Then $x \in X$ and $x \notin \bigcup_{\alpha \in I} A_\alpha$. So $x \notin A_\alpha$ for all $\alpha \in I$ by the definition of union. Hence, for each $\alpha \in I$, $x \in X \setminus A_\alpha$ by the definition of complement. Thus, $x \in \bigcap_{\alpha \in I}(X \setminus A_\alpha)$.

   To show (9), let $x \in \bigcap_{\alpha \in I}(X \setminus A_\alpha)$. Then $x \in X \setminus A_\alpha$ for each $\alpha \in I$. So $x \in X$ and $x \notin A_\alpha$ for each $\alpha \in I$. So $x \notin \bigcup_{\alpha \in I} A_\alpha$. Therefore, $x \in X \setminus \bigcup_{\alpha \in I} A_\alpha$.   ■

Cartesian Product

Definition 1.6   Let A and B be sets. The *Cartesian product* of A and B, denoted by $A \times B$, is the set of all ordered pairs (a, b), where a is in A and b is in B. Thus

$$A \times B = \{(a, b) : a \in A \text{ and } b \in B\}.$$

For example, $\mathbb{R} \times \mathbb{R}$ is the Cartesian plane. For real numbers a and b, the notation (a, b) has two different meanings: it may mean the ordered pair or it may mean the open interval $\{x \in \mathbb{R} : a < x < b\}$. The context should determine which meaning is appropriate.

Proposition 1.9   Let A, B, and C be sets. Then

$$A \times (B \cap C) = (A \times B) \cap (A \times C).$$

Proof   To show that $A \times (B \cap C) \subset (A \times B) \cap (A \times C)$, let $(x, y) \in A \times (B \cap C)$. Then $x \in A$ and $y \in B \cap C$. Thus, $y \in B$ and $y \in C$. So, $x \in A$ and $y \in B$ imply $(x, y) \in A \times B$ while $x \in A$ and $y \in C$ imply $(x, y) \in A \times C$. Therefore, $(x, y) \in (A \times B) \cap (A \times C)$.

   To show the other containment, reverse the above steps.   ■

Exercises

1. Finish the proof of Proposition 1.6.

2. Finish the proof of Proposition 1.7.

3. Finish the proof of Proposition 1.8.

4. Let A and B be sets. The *symmetric difference* of A and B is $(A \cup B) \setminus (A \cap B)$. Show that $(A \cup B) \setminus (A \cap B) = (A \setminus B) \cup (B \setminus A)$.

5. Show that if $A \subset B$, then $A = B \setminus (B \setminus A)$.

6. Show that if $A \subset B$, then $A \cup (B \setminus A) = B$.

7. Show, by example, that for sets A, B, and C, $A \cap B = A \cap C$ does not imply $B = C$.

8. Let $A_n = (n, \infty)$ for each $n \in \mathbb{N}$. Find $\bigcup_{n=1}^{\infty} A_n$ and $\bigcap_{n=1}^{\infty} A_n$.

9. Let $A_n = [0, 1/n]$ for each $n \in \mathbb{N}$. Find $\bigcup_{n=1}^{\infty} A_n$ and $\bigcap_{n=1}^{\infty} A_n$.

10. Let X be a set and let A_α be a set for each α in a nonempty index set I. Prove the distributive properties:

$$X \cap \left(\bigcup_{\alpha \in I} A_\alpha \right) = \bigcup_{\alpha \in I} (X \cap A_\alpha)$$

and

$$X \cup \left(\bigcap_{\alpha \in I} A_\alpha \right) = \bigcap_{\alpha \in I} (X \cup A_\alpha).$$

11. Let A, B, and C be sets. Prove that
$$A \times (B \cup C) = (A \times B) \cup (A \times C).$$

1.3 Functions

The concept of a function is central to mathematics. In this section we consider basic results about functions that we will need throughout the text.

Basic Definitions

Definition 1.7 Let X and Y be sets. A *function* (or *map*) from X into Y is a rule f that assigns to each element x in the set X a unique element $f(x)$ in the set Y. The set X is called the *domain* of the function. The set $\{f(x) : x \in X\}$ is called the *range* of the function.

Definition 1.7 is somewhat vague in the sense that the term rule is never defined. Because of this, an alternative definition is desirable. This form of the definition identifies a function and its graph.

Definition 1.8 A *function* from a set X into a set Y is a subset, denoted by f, of the Cartesian product $X \times Y$ such that if $(x, y_1) \in f$ and $(x, y_2) \in f$, then $y_1 = y_2$.

Notationally, we write $f : X \to Y$ to denote that f is a function from X into Y, and we often denote the ordered pair $(x, y) \in f$ by $y = f(x)$.

Two functions f and g are *equal*, denoted by $f = g$, provided that they have the same domain and that for each domain element x, $f(x) = g(x)$.

Definition 1.9 Let f be a function from X into Y. Let $S \subset X$. The *direct image* of S, denoted by $f(S)$, is defined as

$$f(S) = \{f(s) : s \in S\}.$$

Let $T \subset Y$. The *inverse image* of T, denoted by $f^{-1}(T)$, is defined as

$$f^{-1}(T) = \{x \in X : f(x) \in T\}.$$

For the following examples the reader should graph the functions to help verify the claims made about the direct and inverse images.

Example 1.5 Let $f : \mathbb{R} \to \mathbb{R}$ be defined by $f(x) = x^2$.

1. Let S be the set of integers. Then $f(S) = \{0, 1, 4, 9, 16, \ldots\}$.
2. Let $T = \{64\}$. Then $f^{-1}(T) = \{\pm 8\}$.
3. Let $T = \{y : -3 < y < 4\}$. Then $f^{-1}(T) = \{x : -2 < x < 2\}$. Observe that for an element in T less than 0, there are no real numbers that are mapped to that element. In other words, $f^{-1}((-3, 0)) = \emptyset$.

Example 1.6 Let $f : \mathbb{R} \to \mathbb{R}$ be defined by $f(x) = \sin x$.

1. Let $S = \{x : 0 \le x \le \pi/6\}$. Then $f(S) = \{y : 0 \le y \le \frac{1}{2}\} = [0, \frac{1}{2}]$.
2. Let $T = \{1\}$. Then $f^{-1}(T) = \{x : f(x) = 1\} = \{\pi/2 + 2k\pi : k$ is an integer$\}$.
3. Let $T = \{y : 0 \le y \le 1\}$. Then $f^{-1}(T) = \bigcup_{n=-\infty}^{\infty} [2n\pi, (2n+1)\pi]$.
4. Let $T = \{y : \pi/2 < y < \pi\}$. Then $f^{-1}(T) = \emptyset$.

Proposition 1.10 Let $f : X \to Y$. Let S and T be subsets of Y. Then $f^{-1}(S \cup T) = f^{-1}(S) \cup f^{-1}(T)$.

Proof We will show that $f^{-1}(S \cup T) \subset f^{-1}(S) \cup f^{-1}(T)$ and $f^{-1}(S) \cup f^{-1}(T) \subset f^{-1}(S \cup T)$. To begin, let $x \in f^{-1}(S \cup T)$. Then, by definition of $f^{-1}(S \cup T)$, $f(x) \in S \cup T$. So $f(x) \in S$ or $f(x) \in T$. If $f(x) \in S$, then $x \in f^{-1}(S)$. If $f(x) \in T$, then $x \in f^{-1}(T)$. So $x \in f^{-1}(S) \cup f^{-1}(T)$. Thus, $f^{-1}(S \cup T) \subset f^{-1}(S) \cup f^{-1}(T)$.

To prove the reverse inclusion, let $x \in f^{-1}(S) \cup f^{-1}(T)$. Then either $x \in f^{-1}(S)$ or $x \in f^{-1}(T)$. If $x \in f^{-1}(S)$, then $f(x) \in S$. If $x \in f^{-1}(T)$, then $f(x) \in T$. Therefore, $f(x) \in S \cup T$ and so $x \in f^{-1}(S \cup T)$. Hence, $f^{-1}(S) \cup f^{-1}(T) \subset f^{-1}(S \cup T)$. ∎

Definition 1.10 A function f from X into Y is a *one-to-one* function (or a *1-1* function) if for any pair of distinct points x_1 and x_2 in X, $f(x_1)$ and $f(x_2)$ are distinct points in Y [that is, if $x_1 \neq x_2$ in X, then $f(x_1) \neq f(x_2)$ in Y]. Equivalently, using the contrapositive, f is one-to-one if and only if for each x_1 and x_2 in X, if $f(x_1) = f(x_2)$, then $x_1 = x_2$.

Another way to classify a function between two sets X and Y is to look at how much of Y is taken up by the image of X.

Definition 1.11 Let f be a function from X into Y. If $f(X) = Y$, then f is *onto* Y. Equivalently, a function f from X into Y is onto Y if for each $y \in Y$, there is an $x \in X$ with $f(x) = y$.
  A function from X into Y is a *bijection* of X onto Y if it is both one-to-one and onto Y.

Example 1.7 The function f in Example 1.5 is not one-to-one since $f(-8) = f(8)$. Since no negative number is in the range of f, f fails to be onto $\mathbb{R}$.

Example 1.8 Define $g : \mathbb{R} \to \mathbb{R}$ by
$$g(x) = \begin{cases} x - 1 & \text{if} \quad x \geq 0 \\ x + 1 & \text{if} \quad x < 0. \end{cases}$$
This function is onto $\mathbb{R}$ but fails to be one-to-one since $g(-1) = g(1)$.

Example 1.9 Define $h : \mathbb{R} \to \mathbb{R}$ by $h(x) = 2x + 7$. Then

1. h is a one-to-one function. Suppose $h(x_1) = h(x_2)$. Then $2x_1 + 7 = 2x_2 + 7$. This implies that $x_1 = x_2$.

2. h is onto $\mathbb{R}$. Suppose $y \in \mathbb{R}$. Then $h((y - 7)/2) = 2[(y - 7)/2] + 7 = y$.

Example 1.10 Define $f : \mathbb{R} \to \mathbb{R}$ by $f(x) = e^x$. We leave it to the reader to verify that f is one-to-one but not onto $\mathbb{R}$. [*Hint:* Graph the function.]

Examples 1.7 to 1.10 show that all combinations of one-to-one and onto are possible. In a sense these ideas are independent of one another. Some consequences of these properties appear in the next proposition and in the exercises.

Proposition 1.11 Let f be a function from X into Y.

1. For any subsets A and B of X, $f(A \cap B) \subset f(A) \cap f(B)$.

2. If f is a one-to-one function, then $f(A \cap B) = f(A) \cap f(B)$.

Proof

1. Recall that $f(A \cap B) = \{f(x) : x \in A \cap B\}$. Let $y \in f(A \cap B)$. Then there is some $x \in A \cap B$ with $f(x) = y$. Since $x \in A \cap B$, $x \in A$ and $x \in B$. So $f(x) \in f(A)$ and $f(x) \in f(B)$. Thus $y = f(x) \in f(A) \cap f(B)$. Hence $f(A \cap B) \subset f(A) \cap f(B)$.

2. We must show that if f is a one-to-one function, then $f(A) \cap f(B) \subset f(A \cap B)$. If this holds, then combining this with the first part yields the desired result.

   Let $y \in f(A) \cap f(B)$. Then $y \in f(A)$ and $y \in f(B)$. Since $y \in f(A)$, there is an $a \in A$ with $f(a) = y$. Similarly, there is a $b \in B$ with $f(b) = y$. Thus $f(a) = f(b)$. Since f is a one-to-one function, $a = b$. That is, $a \in A \cap B$. Therefore, $y = f(a) \in f(A \cap B)$ and so $f(A) \cap f(B) \subset f(A \cap B)$. ∎

In Example 1.5 we defined a function $f : \mathbb{R} \to \mathbb{R}$ by $f(x) = x^2$. If we restrict our attention to the subset of $\mathbb{R}$ consisting of the nonnegative real numbers, we obtain a function that is one-to-one. This motivates the following definition.

Definition 1.12   Let f be a function from X into Y. Let $A \subset X$. The *restriction* of f to A, denoted by $f|_A$, is defined by $f|_A(x) = f(x)$ for all x in A. In a similar vein, let g be a function from A into Y. A function $f : X \to Y$ that satisfies $f|_A = g$ is called an *extension* of g to X.

Operations

We assume that the reader is familiar with the basic algebraic operations on functions (the sum, difference, product, and quotient of two functions). An important operation on functions is that of composition.

Definition 1.13   Let f be a function from X into Y. Let g be a function from Y into Z. The *composition* of f and g, denoted by $g \circ f$, is a function from X into Z defined by $(g \circ f)(x) = g(f(x))$ for all x in X.

This definition can be extended to any finite number of functions.

Proposition 1.12   Let f be a function from X into Y and let g be a function from Y into Z.

1. If both f and g are one-to-one, then so is $g \circ f$.

2. If both f and g are onto functions, then so is $g \circ f$.

3. If both f and g are bijections, then so is $g \circ f$.

Proof   See Exercise 6. ∎

Example 1.11   We construct a bijection from $\mathbb{R}$ onto the open interval $(0, 1)$. Define

$$f : \mathbb{R} \to (0, \infty) \qquad \text{by} \quad f(x) = e^x$$
$$g : (0, \infty) \to (1, \infty) \quad \text{by} \quad g(x) = x + 1$$
$$h : (1, \infty) \to (0, 1) \quad \text{by} \quad h(x) = 1/x.$$

The reader should draw the graphs of these functions to verify that each function is one-to-one and onto the appropriate set. By Proposition 1.12, the composition $h \circ g \circ f$ is a one-to-one map of $\mathbb{R}$ onto $(0, 1)$.

We conclude this section by showing that a bijection has a natural function associated with it and that this function in some sense reverses what the original function does.

Proposition 1.13   Let f be a function from X into Y. Then f is a bijection from X onto Y if and only if there is a function g from Y into X such that $(g \circ f)(x) = x$ for all x in X and $(f \circ g)(y) = y$ for all y in Y.

Proof   First assume that f is a bijection from X onto Y. We show how to construct a function with the stated properties.

For y in Y we define $g(y) = x$ if $f(x) = y$. Since f is onto Y, for any y in Y there is an element x in X with $f(x) = y$. This shows that g is defined for all elements of the set Y. To show that g is well-defined, let y be an element of Y and assume that $g(y) = x_1$ and $g(y) = x_2$. Then, by definition of g, $f(x_1) = y = f(x_2)$. Since f is a one-to-one function, we have $x_1 = x_2$ and so g is a well-defined function from Y into X. The properties of the composites follow at once.

Next, assume that such a function g exists. We must show that f is a bijection. Suppose that $f(x_1) = f(x_2)$. Since both $f(x_1)$ and $f(x_2)$ are in Y, we can apply g to them and obtain $g(f(x_1)) = g(f(x_2))$. Thus, $x_1 = g(f(x_1)) = g(f(x_2)) = x_2$, and so f is one-to-one. To show that f is onto Y, let y be in Y. Then $g(y)$ is in X. So, applying f, we get $f(g(y)) = (f \circ g)(y) = y$ by assumption. Hence, we have found an element x in X—namely, $x = g(y)$—with $f(x) = y$. This shows that f is onto Y. ∎

The function g that was constructed in Proposition 1.13 is called the *inverse* of f, and we write $f^{-1} = g$. Restating part of Proposition 1.13 (if f is a bijection), $(f^{-1} \circ f)(x) = x$ for all x in X and $(f \circ f^{-1})(y) = y$ for all y in Y.

Example 1.12   The function $f : \mathbb{R} \to \mathbb{R}$ defined by $f(x) = x^2$ has no inverse. (Why?) However, if we restrict the domain of f to the set of nonnegative real numbers $[0, \infty)$, then $f|_{[0,\infty)}$ is a bijection onto $[0, \infty)$. By Proposition 1.13, $f|_{[0,\infty)}$ has an inverse. It should come as no surprise that the inverse of $f|_{[0,\infty)}$ is the square root function.

≡ Exercises ≡

In the exercises below, X, Y, and Z are sets.

1. Let f be a function from X into Y. Let A and B be subsets of X. Prove that $f(A \cup B) = f(A) \cup f(B)$.

2. Define a function f from $\mathbb{R}$ into $\mathbb{R}$ by

$$f(x) = \begin{cases} -1 & \text{if} \quad x < 0 \\ x & \text{if} \quad x \geq 0. \end{cases}$$

Find each of the following.

(a) $f([-1, 1])$ (d) $f^{-1}(\{-2\})$

(b) $f^{-1}([-1, 1])$ (e) $f^{-1}(f((-\infty, 0)))$

(c) $f^{-1}(\{-1\})$ (f) $f(f^{-1}((-\infty, 0)))$

3. Let f be a function from X into Y. Let S and T be subsets of Y. Show that $f^{-1}(S \cap T) = f^{-1}(S) \cap f^{-1}(T)$.

4. Let f be a one-to-one function from X into Y and let $A \subset X$. Show that $f^{-1}(f(A)) = A$. Give an example showing that equality need not hold if f is not one-to-one.

5. Let f be a function from X onto Y and let $B \subset Y$. Show that $f(f^{-1}(B)) = B$. Give an example showing that equality need not hold if f is not onto Y.

6. Prove Proposition 1.12.

7. Let f be a function from X into Y. Let g be a function from Y into Z. Assume that $g \circ f$ is one-to-one. Prove that f is one-to-one. Give an example showing that g need not be one-to-one.

8. Let f be a function from X into Y. Let g be a function from Y into Z. Assume that $g \circ f$ is onto Z. Prove that g is onto Z. Give an example showing that f need not be onto Y.

9. Find a bijection from $(0, \infty)$ onto $(0, 1)$.

10. Let a and b be real numbers with $a < b$. Find a bijection from $(0, 1)$ onto (a, b). Combine this result with that of Exercise 9 to produce a bijection from $(0, \infty)$ onto any open interval (a, b).

11. Let f be a bijection from X onto Y. Prove that f^{-1} is a bijection from Y onto X.

12. Let f be a function from X into Y. Let g be a function from Y into Z. Let B be a subset of Z. Show that $(g \circ f)^{-1}(B) = f^{-1}(g^{-1}(B))$.

1.4 Mathematical Induction

Mathematical induction will be used often throughout the text. For instance, in Chapter 3 we will use it to construct sequences inductively. The examples following the proof of Theorem 1.3 illustrate the method of mathematical induction. Recall from Section 1.1 that statements are either true or false, but not both.

Mathematical Induction. Let $p(n)$ be a statement for each n in $\mathbb{N}$. Assume that

1. $p(1)$ is true

and

2. for each $k \geq 1$, if $p(k)$ is true, then $p(k + 1)$ is true.

Then $p(n)$ is true for all n in $\mathbb{N}$.

In assumption 2, " $p(k)$ is true" is called the *induction hypothesis*. Intuitively, $p(1)$ true implies that $p(2)$ is true by assumption 2; $p(2)$ true implies that $p(3)$ is true by assumption 2; and so on.

Before illustrating the method of mathematical induction, we examine the relationship between mathematical induction and the following concept, which will be assumed as an axiom in Section 2.1.

$\mathbb{N}$ *is Well-Ordered*. Every nonempty subset of $\mathbb{N}$ has a least element. That is, if $A \subset \mathbb{N}$ and $A \neq \emptyset$, then there is an a_0 in A such that $a_0 \leq a$ for all a in A.

Theorem 1.3 $\mathbb{N}$ is well-ordered if and only if mathematical induction is true.

Proof   First assume that $\mathbb{N}$ is well-ordered. For each positive integer n, let $p(n)$ be a statement satisfying assumptions 1 and 2 of mathematical induction. We want to show that $p(n)$ is true for all positive integers n. Suppose this is false. Since $\mathbb{N}$ is well-ordered, let n_0 be the least positive integer such that $p(n_0)$ is false. By assumption 1, $n_0 > 1$. Therefore $n_0 - 1$ is a positive integer and $p(n_0-1)$ is true (since n_0 is the least positive integer for which the corresponding statement is false). By assumption 2, $p(n_0) = p[(n_0 - 1) + 1]$ is true, which is a contradiction. Therefore, $p(n)$ is true for every positive integer n.

Next, suppose mathematical induction holds. We wish to show that $\mathbb{N}$ is well-ordered. Let A be a nonempty subset of $\mathbb{N}$. Suppose A has no least element. For each n in $\mathbb{N}$, let $p(n)$ be the statement

$$A \cap \{1, 2, \ldots, n\} = \emptyset.$$

Suppose $p(1)$ is false. Then $A \cap \{1\} \neq \emptyset$ and A has a least element—namely, 1—which is a contradiction to A having no least element. Therefore, $p(1)$ is true. Let $k \geq 1$ and assume that $p(k)$ is true. Thus, $A \cap \{1, 2, \ldots, k\} = \emptyset$. Suppose that $p(k + 1)$ is false. Then $A \cap \{1, 2, \ldots, k, k+1\} \neq \emptyset$. Thus, A has a least element—namely, $k + 1$—which again is a contradiction to A having no least element. Therefore, $p(k + 1)$ is true.

By mathematical induction, $p(n)$ is true for all n in $\mathbb{N}$. Therefore $A = \emptyset$, which is a contradiction to A being nonempty. Therefore, A has a least element.   ∎

Example 1.13   For each n in $\mathbb{N}$, $1^2 + 2^2 + \cdots + n^2 = \frac{1}{6}n(n + 1)(2n + 1)$.

For each n in $\mathbb{N}$, let $p(n)$ be the statement

$$1^2 + 2^2 + \cdots + n^2 = \frac{1}{6}n(n + 1)(2n + 1).$$

$p(1)$ is true since $1^2 = \frac{1}{6}(1)(2)(3)$. Let $k \geq 1$ and assume that $p(k)$ is true. That is, assume $1^2+2^2+\cdots+k^2 = \frac{1}{6}k(k+1)(2k+1)$. We need to show that $p(k+1)$ is true. We need to show that $1^2+2^2+\cdots+k^2+(k+1)^2 = \frac{1}{6}(k+1)(k+2)(2k+3)$. By the induction hypothesis,

$$1^2 + 2^2 + \cdots + k^2 + (k + 1)^2 = \frac{1}{6}k(k + 1)(2k + 1) + (k + 1)^2$$

$$= \frac{1}{6}(k + 1)[k(2k + 1) + 6(k + 1)]$$

$$= \frac{1}{6}(k + 1)(2k^2 + 7k + 6)$$

$$= \frac{1}{6}(k + 1)(k + 2)(2k + 3).$$

Thus, $p(k + 1)$ is true.   By mathematical induction, $p(n)$ is true for all n in $\mathbb{N}$.

Example 1.14   For each n in $\mathbb{N}$, $2^n \geq n + 1$.

For each n in $\mathbb{N}$, $p(n)$ is the statement: $2^n \geq n + 1$. $p(1)$ is true since $2^1 \geq 2 = 1 + 1$. Let $k \geq 1$ and assume that $p(k)$ is true. That is, assume $2^k \geq k + 1$. We want to show that $p(k + 1)$ is true. We need to show that $2^{k+1} \geq (k + 1) + 1 = k + 2$. Observe that

$$2^{k+1} = 2^k \cdot 2$$
$$\geq (k + 1) \cdot 2 \qquad \text{(by the induction hypothesis)}$$
$$= 2k + 2$$
$$\geq k + 2 \qquad \text{(since } 2k \geq k).$$

Thus, $p(k + 1)$ is true and so $p(n)$ is true for each n in $\mathbb{N}$.

Example 1.15   For each n in $\mathbb{N}$, 9 divides $n^3 + (n + 1)^3 + (n + 2)^3$ (where "divides" means with 0 remainder).

For each n in $\mathbb{N}$, let $p(n)$ be the statement: 9 divides $n^3 + (n+1)^3 + (n+2)^3$. Since $1^3 + 2^3 + 3^3 = 36$, $p(1)$ is true. Let $k \geq 1$ and assume that $p(k)$ is true. That is, assume that 9 divides $k^3 + (k + 1)^3 + (k + 2)^3$. We must show that 9 divides $(k + 1)^3 + (k + 2)^3 + (k + 3)^3$. Observe that

$$(k+1)^3 + (k+2)^3 + (k+3)^3 = (k+1)^3 + (k+2)^3 + k^3 + 9k^2 + 27k + 27$$
$$= k^3 + (k+1)^3 + (k+2)^3 + 9(k^2 + 3k + 3).$$

By the induction hypothesis, 9 divides $k^3 + (k + 1)^3 + (k + 2)^3$. Since $9(k^2 + 3k + 3)$ is a multiple of 9, 9 divides it. Thus $p(k + 1)$ is true and so $p(n)$ is true for all n in $\mathbb{N}$.

Remark   Sometimes mathematical induction is stated as follows. Let A be a subset of $\mathbb{N}$ satisfying

    1. $1 \in A$

and

    2. if $k \geq 1$ and $k \in A$, then $k + 1 \in A$.

Then $A = \mathbb{N}$. That this is equivalent to the previous version of mathematical induction can be seen be letting $A = \{n \in \mathbb{N} : p(n) \text{ is true}\}$.

Remark   In mathematical induction, one need not start with 1. Let n_0 be an integer and suppose that $p(n)$ is a statement for each $n \geq n_0$. Assume that (1) $p(n_0)$ is true and (2) for each $k \geq n_0$, if $p(k)$ is true, then $p(k + 1)$ is true. Then $p(n)$ is true for all $n \geq n_0$. To see this, for each n in $\mathbb{N}$, let $q(n)$ be the statement: $p(n_0 + n - 1)$. Note that $q(1) = p(n_0)$. A little thought shows that assumptions 1 and 2 imply by mathematical induction that $q(n)$ is true for all n in $\mathbb{N}$. Equivalently, $p(n)$ is true for all $n \geq n_0$.

Remark   The following represents a misuse of mathematical induction. Find the error. For which k does the argument fail?

For each n in $\mathbb{N}$, let $p(n)$ be the statement: any set of n horses are all of the same color. $p(1)$ is true since we have only one horse. Let $k \geq 1$ and assume that $p(k)$ is true. That is, assume that any set of k horses are all of the same color. We want to show that $p(k + 1)$ is true. Let $X = \{x_1, x_2, \ldots, x_{k+1}\}$ be a set of $k + 1$ horses. We want to show that all $k + 1$ horses are of the same color. Since $\{x_1, x_2, \ldots, x_k\}$ is a set of k horses, by the induction hypothesis, these are all of the same color. Since $\{x_2, x_3, \ldots, x_{k+1}\}$ is a set of k horses, by the induction hypothesis, these are all of the same color. Thus, all $k + 1$ horses in X are of the same color. Therefore, $p(n)$ is true for all n in $\mathbb{N}$.

Exercises

1. Prove that for each n in $\mathbb{N}$, $1 + 2 + \cdots + n = n(n + 1)/2$.

2. Prove that for each n in $\mathbb{N}$, $1^3 + 2^3 + \cdots + n^3 = [n(n + 1)/2]^2$.

3. Prove that for each n in $\mathbb{N}$, $1 + 3 + 5 + \cdots + (2n - 1) = n^2$.

4. Prove that for each $n \geq 4$, $2^n < n!$.

5. Prove that for each n in $\mathbb{N}$, 4 divides $7^n - 3^n$.

6. Prove that for each n in $\mathbb{N}$, 5 divides $n^5 - n$.

7. Prove that for each n in $\mathbb{N}$, $\sum_{j=1}^{n} (-1)^{j+1} j^2 = (-1)^{n+1} \sum_{j=1}^{n} j$.

8. Prove that if $x_1, x_2, \ldots, x_n$ are n real numbers in the closed interval $[a, b]$, then $\frac{x_1 + \cdots + x_n}{n}$ is in the closed interval $[a, b]$.

9. Prove that a set with n elements has 2^n subsets for each $n \in \mathbb{N} \cup \{0\}$.

10. Prove Bernoulli's inequality: If $x > -1$, then $(1 + x)^n \geq 1 + nx$ for each n in $\mathbb{N}$.

11. Prove DeMoivre's Theorem: for t a real number,

    $$(\cos t + i \sin t)^n = \cos nt + i \sin nt$$

    for each n in $\mathbb{N}$, where $i = \sqrt{-1}$.

2 The Structure of $\mathbb{R}$

By the end of this chapter, the reader should understand the difference between the real numbers and the rational numbers. In Section 2.1, we note that both are fields with an order relation. However, in Sections 2.2 and 2.3, we show that the real numbers are "complete" whereas the rational numbers are not complete. In Section 2.4, we show that the rational numbers are "countable" but that the real numbers are "uncountable."

2.1 Algebraic and Order Properties of $\mathbb{R}$

Throughout the text we use the following notation.

$\mathbb{R}$ is the set of real numbers.

$\mathbb{N} = \{1, 2, 3, \ldots\}$ is the set of positive integers (or natural numbers).

$\mathbb{Z} = \{\ldots, -2, -1, 0, 1, 2, \ldots\}$ is the set of integers.

$\mathbb{Q} = \{m/n : m, n \in \mathbb{Z}, n \neq 0\} = \{m/n : m \in \mathbb{Z}, n \in \mathbb{N}\}$ is the set of rational numbers.

$\mathbb{R} \setminus \mathbb{Q}$, the complement of $\mathbb{Q}$ in $\mathbb{R}$, is the set of irrational numbers.

In this text we are not going to construct the set of real numbers. Rather, we assume that the reader is familiar with many of the properties of the real numbers discussed in this section.

Field Axioms for $\mathbb{R}$

To each pair of real numbers a and b there correspond unique real numbers $a + b$ and $a \cdot b$ that satisfy:

Axiom 2.1

$$a + b = b + a$$
$$a \cdot b = b \cdot a \qquad \text{(commutative laws)}$$

Axiom 2.2 For all c in $\mathbb{R}$,

$$a + (b + c) = (a + b) + c$$
$$a \cdot (b \cdot c) = (a \cdot b) \cdot c \qquad \text{(associative laws)}$$

Axiom 2.3 For all c in $\mathbb{R}$,

$$a \cdot (b + c) = a \cdot b + a \cdot c \qquad \text{(distributive law)}$$

Axiom 2.4 There exist distinct real numbers 0 and 1 such that for all a in $\mathbb{R}$,

$$a + 0 = a$$
$$a \cdot 1 = a \qquad \text{(identity elements)}$$

Axiom 2.5 For each a in $\mathbb{R}$, there is an element $-a$ in $\mathbb{R}$ such that

$$a + (-a) = 0$$

and for each b in $\mathbb{R}$, $b \neq 0$, there is an element $b^{-1} = 1/b$ in $\mathbb{R}$ such that

$$b \cdot \frac{1}{b} = 1. \qquad \text{(inverse elements)}$$

Note that $(a, b) \to a + b$ and $(a, b) \to a \cdot b$ actually define functions from $\mathbb{R} \times \mathbb{R} \to \mathbb{R}$. We usually write ab instead of $a \cdot b$. From these axioms we can derive the usual laws of arithmetic; for example, $a \cdot 0 = 0$, $-(-a) = a$, and so on.

The axioms above make $(\mathbb{R}, +, \cdot)$ into a field. However, $(\mathbb{Q}, +, \cdot)$ is also a field. This amounts to showing that the sum and the product of two fractions are again fractions.

Order Axioms for $\mathbb{R}$

On $\mathbb{R}$ there is an order relation, denoted by $<$, that satisfies:

Axiom 2.6 For all a and b in $\mathbb{R}$, exactly one of the following holds:

$$a = b, \; a < b, \; b < a \qquad \text{(trichotomy)}.$$

Axiom 2.7 For all a, b, and c in $\mathbb{R}$, if $a < b$, then $a + c < b + c$.

Axiom 2.8 For all a, b, and c in $\mathbb{R}$, if $a < b$ and $0 < c$, then $ac < bc$.

Axiom 2.9 For all a, b, and c in $\mathbb{R}$, if $a < b$ and $b < c$, then $a < c$ (transitivity).

We write $b > a$ if $a < b$ and we define $a \leq b$ (or $b \geq a$) if either $a < b$ or $a = b$. From these axioms we can derive the usual properties of inequalities; for example, if $a < b$ and $c < 0$, then $ac > bc$. The reader will find it instructive to show from the axioms above that $0 < 1$ (see Exercises 4 and 5).

We will assume two more axioms on the set of real numbers. The second one will be introduced in the next section. The first one, which was defined in Section 1.4, we now state:

Axiom 2.10 The positive integers are well-ordered.

Decimal Representations

The following is essentially given in Apostol, pp. 2–3. Our purpose in discussing decimals is in anticipation of Theorem 2.5.

Every real number x has a decimal representation of the form

$$x = n.a_1 a_2 a_3 \cdots$$

where n is an integer and each a_i is one of the digits from 0 to 9. The notation $.a_1 a_2 a_3 \cdots$ is an abbreviation for the infinite series

$$\frac{a_1}{10} + \frac{a_2}{100} + \frac{a_3}{1000} + \cdots.$$

For example

$$0.499999\ldots = \frac{4}{10} + \frac{9}{100} + \frac{9}{1000} + \cdots$$

$$= \frac{4}{10} + \frac{9}{100}\left[1 + \frac{1}{10} + \left(\frac{1}{10}\right)^2 + \cdots\right]$$

$$= \frac{4}{10} + \frac{9}{100}\left(\frac{1}{1 - \frac{1}{10}}\right) \qquad \text{(the sum of a geometric series)}$$

$$= \frac{4}{10} + \frac{9}{100}\left(\frac{10}{9}\right)$$

$$= \frac{5}{10}$$

$$= 0.5.$$

Thus, $\frac{1}{2}$ has two decimal expansions. More generally, if a number has a decimal expansion ending in zeros, it also has a decimal expansion ending in nines. For example, $\frac{1}{4} = 0.2499999\cdots = 0.250000\cdots$. Except in situations such as this, decimal expansions are unique.

A real number is rational if and only if its decimal expansion is repeating. For example, $\frac{1}{3} = 0.33333\cdots$, $\frac{1}{8} = 0.125000\cdots$, and $1 = 0.9999999\cdots$.

Given the positive rational number m/n, in lowest terms with $m < n$, the process of long division produces the decimal expansion of m/n. Since at each step of the division a remainder that is an integer in the set $\{0, 1, 2, \ldots, n-1\}$ occurs, after at most n steps, the quotient will repeat. Conversely, as with $0.49999\cdots$ above, a decimal that repeats forms a geometric series of the form

$$a + b(1 + r + r^2 + r^3 + \cdots) = a + b\left(\frac{1}{1-r}\right)$$

where a, b, and r are rational, with r being a power of $\frac{1}{10}$. Since $\mathbb{Q}$ is a field, $a + b[1/(1-r)]$ is rational.

Absolute Value

Definition 2.1   For x in $\mathbb{R}$, the *absolute value* of x, denoted by $|x|$, is defined by

$$|x| = \begin{cases} x & \text{if} \quad x \geq 0 \\ -x & \text{if} \quad x < 0. \end{cases}$$

Note that $|x| \geq 0$ for all x in $\mathbb{R}$. Geometrically, $|x|$ is the distance from x to 0.

Proposition 2.1   Let a, b, and c be in $\mathbb{R}$. Then

1. $|a| = 0$ if and only if $a = 0$;
2. $|-a| = |a|$;
3. $|ab| = |a||b|$;
4. if $c \geq 0$, then $|a| \leq c$ if and only if $-c \leq a \leq c$.

Proof   The proofs of parts 1 and 2 are left as exercises.

Proof of part 3. If either a or b is 0, then both $|ab|$ and $|a||b|$ are 0. If $a > 0$ and $b > 0$, then $ab > 0$ and so $|ab| = ab = |a||b|$. If $a > 0$ and $b < 0$, then $ab < 0$ and so $|ab| = -(ab) = a(-b) = |a||b|$. If $a < 0$ and $b < 0$, then $ab > 0$ and so $|ab| = ab = (-a)(-b) = |a||b|$.

Proof of part 4. Suppose that $-c \le a \le c$. If $a \ge 0$, then $|a| = a \le c$. If $a < 0$, then $|a| = -a \le c$ since $-c \le a$. In either case, $|a| \le c$. Conversely, assume that $|a| \le c$. If $a \ge 0$, then $-c \le 0 \le a = |a| \le c$. If $a < 0$, then $-a = |a| \le c$ implies that $a \ge -c$ and so $-c \le a < 0 \le c$.   ∎

Proposition 2.2   (triangle inequality) For a and b in $\mathbb{R}$,
$$|a + b| \le |a| + |b|.$$

Proof   One could make a case-by-case argument as in Proposition 2.1. However, we prefer the following:
$$\begin{aligned}
0 &\le (a + b)^2 \\
&= a^2 + 2ab + b^2 \\
&= |a|^2 + 2ab + |b|^2 \\
&\le |a|^2 + 2|ab| + |b|^2 \\
&= |a|^2 + 2|a||b| + |b|^2 \\
&= (|a| + |b|)^2.
\end{aligned}$$
Taking square roots (see Exercises 8 and 9),
$$|a + b| \le |a| + |b|.$$ ∎

By mathematical induction, the triangle inequality can be extended to any finite number of elements in $\mathbb{R}$. That is, for any $a_1, a_2, \ldots, a_n$ in $\mathbb{R}$,
$$|a_1 + a_2 + \cdots + a_n| \le |a_1| + |a_2| + \cdots + |a_n|.$$

The following simple result, especially the second part, will be used often in Chapters 3 and 4.

Corollary 2.1   For a and b in $\mathbb{R}$,
1. $|a - b| \le |a| + |b|$

and

2. $|\,|a| - |b|\,| \le |a - b|$.

Proof   For part 1, $|a - b| = |a + (-b)| \le |a| + |-b| = |a| + |b|$. For part 2, we also use the triangle inequality:
$$|a| = |(a - b) + b| \le |a - b| + |b|.$$
Therefore, $|a| - |b| \le |a - b|$. Reversing the roles of a and b, we get
$$|b| - |a| \le |b - a| = |a - b|.$$
Thus, $|\,|a| - |b|\,| \le |a - b|$.   ∎

≡ Exercises ≡

1. Let x be irrational. Show that if r is rational, then $x + r$ is irrational. Also show that if r is a nonzero rational, then rx is irrational.

2. Show, by example, that if x and y are irrational, then $x + y$ and xy may be rational.

3. Show that $\sqrt{2} + \sqrt{3}$ is irrational. [*Hint:* Use the fact that $\sqrt{6}$ is irrational.]

4. From the order axioms for $\mathbb{R}$, show that the set of positive real numbers, $\{x \in \mathbb{R} : x > 0\}$, is closed under addition and multiplication.

5. From the order axioms for $\mathbb{R}$, show that $0 < 1$. [*Hint:* From the field axioms, $0 \neq 1$. By the trichotomy property, either $0 < 1$ or $1 < 0$. Assuming $1 < 0$, get $0 < -1$. Now use Exercise 4.]

6. Write $0.33474747 \cdots$ as a fraction.

7. Prove parts 1 and 2 of Proposition 2.1.

8. For a in $\mathbb{R}$, show that $|a| = \sqrt{a^2}$. (Note that for $b \geq 0$, $\sqrt{b}$ denotes the nonnegative square root of b.)

9. Let $a \geq 0$ and $b \geq 0$. Show that $a \leq b$ if and only if $a^2 \leq b^2$.

10. Show that equality holds in the triangle inequality if and only if $ab \geq 0$.

11. For x in $\mathbb{R}$ and $\varepsilon > 0$, let $(x - \varepsilon, x + \varepsilon)$ be the open interval centered at x of radius ε. That is,

$$(x - \varepsilon, x + \varepsilon) = \{y \in \mathbb{R} : x - \varepsilon < y < x + \varepsilon\}$$
$$= \{y \in \mathbb{R} : |y - x| < \varepsilon\}.$$

Use the triangle inequality to show that if a and b are in $\mathbb{R}$ with $a \neq b$, then there exist open intervals U centered at a and V centered at b, both of radius $\varepsilon = \frac{1}{2}|a - b|$, with $U \cap V = \emptyset$.

2.2 The Completeness Axiom

Boundedness

Definition 2.2   The *extended real number system*, denoted by $\mathbb{R}^{\#}$, consists of the real number system together with two distinct symbols, ∞ and $-\infty$, neither of which is a real number. That is,

$$\mathbb{R}^{\#} = \mathbb{R} \cup \{\infty\} \cup \{-\infty\}.$$

We extend the usual ordering on $\mathbb{R}$ to $\mathbb{R}^{\#}$ by

$$-\infty < \infty, \quad -\infty < x < \infty \text{ for all } x \text{ in } \mathbb{R},$$

and $u \leq v$ if either $u < v$ or $u = v$ for all u and v in $\mathbb{R}^{\#}$.

We also use the symbol $+\infty$ for ∞ where convenient.

Definition 2.3   Let S be a subset of $\mathbb{R}^{\#}$. Let u and v be in $\mathbb{R}^{\#}$.

1. u is an *upper bound* of S if $s \le u$ for all s in S.
2. v is a *lower bound* of S if $v \le s$ for all s in S.
3. S is *bounded above* if there is a real number that is an upper bound of S.
4. S is *bounded below* if there is a real number that is a lower bound of S.
5. S is *bounded* if S is both bounded above and bounded below.

Obviously, ∞ is an upper bound and $-\infty$ is a lower bound of every subset of $\mathbb{R}^{\#}$. However, the positive real numbers $\{x \in \mathbb{R} : x > 0\}$ are bounded below but not bounded above; the negative real numbers $\{x \in \mathbb{R} : x < 0\}$ are bounded above but not bounded below; and the closed interval $[0, 1] = \{x \in \mathbb{R} : 0 \le x \le 1\}$ is bounded.

Definition 2.4   Let S be a subset of $\mathbb{R}^{\#}$ and let α and β be elements of $\mathbb{R}^{\#}$.

1. if α is the *least upper bound* of S or the *supremum* of S
   (a) if α is an upper bound of S
   and
   (b) whenever γ is an upper bound of S, $\alpha \le \gamma$.
      Notation: $\alpha = \text{lub } S = \sup S$.
2. β is the *greatest lower bound* of S or the *infimum* of S
   (a) if β is a lower bound of S
   and
   (b) whenever γ is a lower bound of S, $\gamma \le \beta$.
      Notation: $\beta = \text{glb } S = \inf S$.

As a simple exercise, the reader should show that $\sup S$ and $\inf S$ are unique.

Example 2.1   The following illustrate the basic ideas.

1. $\sup \mathbb{R} = +\infty$ and $\inf \mathbb{R} = -\infty$.
2. $\sup[0, 1] = \sup(0, 1) = 1$ and $\inf[0, 1] = \inf(0, 1) = 0$, where $(0, 1) = \{x \in \mathbb{R} : 0 < x < 1\}$.
3. $\sup\{x \in \mathbb{R} : x < \pi\} = \pi$ and $\inf\{x \in \mathbb{R} : x < \pi\} = -\infty$.
4. Pathologically, $\sup \emptyset = -\infty$ and $\inf \emptyset = +\infty$.

Observe that the supremum and infimum of a set may or may not be in that set.

Definition 2.5   Let S be a subset of $\mathbb{R}^{\#}$ and let s_0 and s_1 be elements of S.

1. s_0 is the *smallest (least, minimum) element* of S if $s_0 \leq x$ for all x in S.

2. s_1 is the *greatest (largest, maximum) element* of S if $x \leq s_1$ for all x in S.

   Notation: $\min S = s_0$ and $\max S = s_1$.

Proposition 2.3   Let S be a subset of $\mathbb{R}^{\#}$.

1. If S has a smallest element, then this smallest element is the infimum of S.

2. If S has a greatest element, then this greatest element is the supremum of S.

Proof   We prove part 1, leaving part 2 as an exercise.

Let s_0 be the smallest element of S. Then s_0 is a lower bound of S since $s_0 \leq x$ for all x in S. Suppose γ is a lower bound of S. Then $\gamma \leq x$ for all x in S. In particular, $\gamma \leq s_0$. Thus, $s_0 = \inf S$.   ■

The Completeness Axiom

The next axiom is our final assumption about the set of real numbers. We will see in the next section that this axiom distinguishes the real numbers from the rational numbers. Recall that bounded above (or below) means bounded above (or below) by a real number.

Completeness Axiom for $\mathbb{R}$ Every nonempty subset of $\mathbb{R}$ that is bounded above has a supremum in $\mathbb{R}$.

Proposition 2.4   Every nonempty subset of $\mathbb{R}$ that is bounded below has an *inf*imum in $\mathbb{R}$.

Proof   Let S be a nonempty subset of $\mathbb{R}$ that is bounded below. Let

$$A = \{x \in \mathbb{R} : x \text{ is a lower bound of } S\}.$$

Then A is nonempty and A is bounded above by each point in S. By the Completeness Axiom for $\mathbb{R}$, $\alpha = \sup A$ is a real number. We will show that $\alpha = \inf S$. Let s be in S. Then s is an upper bound of A. Since α is the least upper bound of A, $\alpha \leq s$. Thus, α is a lower bound of S.

To show that α is the greatest lower bound of S, let γ be a real lower bound of S. (If $\gamma = -\infty$, then clearly $\gamma \leq \alpha$.) We need to show that $\gamma \leq \alpha$. Since γ is a lower bound of S, γ is an element of A. Since α is an upper bound of A, $\gamma \leq \alpha$.   ■

Let S be a subset of $\mathbb{R}^{\#}$. If S is not bounded above (that is, for each real number $x > 0$, there is an s in S such that $x < s$), then $\sup S = +\infty$. Also, if S is not bounded below, then $\inf S = -\infty$. Thus, every subset of $\mathbb{R}^{\#}$ has a supremum and an infimum in $\mathbb{R}^{\#}$, whereas a subset of $\mathbb{R}$ need not have a supremum or an infimum in $\mathbb{R}$. This is why we chose to work in $\mathbb{R}^{\#}$.

We end this section with an example showing how to work with the supremum and the infimum. For a subset S of $\mathbb{R}$ and a point a in $\mathbb{R}$, we define $aS = \{as : s \in S\}$.

Example 2.2   Let S be a nonempty bounded subset of $\mathbb{R}$ and let $a > 0$. Then

1. $\sup(aS) = a \sup S$

and

2. $\inf(aS) = a \inf S$.

To show part 1, let $\alpha = \sup S$. By the Completeness Axiom for $\mathbb{R}$, α is a real number. We wish to show that $a\alpha = \sup(aS)$. First note that for each s in S, $s \leq \alpha$ since α is an upper bound of S. So, $as \leq a\alpha$ for each s in S and, therefore, $a\alpha$ is an upper bound of aS.

To show that $a\alpha$ is the least upper bound of aS, let γ be a real upper bound of aS. (If $\gamma = \infty$, then clearly $\gamma \geq a\alpha$.) We need to show that $a\alpha \leq \gamma$. For each s in S, since γ is an upper bound of aS, $as \leq \gamma$. So, for each s in S, $s \leq \gamma/a$ and, hence, γ/a is an upper bound of S. Since the least upper bound of S is α, $\alpha \leq \gamma/a$. Thus, $a\alpha \leq \gamma$.

To show part 2, let $\beta = \inf S$. By Proposition 2.4, β is a real number. We wish to show that $a\beta = \inf(aS)$. For each s in S, $\beta \leq s$ since β is a lower bound of S. So $a\beta \leq as$ for each s in S and, therefore, $a\beta$ is a lower bound of aS.

To show that $a\beta$ is the greatest lower bound of aS, let γ be a real lower bound of aS. We need to show that $\gamma \leq a\beta$. For each s in S, since γ is a lower bound of aS, $\gamma \leq as$. So, for each s in S, $\gamma/a \leq s$ and, hence, γ/a is a lower bound of S. Since β is the greatest lower bound of S, $\gamma/a \leq \beta$. Thus, $\gamma \leq a\beta$.

Exercises

1. Find the supremum and the infimum of each of the following sets.

    (a) $\{x \in \mathbb{R} : 0 < x^2 < 2\}$ (c) $\{x \in \mathbb{R} : 0 < x \text{ and } x^2 > 2\}$

    (b) $\{x \in \mathbb{R} : x^2 < 2\}$ (d) $\{x \in \mathbb{R} : x^2 > 2\}$

2. Prove part 2 of Proposition 2.3.

3. Let A and B be nonempty subsets of $\mathbb{R}$ with $A \subset B$. Show that

    $$\inf B \leq \inf A \leq \sup A \leq \sup B.$$

4. Let S be a nonempty bounded subset of $\mathbb{R}$ and let $b < 0$. Show that $\inf(bS) = b \sup S$ and $\sup(bS) = b \inf S$.

5. Let S be a nonempty subset of $\mathbb{R}$ that is bounded above. Prove that $\inf(-S) = -\sup S$, where $-S = (-1)S = \{-s : s \in S\}$.

6. Let S be a subset of $\mathbb{R}$ and let a be an element of $\mathbb{R}$. Define $a + S = \{a + s : s \in S\}$. Assume that S is a nonempty bounded subset of $\mathbb{R}$. Show that $\sup(a + S) = a + \sup S$ and $\inf(a + S) = a + \inf S$.

7. Let A and B be subsets of $\mathbb{R}$. Show that

$$\sup(A \cup B) = \max\{\sup A, \sup B\}$$

and

$$\inf(A \cup B) = \min\{\inf A, \inf B\}.$$

8. Show that a nonempty finite subset of $\mathbb{R}$ contains both a maximum and a minimum element. [*Hint:* Use induction.]

9. Let A and B be nonempty bounded subsets of $\mathbb{R}$, let $\alpha = \sup A$, and let $\beta = \sup B$. Let $C = \{ab : a \in A \text{ and } b \in B\}$. Show, by example, that $\alpha\beta \neq \sup C$ in general.

2.3 The Rational Numbers Are Dense in $\mathbb{R}$

In this section we establish three main results: first, that the Archimedean Principle holds; second, that the rationals are "dense" in the reals; and third, that the rationals are not "complete."

We also expand our techniques for working with the supremum and infimum. The next proposition states that we can approximate a real supremum or infimum by an element of the set. That is, we can find an element of the set that is as close as we would like to a real supremum or infimum.

The Archimedean Principle

Proposition 2.5   Let S be a nonempty bounded subset of $\mathbb{R}$. Let $\alpha = \sup S$ and let $\beta = \inf S$. Let $\varepsilon > 0$. Then

1. there exists an s_0 in S such that $\alpha - \varepsilon < s_0$

and

2. there exists an s_1 in S such that $s_1 < \beta + \varepsilon$.

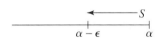

Figure 2.1

Proof   We prove part 1, leaving part 2 as an exercise. Figure 2.1 illustrates the fact that α is an upper bound of S and α is in $\mathbb{R}$. Suppose $s \leq \alpha - \varepsilon$ for all s in S. Then $\alpha - \varepsilon$ is an upper bound of S, contradicting the fact that α is the least upper bound of S.   ∎

Theorem 2.1   The set $\mathbb{N}$ of positive integers is unbounded above.

Proof   Suppose $\mathbb{N}$ is bounded above. By the Completeness Axiom for $\mathbb{R}$, $\alpha = \sup \mathbb{N}$ is a real number. By Proposition 2.5, there is an n_0 in $\mathbb{N}$ such that $\alpha - 1 < n_0 \leq \alpha$. Since $n_0 + 1$ is in $\mathbb{N}$ and $\alpha < n_0 + 1$, this contradicts the fact that α is an upper bound of $\mathbb{N}$.   ∎

We now state the Archimedean Principle, whose truth follows from Proposition 1.2.

Archimedean Principle If $x > 0$, then there exists a positive integer n such that $1/n < x$.

It follows from the Archimedean Principle that $\inf\{1/n : n \in \mathbb{N}\} = 0$.

The Density of ℚ in ℝ

Definition 2.6 Let A be a subset of ℝ. Then A is *dense* in ℝ if between every two real numbers there exists an element of A.

Equivalently, A is dense in ℝ if and only if for each x and y in ℝ with $x < y$, one has that $A \cap (x, y) \neq \emptyset$. Here (x, y) is the open interval $\{z \in \mathbb{R} : x < z < y\}$ in ℝ.

Theorem 2.2 ℚ is dense in ℝ.

Proof   We consider various cases.

Case 1   If x is negative and y is positive, then $x < 0 < y$ and 0 is rational.

Case 2 $0 < x$. By the Archimedean Principle, there is an n in ℕ such that $0 < 1/n < x$, and $1/n$ is a rational number.

Figure 2.2

Case 3 $0 < x < y$. Refer to Figure 2.2. Since $y - x$ is positive, by the Archimedean Principle, there is an n in ℕ with $0 < 1/n < y - x$. [Informally, if we add $1/n$ to itself a sufficient number of times, say m times, then $x < m/n < y$, because we cannot "jump" over the interval (x, y) since $1/n < y - x$.] By Exercise 2, there is an m in ℕ such that $m - 1 \le nx < m$. So, $m/n - 1/n \le x < m/n$. Manipulating this expression and recalling how $1/n$ was chosen, we get

$$x < \frac{m}{n} \le x + \frac{1}{n} < x + (y - x) = y,$$

and m/n is a rational number.

Case 4 $x < 0$. By Case 2, there is an n in ℕ such that $0 < 1/n < -x$, and so $x < -1/n < 0$ and $-1/n$ is a rational number.

Case 5 $x < y < 0$. By Case 3, there is a rational number q with $-y < q < -x$. Then $-q$ is rational and $x < -q < y$.   ■

Definition 2.7 A subset C of ℝ is *complete* if every nonempty bounded subset of C has both a supremum and an infimum in C.

For instance, $[0, 1]$ is complete but $(0, 1)$ is not complete. Also, ℝ is complete by the Completeness Axiom for ℝ.

Corollary 2.2   The rational numbers are not complete.

Proof   Let $A = \{x \in \mathbb{Q} : 0 < x < \sqrt{2}\}$. (Recall that $\sqrt{2}$ is irrational by Theorem 1.1.) Since $\sqrt{2}$ is an upper bound of the set A, sup $A \leq \sqrt{2}$. Suppose sup $A < \sqrt{2}$. Since $\mathbb{Q}$ is dense in $\mathbb{R}$, there is a rational number q such that sup $A < q < \sqrt{2}$. Thus q is in A and $q >$ sup A, which is a contradiction. So, sup $A = \sqrt{2}$. Thus, A is a nonempty bounded subset of $\mathbb{Q}$ that does not have a supremum in $\mathbb{Q}$. Therefore, $\mathbb{Q}$ is not complete.   ■

More on Suprema and Infima

We end this section by expanding our techniques for working with the supremum and the infimum. We give two different arguments for the more difficult part of the next proof because both are elegant.

For subsets A and B of $\mathbb{R}$, we define

$$A + B = \{a + b : a \in A \text{ and } b \in B\}.$$

Proposition 2.6   Let A and B be nonempty subsets of $\mathbb{R}$ that are both bounded above. Then $\sup(A + B) = \sup A + \sup B$.

Proof   Let $\alpha = \sup A$ and $\beta = \sup B$. Both α and β are real numbers by the Completeness Axiom for $\mathbb{R}$. We need to show that $\alpha + \beta = \sup(A + B)$. For each a in A and b in B, $a + b \leq \alpha + \beta$, and so $\alpha + \beta$ is an upper bound of $A + B$.

Let γ be a real upper bound of $A + B$. We must show that $\alpha + \beta \leq \gamma$.

Argument 1   Fix b_0 in B. Then, for each a in A, since γ is an upper bound of $A + B$, $a + b_0 \leq \gamma$. So, for each a in A, $a \leq \gamma - b_0$, and thus $\gamma - b_0$ is an upper bound of A. Since α is the least upper bound of A, $\alpha \leq \gamma - b_0$ or, equivalently, $b_0 \leq \gamma - \alpha$. Since b_0 is an arbitrary element of B, $b \leq \gamma - \alpha$ for all b in B. Hence, $\gamma - \alpha$ is an upper bound of B, and since β is the least upper bound of B, $\beta \leq \gamma - \alpha$. Therefore, $\alpha + \beta \leq \gamma$.

Argument 2   Let $\varepsilon > 0$. By Proposition 2.5, there exist an a_0 in A and a b_0 in B such that $\alpha - \varepsilon/2 < a_0$ and $\beta - \varepsilon/2 < b_0$. Thus, $\alpha + \beta - \varepsilon < a_0 + b_0 \leq \gamma$ since γ is an upper bound of $A + B$. Hence, $\alpha + \beta < \gamma + \varepsilon$. Since ε is an arbitrary positive number, $\alpha + \beta \leq \gamma$ (see Proposition 1.5).   ■

Definition 2.8   Let f be a function from a set X into $\mathbb{R}$. Then f is *bounded on* X if the range of f is a bounded subset of $\mathbb{R}$. For a subset D of X, f is *bounded on* D if the restriction of f to D, $f|_D$, has bounded range in $\mathbb{R}$.

Thus, for $f : X \to \mathbb{R}$ and D a subset of X, f is bounded on D if and only if there exists an $M > 0$ such that $|f(x)| \leq M$ for all x in D.

For example, $f(x) = x^2$ is bounded on $(0, 1)$ while $g(x) = 1/x$ is not bounded on $(0, 1)$. Also, $\inf\{f(x) : x \in (0, 1)\} = 0$, $\sup\{f(x) : x \in (0, 1)\} = 1$, $\inf\{g(x) : x \in (0, 1)\} = 1$, and $\sup\{g(x) : x \in (0, 1)\} = \infty$.

Proposition 2.7 Let f and g be bounded functions from a nonempty set X into $\mathbb{R}$. Then

$$\inf\{f(x) : x \in X\} + \inf\{g(x) : x \in X\} \leq \inf\{f(x) + g(x) : x \in X\}.$$

Proof Let $\alpha = \inf\{f(x) : x \in X\}$ and let $\beta = \inf\{g(x) : x \in X\}$. By Proposition 2.4, both α and β are real numbers. For each x in X,

$$\alpha + \beta \leq f(x) + g(x).$$

So, $\alpha + \beta$ is a lower bound of $\{f(x) + g(x) : x \in X\}$. Since

$$\inf\{f(x) + g(x) : x \in X\}$$

is the greatest lower bound of $\{f(x) + g(x) : x \in X\}$,

$$\alpha + \beta \leq \inf\{f(x) + g(x) : x \in X\}. \qquad \blacksquare$$

The difference between the two previous propositions is that in Proposition 2.7 we have only one variable, x in X, while in Proposition 2.6 we have two variables, a in A and b in B. To make Proposition 2.7 analogous to Proposition 2.6, we would need to consider $\inf\{f(x) + g(y) : x \in X \text{ and } y \in X\}$.

Example 2.3 1. If we let $f(x) = g(x) = 1$ for all x in X, we obtain equality in Proposition 2.7.

2. Define $f : [0, 2] \to \mathbb{R}$ by

$$f(x) = \begin{cases} 1 & \text{if} \quad 0 \leq x \leq 1 \\ -1 & \text{if} \quad 1 < x \leq 2 \end{cases}$$

and define $g : [0, 2] \to \mathbb{R}$ by

$$g(x) = \begin{cases} -1 & \text{if} \quad 0 \leq x \leq 1 \\ 1 & \text{if} \quad 1 < x \leq 2. \end{cases}$$

The reader should graph these functions. Then $\inf\{f(x) : x \in [0, 2]\} = \inf\{g(x) : x \in [0, 2]\} = -1$ and $\inf\{f(x) + g(x) : x \in [0, 2]\} = \inf\{0\} = 0$. Thus, we obtain strict inequality in Proposition 2.7.

Exercises

1. Prove part 2 of Proposition 2.5.

2. Let $0 < x$. Show that there is a unique m in $\mathbb{N}$ such that $m - 1 \leq x < m$. [*Hint:* Consider $\{n \in \mathbb{N} : x < n\}$ and use the well-ordering of $\mathbb{N}$.]

3. Find the sup and inf of
   (a) $\{1 - 1/n : n \in \mathbb{N}\}$ (c) $\{n - 1/n : n \in \mathbb{N}\}$
   (b) $\mathbb{Q}$ (d) $\{x \in \mathbb{Q} : x^2 < 2\}$.

4. Show that the irrational numbers are dense in $\mathbb{R}$. [*Hint:* Use the fact that $\sqrt{2}$ is irrational.]

5. Let A and B be nonempty subsets of $\mathbb{R}$ that are both bounded below. Prove that

$$\inf(A + B) = \inf A + \inf B.$$

6. Let f and g be bounded functions from a nonempty set X into $\mathbb{R}$. Show that
$$\sup\{f(x) + g(x) : x \in X\} \leq \sup\{f(x) : x \in X\} + \sup\{g(x) : x \in X\}.$$
Show by examples that both equality and strict inequality can occur.

7. Let f and g be bounded functions from a nonempty set X into $\mathbb{R}$.
   (a) Prove that if $f(x) \leq g(x)$ for all x in X, then $\inf f(X) \leq \inf g(X)$ and $\sup f(X) \leq \sup g(X)$.
   (b) Prove that if $f(x) \leq g(y)$ for all x and y in X, then $\sup f(X) \leq \inf g(X)$.
   (c) Give an example showing that the hypothesis of part (a) does not imply the conclusion of part (b).

2.4 Cardinality

In this section we distinguish between "finite" and "infinite" sets and then we classify infinite sets as to whether they are "countable" or "uncountable." The main results of this section are Theorems 2.3, 2.4, and 2.5 and Corollaries 2.4 and 2.5. Our goal is to prove that the set of rational numbers is countable and that the set of real numbers is uncountable.

The Cardinality of a Set

Definition 2.9   Two sets A and B have the same *cardinal number*, denoted by $A \sim B$, if there exists a one-to-one function from A onto B (in other words, if there exists a bijection from A onto B).

If $A \sim B$, then A and B are said to be equivalent sets and, intuitively, A and B have the same number of elements. Thus, $\sim$ is read as "is equivalent to." From a previous course, the reader may be familiar with the term "equivalence relation." Such a relation has the properties of being reflexive, symmetric, and transitive. Observe that $\sim$ is an equivalence relation:

1. $\sim$ is reflexive since $A \sim A$ by the identity function;

2. $\sim$ is symmetric, because if $A \sim B$ and if f is a bijection from A onto B, then f^{-1} is a bijection from B onto A and so $B \sim A$;

3. $\sim$ is transitive, because if $A \sim B$ and $B \sim C$, with f and g being bijections from A onto B and from B onto C, respectively, then $g \circ f$ is a bijection from A onto C, and so $A \sim C$.

As examples, note that $\mathbb{N}$ is equivalent to the set of even positive integers by the function $n \to 2n$ and that $\mathbb{N}$ is equivalent to the set of odd positive integers by the function $n \to 2n - 1$. Also, $\mathbb{R} \sim (0, 1)$ by Example 1.11.

Definition 2.10 A set A is *finite* if either A is the empty set (which has cardinal number 0) or there is an n in $\mathbb{N}$ such that $A \sim \{1, 2, \ldots, n\}$ (in which case A has cardinal number n or, equivalently, A has n elements). A set is *infinite* if it is not finite.

Observe that if A has n elements, we can write A as $\{x_1, x_2, \ldots, x_n\}$. From the Pigeonhole Principle (Section 1.1, Exercise 8), which states that if there are m pigeons and n pigeonholes with $m > n$, then at least two pigeons must get in the same hole, it is clear that there cannot be a bijection from a finite set onto a proper subset of itself. From the paragraph preceding Definition 2.10, this is not the case with infinite sets.

Proposition 2.8 Let A and B be sets.

1. If A is finite and $A \sim B$, then B is finite.

2. If B is infinite and $A \sim B$, then A is infinite.

3. If A is finite and $B \subset A$, then B is finite.

4. If B is infinite and $B \subset A$, then A is infinite.

Proof For part 1, first note that if A is empty, then so is B. Otherwise, $A \sim \{1, 2, \ldots, n\}$ for some n in $\mathbb{N}$. Since $\sim$ is transitive, $B \sim \{1, 2, \ldots, n\}$.

For part 4, note that it is the contrapositive of part 3. The rest of the proof is left as an exercise. ∎

Proposition 2.9 Let A and B be sets.

1. If A is finite and there exists a function f from A onto B, then B is finite.

2. If A is infinite and there exists a one-to-one function from A into B, then B is infinite.

Proof We first prove part 1. We can assume that B is nonempty. For each b in B, $f^{-1}(\{b\})$ is nonempty since f maps A onto B. Choose one a_b in $f^{-1}(\{b\})$ for each b in B. Then the function $b \to a_b$ is a bijection from B onto a subset of A, which is finite by Proposition 2.8.

To prove part 2, note that A is equivalent to a subset of B. Again, by Proposition 2.8, this subset of B, and hence B, is infinite. ∎

In reference to Proposition 2.9, since a function is single-valued, the range of a function cannot be larger than the domain with respect to cardinality. In the proof of part 1, the reader who wonders, or even questions, why we can choose an element from each member of a collection of sets indexed by the set B (when, at the time of doing this, all we know about B is that B is nonempty) should see the end of this section.

Proposition 2.10 $\mathbb{N}$ is infinite.

Proof Suppose that $\mathbb{N}$ is not infinite. Then, by Exercise 8 in Section 2.2, $\mathbb{N}$ is bounded, which contradicts Theorem 2.1. ∎

Sequences and Infinite Sets

Definition 2.11   A *sequence* in a set X is a function from $\mathbb{N}$ into X.

For a sequence it is customary to use a letter such as x for the function and to denote the value $x(n)$ as x_n for each n in $\mathbb{N}$. Thus, we think of the sequence as $(x_1, x_2, x_3, \ldots)$ or as $(x_n)_{n \in \mathbb{N}}$ or as $(x_n)_{n=1}^{\infty}$.

Example 2.4   The following are examples of sequences.

1. $(n)_{n=1}^{\infty} = (1, 2, 3, \ldots)$
2. $\left(\frac{1}{n}\right)_{n=1}^{\infty} = (1, \frac{1}{2}, \frac{1}{3}, \ldots)$
3. $((-1)^n)_{n=1}^{\infty} = (-1, 1, -1, 1, \ldots)$

The first two examples are sequences of distinct points, whereas the third is not.

Theorem 2.3   A set X is infinite if and only if X contains a sequence of distinct points.

Proof   If X contains the sequence $(x_n)_{n=1}^{\infty}$ of distinct points, then $\mathbb{N} \sim \{x_n : n \in \mathbb{N}\}$ by the function $n \to x_n$. Hence, X has an infinite subset and so X is infinite.

Let X be an infinite set. We want to construct a sequence of distinct points in X. Since X is nonempty, there is some element x_1 in X. Since $X \setminus \{x_1\}$ is nonempty, there is an element x_2 in $X \setminus \{x_1\}$. Suppose that distinct points $x_1, x_2, x_3, \ldots, x_k$ have been chosen in X. Since $X \setminus \{x_1, x_2, x_3, \ldots, x_k\}$ is nonempty, there exists an element x_{k+1} in $X \setminus \{x_1, x_2, x_3, \ldots, x_k\}$. Then $(x_n)_{n=1}^{\infty}$ is a sequence of distinct points in X.   ∎

In reference to the proof above, we constructed the sequence inductively. Once we acquire facility with this type of construction, we need only generate the first two or three terms and then say "continuing by induction, we obtain the rest."

Corollary 2.3   A set X is infinite if and only if there exists a one-to-one function from X onto a proper subset of X.

Proof   If X is finite, then no such bijection can exist by the Pigeonhole Principle.

Let X be infinite. By Theorem 2.3, let $(x_n)_{n=1}^{\infty}$ be a sequence of distinct points in X. Define .

$$f : X \overset{1-1}{\underset{\text{onto}}{\to}} X \setminus \{x_1\}$$

by

$$f(x_n) = x_{n+1} \text{ for all } n \text{ in } \mathbb{N}$$
$$f(x) = x \text{ otherwise.}$$

The reader should verify that this function is one-to-one and onto the desired set. ■

The proof above provides a standard way to construct one-to-one functions on an infinite set. The function on the sequence of distinct points is really a function on $\mathbb{N}$ since you manipulate the subscripts, and the function on the rest of the set is the identity. There are obviously many more such functions. For example, we could have eliminated all the x_n's with n odd by using $f(x_n) = x_{2n}$.

Countable and Uncountable Sets

So far, we have split sets into finite and infinite. Next, we divide the infinite sets into two classifications.

> **Definition 2.12**   Let A be a set.
>
> 1. A is *countably infinite* (or *denumerable*) if $A \sim \mathbb{N}$.
>
> 2. A is *countable* if either A is finite or A is countably infinite.
>
> 3. A is *uncountable* if A is not countable.

Note that $\mathbb{N}$, the set of even positive integers, and the set of odd positive integers are all countably infinite. Informally, Theorem 2.3 states that the countably infinite sets are the "smallest" infinite sets. Also note that if A is countably infinite, we can write A as a sequence of distinct points.

Proposition 2.11   Let A and B be sets.

1. If A is countable and $A \sim B$, then B is countable.

2. If B is uncountable and $A \sim B$, then A is uncountable.

3. If A is countable and $B \subset A$, then B is countable.

4. If B is uncountable and $B \subset A$, then A is uncountable.

Proof   Note that part 4 is the contrapositive of part 3. We prove part 3, leaving parts 1 and 2 as exercises. We have that A is countable with $B \subset A$ and we want to show that B is countable. If B is finite, then B is countable. So we can assume that B is infinite. Hence, A is countably infinite. So we enumerate A as a sequence of distinct points $(x_n)_{n=1}^{\infty}$. Recall that $\mathbb{N}$ is well-ordered. Let n_1 be the smallest positive integer such that x_{n_1} is in B. Let n_2 be the smallest positive integer greater than n_1 such that x_{n_2} is in B. Continuing by induction, we get $B = \{x_{n_1}, x_{n_2}, x_{n_3}, \ldots\}$. Hence, $\mathbb{N} \sim B$ by the function $k \to x_{n_k}$. ■

Proposition 2.12   Let A and B be sets.

1. If A is countable and there exists a function from A onto B, then B is countable.

2. If A is uncountable and there exists a one-to-one function from A into B, then B is uncountable.

Proof   We first prove part 1. If B is finite, then B is countable; so we can assume that B is infinite. For each b in B, choose one a_b in $f^{-1}(\{b\})$. Then the function $b \rightarrow a_b$ is a bijection from B onto a subset of A. Thus, B is equivalent to a subset of A, which is countable by Proposition 2.11.

To prove part 2, note that A is equivalent to a subset of B. Again, by Proposition 2.11, this subset of B, and hence B, is uncountable. ∎

In reference to the proof of part 1, the comment after Proposition 2.9 also applies here.

Lemma 2.1 $\mathbb{N} \times \mathbb{N}$ is countable.

Proof   Define $f : \mathbb{N} \times \mathbb{N} \rightarrow \mathbb{N}$ by $f(n, m) = 2^n 3^m$. (Any two distinct primes will work.) We claim that f is one-to-one and hence $\mathbb{N} \times \mathbb{N} \sim$ a subset of $\mathbb{N}$ and therefore $\mathbb{N} \times \mathbb{N}$ is countable. To see that f is one-to-one, suppose that $f(n, m) = f(r, s)$. Then $2^n 3^m = 2^r 3^s$. If $n > r$, then $2^{n-r} = 3^{s-m}$. Since 2 divides the left side, 2 divides 3^{s-m} and so 2 divides 3, which is a contradiction. The case $n < r$ is handled similarly. Thus, $n = r$ and hence $m = s$. ∎

Note that if A is a countably infinite set, then there is a bijection from $\mathbb{N}$ onto A. Also, if A is finite and nonempty, there is a function from $\mathbb{N}$ onto A, because if $A = \{x_1, x_2, \ldots, x_n\}$, we can define a function f from $\mathbb{N}$ onto A by

$$f(k) = \begin{cases} x_k & \text{if} \quad k < n \\ x_n & \text{if} \quad k \geq n. \end{cases}$$

So, if A is countable and nonempty, there is a function from $\mathbb{N}$ onto A.

Theorem 2.4   The countable union of countable sets is countable. That is, if I is a countable index set and A_α is a countable set for each α in I, then $\bigcup_{\alpha \in I} A_\alpha$ is countable.

Proof   We can assume that $I \neq \emptyset$ since $\bigcup_{\alpha \in \emptyset} A_\alpha = \emptyset$, and we can assume that each A_α is nonempty since empty A_α's add nothing to the union. Since I is countable, there exists a function f from $\mathbb{N}$ onto I. Since each A_α is countable, for each α in I there exists a function g_α from $\mathbb{N}$ onto A_α. Define

$$h : \mathbb{N} \times \mathbb{N} \rightarrow \bigcup_{\alpha \in I} A_\alpha$$

by

$$h(n, m) = g_{f(n)}(m).$$

We claim that h maps $\mathbb{N} \times \mathbb{N}$ onto $\bigcup_{\alpha \in I} A_\alpha$ and hence, by Lemma 2.1 and Proposition 2.12, $\bigcup_{\alpha \in I} A_\alpha$ is countable. Let x be in $\bigcup_{\alpha \in I} A_\alpha$. Then there exists an α_0 in I such that x is in A_{α_0}. Since g_{α_0} maps $\mathbb{N}$ onto A_{α_0}, there exists an m in $\mathbb{N}$ such that $g_{\alpha_0}(m) = x$. Since f maps $\mathbb{N}$ onto I, there is an n in $\mathbb{N}$ with $f(n) = \alpha_0$. Then, $h(n, m) = g_{f(n)}(m) = g_{\alpha_0}(m) = x$. ∎

Corollary 2.4   The sets $\mathbb{Z}$, $\mathbb{Q}$, and $\mathbb{Q} \times \mathbb{Q}$ are countable.

Proof   We write each set as the countable union of countable sets.

$$\mathbb{Z} = \{\ldots, -3, -2, -1\} \cup \{0\} \cup \{1, 2, 3, \ldots\}.$$
$$\mathbb{Q} = \bigcup_{n=1}^{\infty} \{m/n : m \in \mathbb{Z}\}.$$
$$\mathbb{Q} \times \mathbb{Q} = \bigcup_{q \in \mathbb{Q}} \{(q, p) : p \in \mathbb{Q}\}.$$
∎

If we define $\mathbb{Q}^n = \{(q_1, q_2, \ldots, q_n) : q_i \in \mathbb{Q}$ for $i = 1, 2, \ldots, n\}$, we can show that $\mathbb{Q}^n$ is countable by induction and by writing $\mathbb{Q}^n = \bigcup_{q \in \mathbb{Q}}(\mathbb{Q}^{n-1} \times \{q\})$.

Theorem 2.5   The closed interval $[0, 1]$ is uncountable.

Proof   The technique used here is called the Cantor diagonalization argument. Suppose that $[0, 1]$ is countable. Then there exists a bijection f from $\mathbb{N}$ onto $[0, 1]$. We list the elements of $[0, 1]$ in their decimal expansion as an infinite matrix:

$$f(1) = .a_{11}a_{12}a_{13}\cdots$$
$$f(2) = .a_{21}a_{22}a_{23}\cdots$$
$$f(3) = .a_{31}a_{32}a_{33}\cdots$$
$$\vdots$$
$$f(n) = .a_{n1}a_{n2}a_{n3}\cdots$$
$$\vdots$$

where each a_{ij} is a digit from 0 to 9.

We now construct an element of the interval $[0, 1]$ that is not in the range of f by "going down" the diagonal of the matrix. For each n in $\mathbb{N}$, let

$$b_n = \begin{cases} 3 & \text{if} \quad a_{nn} \neq 3 \\ 4 & \text{if} \quad a_{nn} = 3. \end{cases}$$

Then $x = .b_1b_2b_3\cdots$ is in $[0, 1]$, but for all n in $\mathbb{N}$, $x \neq f(n)$ since $b_n \neq a_{nn}$. (The choice of 3 and 4 for b_n is done simply to avoid duplicate representations, which occur only with tails of nines or zeros.)   ∎

Corollary 2.5   The real numbers and the irrational numbers are uncountable.

Proof   That $\mathbb{R}$ is uncountable follows from part 4 of Proposition 2.11. Since $\mathbb{R} = \mathbb{Q} \cup (\mathbb{R} \setminus \mathbb{Q})$, $\mathbb{R} \setminus \mathbb{Q}$ is uncountable by Theorem 2.4.   ∎

We make one final comment. The following axiom is independent of the other axioms of set theory.

Axiom of Choice   If I is a nonempty set and if A_α is a nonempty set for each α in I, we may choose an x_α in A_α for each α in I.

If I is countable, then we can use mathematical induction to choose each x_α. The strength of the Axiom of Choice comes when I is uncountable. The reason we mention the Axiom of Choice is because we used the Axiom of Choice in the proofs of Propositions 2.9 and 2.12. We could have given alternative proofs using the fact that $\mathbb{N}$ is well-ordered, but we felt that this was a needless complication.

Exercises

1. Complete the proof of Proposition 2.8.

2. Let f be a one-to-one function from A into B with B finite. Show that A is finite.

3. If A and B are finite sets, show that $A \cup B$ is a finite set. Conclude that the finite union of finite sets is finite.

4. If X is an infinite set and x is in X, show that $X \sim X \setminus \{x\}$.

5. Define explicitly a bijection from $[0, 1]$ onto $(0, 1)$.

6. Complete the proof of Proposition 2.11.

7. Let f be a one-to-one function from A into B with B countable. Prove that A is countable.

8. For m in $\mathbb{N}$, show that $\mathbb{N} \sim \mathbb{N} \setminus \{1, 2, \ldots, m\}$.

9. If A and B are countable sets, show that $A \times B$ is countable.

10. Let A be an uncountable set and let B be a countable subset of A. Show that $A \setminus B$ is uncountable.

11. Let $\mathfrak{A}$ be a collection of pairwise disjoint open intervals. That is, members of $\mathfrak{A}$ are open intervals in $\mathbb{R}$ and any two distinct members of $\mathfrak{A}$ are disjoint. Show that $\mathfrak{A}$ is countable.

12. Let $\mathfrak{S}$ be the set of all open intervals with rational endpoints. Show that $\mathfrak{S}$ is countable.

13. Let A be the set of all sequences whose terms are the digits 0 and 1. Show that A is uncountable.

14. The purpose of this exercise is to show that given any set, there is always a larger set with respect to cardinality. For a set X, the *power set* of X, denoted by $\mathcal{P}(X)$, is the collection of all subsets of X. Recall from Exercise 9 in Section 1.4 that a set with n elements has 2^n subsets for n in $\mathbb{N} \cup \{0\}$. The map $x \to \{x\}$ is a one-to-one function from X into $\mathcal{P}(X)$. Show that there does not exist a function from X onto $\mathcal{P}(X)$. [*Hint:* Suppose f is a function from X onto $\mathcal{P}(X)$. Let $A = \{x \in X : x \notin f(x)\}$. Then A is in $\mathcal{P}(X)$ and so $A = f(x_0)$ for some x_0 in X. Ask yourself "where is x_0?" to obtain a contradiction.]

15. Let A be the set of all real-valued functions on $[0, 1]$. Show that there does not exist a function from $[0, 1]$ onto A.

3 Sequences

This is an important chapter because we will use sequences throughout the text. The main theorem is the Bolzano-Weierstrass Theorem for sequences in Section 3.5. Other very important theorems are the Monotone Convergence Theorem (Section 3.4), Theorem 3.5 (Section 3.2), and Theorem 3.12 (Section 3.6).

3.1 Convergence

Limit of a Sequence

Recall that $\mathbb{N} = \{1, 2, 3, \ldots\}$ and that $\mathbb{R}$ is the set of real numbers.

> **Definition 3.1** A *sequence of real numbers* (or a *sequence in $\mathbb{R}$*) is a function whose domain is $\mathbb{N}$ and whose range is a subset of $\mathbb{R}$.

Thus, a sequence in $\mathbb{R}$ is a function from $\mathbb{N}$ into $\mathbb{R}$.

Notation When we use the word "sequence" without qualification, we mean a sequence in $\mathbb{R}$. For a sequence it is customary to use a letter such as x for the function and to denote the value $x(n)$ as x_n for each n in $\mathbb{N}$. Thus, we think of a sequence as $x : \mathbb{N} \to \mathbb{R}$ or as $(x_1, x_2, x_3, \ldots)$ or as $(x_n)_{n \in \mathbb{N}}$ or as $(x_n)_{n=1}^{\infty}$. Each x_n is a *term* of the sequence. Following Bartle and Sherbert, we distinguish between the sequence $(x_n)_{n \in \mathbb{N}}$, whose terms have an order induced by $\mathbb{N}$, and the range $\{x_n : n \in \mathbb{N}\}$ of the sequence, which is not ordered. For example, the sequence $\left((-1)^{n+1}\right)_{n \in \mathbb{N}} = (1, -1, 1, -1, \ldots)$ has range $\{-1, 1\}$. Some books use the notations $(x_n)_{n \in \mathbb{N}}$ and $\{x_n : n \in \mathbb{N}\}$ interchangeably, whereas we will reserve the latter notation for the range of the sequence.

Sequences may be given explicitly, as in $(1/n)_{n=1}^{\infty} = (1, \frac{1}{2}, \frac{1}{3}, \ldots)$, or recursively, as in the *Fibonacci sequence*:

let $x_1 = x_2 = 1$

let $x_n = x_{n-1} + x_{n-2}$ for $n \geq 3$.

> **Definition 3.2** A sequence $(x_n)_{n \in \mathbb{N}}$ *eventually* has a certain property if there exists an n_0 in $\mathbb{N}$ such that
> $$(x_n)_{n \geq n_0} = (x_{n_0}, x_{n_0+1}, x_{n_0+2}, \ldots)$$
> has this property.

For example, a *constant sequence* is a sequence whose range consists of a single number, whereas the sequence $(1, 2, 3, 4, 4, 4, \ldots)$ is eventually constant.

Terminology   In Definition 3.2, $(x_n)_{n \geq n_0}$ is a *tail* of the sequence $(x_n)_{n \in \mathbb{N}}$. This tail is also a sequence. To see this, define a bijection f from $\mathbb{N}$ onto $\{n_0, n_0 + 1, n_0 + 2, \ldots\}$ by $f(n) = n_0 + n - 1$ for each n in $\mathbb{N}$. Then $x \circ f$ is a sequence and $x_n = (x \circ f)(n + 1 - n_0)$ for $n \geq n_0$.

A useful and important concept is that of a *neighborhood* of a point.

Definition 3.3   For x in $\mathbb{R}$ and $\varepsilon > 0$, the open interval

$$(x - \varepsilon, x + \varepsilon) = \{y \in \mathbb{R} : |y - x| < \varepsilon\}$$

centered at x of radius ε is a *neighborhood of x*.

Definition 3.4   Let $(x_n)_{n \in \mathbb{N}}$ be a sequence in $\mathbb{R}$ and let x be in $\mathbb{R}$. The sequence $(x_n)_{n \in \mathbb{N}}$ *converges* to x (or has *limit x*) if for every neighborhood U of x the sequence $(x_n)_{n \in \mathbb{N}}$ is eventually in U.

Notation and Terminology   If the sequence $(x_n)_{n \in \mathbb{N}}$ converges to x, we write $\lim_{n \to \infty} x_n = x$ or $\lim_n x_n = x$ or $x_n \to x$ or simply $x_n \to x$, and we call $(x_n)_{n \in \mathbb{N}}$ a *convergent sequence*. A sequence that is not convergent is *divergent*.

Paraphrasing Definitions 3.2 through 3.4, we obtain the following proposition.

Proposition 3.1   Let $(x_n)_{n \in \mathbb{N}}$ be a sequence in $\mathbb{R}$ and let x be in $\mathbb{R}$. Then $(x_n)_{n \in \mathbb{N}}$ converges to x if and only if for all $\varepsilon > 0$ there exists an n_0 in $\mathbb{N}$ such that if $n \geq n_0$, then $|x_n - x| < \varepsilon$.

Proof   Combining Definitions 3.2 through 3.4, we have that $x_n \to x \iff (x_n)_{n \in \mathbb{N}}$ is eventually in every neighborhood of $x \iff$ for all $\varepsilon > 0$, $(x_n)_{n \in \mathbb{N}}$ is eventually in $(x - \varepsilon, x + \varepsilon) \iff$ for all $\varepsilon > 0$, eventually, $|x_n - x| < \varepsilon$. $\blacksquare$

In general, n_0 depends on ε. Figure 3.1 graphically depicts Definition 3.4 and Proposition 3.1. Usually, the smaller ε becomes, the larger n_0 must be in order for the distance from x_n to x to be less than ε for all $n \geq n_0$.

Example 3.1   Let $(x_n)_{n \in \mathbb{N}}$ be the constant sequence c—that is, $x_n = c$ for all n in $\mathbb{N}$. Then, $x_n \to c$. To see this, let $\varepsilon > 0$. Let $n_0 = 17$. (Here n_0 does not depend on ε. Any other n_0 would also work.) Let $n \geq n_0$. Then

$$|x_n - c| = |c - c| = 0 < \varepsilon.$$

Example 3.2 $\lim_{n \to \infty} (1/n) = 0$. Let $\varepsilon > 0$. We want to determine a value of n_0 as specified in Proposition 3.1. Right now we do not know how to choose n_0, and so we proceed as if we know n_0 with the hope that the calculations below

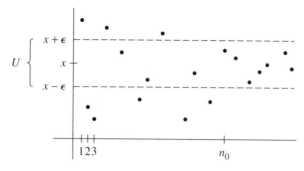

Figure 3.1

will indicate how to choose n_0. Let $n \geq n_0$. Then

$$\left| \frac{1}{n} - 0 \right| = \frac{1}{n} \leq \frac{1}{n_0} < \varepsilon$$

if we choose n_0 in $\mathbb{N}$ such that $1/n_0 < \varepsilon$ (by the Archimedean Principle), or equivalently, choose n_0 in $\mathbb{N}$ with $n_0 > 1/\varepsilon$ since the positive integers are unbounded above (Theorem 2.1).

Example 3.3 $\lim\limits_{n \to \infty} (2n + 3)/(3n + 5) = 2/3$. Let $\varepsilon > 0$. Let $n_0 = ?$ Pretend that you know n_0. For $n \geq n_0$,

$$\left| \frac{2n + 3}{3n + 5} - \frac{2}{3} \right| = \left| \frac{3(2n + 3) - 2(3n + 5)}{3(3n + 5)} \right|$$

$$= \frac{1}{9n + 15}$$

$$< \frac{1}{9n}$$

$$\leq \frac{1}{9n_0}.$$

Now we know how to choose n_0. By the Archimedean Principle, choose n_0 in $\mathbb{N}$ such that $1/n_0 < 9\varepsilon$. Then, for $n \geq n_0$,

$$\left| \frac{2n + 3}{3n + 5} - \frac{2}{3} \right| < \frac{1}{9n_0} < \frac{1}{9}(9\varepsilon) = \varepsilon.$$

An important property of the limit of a sequence is uniqueness.

Lemma 3.1 Distinct points in $\mathbb{R}$ can be separated by disjoint neighborhoods. That is, if x and y are in $\mathbb{R}$ with $x \neq y$, then there exist neighborhoods U of x and V of y such that $U \cap V = \emptyset$.

Proof Let x and y be in $\mathbb{R}$ with $x \neq y$. Let $\varepsilon = \frac{1}{2}|x - y|$. Let $U = (x - \varepsilon, x + \varepsilon)$ and $V = (y - \varepsilon, y + \varepsilon)$. We claim that $U \cap V = \emptyset$. Suppose that z is in $U \cap V$. Then, by the triangle inequality,

$$|x - y| \leq |x - z| + |z - y| < \varepsilon + \varepsilon = 2\varepsilon = |x - y|.$$

This is a contradiction, because a real number cannot be less than itself. Therefore, U and V are disjoint neighborhoods of x and y. (The reader should compare this with Exercise 11 in Section 2.1.) ∎

Theorem 3.1 Limits of sequences are unique.

Proof Let $(x_n)_{n \in \mathbb{N}}$ be a sequence in $\mathbb{R}$. Suppose that x and y are in $\mathbb{R}$ with $x_n \to x$ and $x_n \to y$. We wish to show that $x = y$. Suppose that $x \neq y$. By Lemma 3.1, let U and V be disjoint neighborhoods of x and y, respectively. Since $x_n \to x$, there is an n_1 in $\mathbb{N}$ such that x_n is in U for all $n \geq n_1$. Since $x_n \to y$, there is an n_2 in $\mathbb{N}$ such that x_n is in V for all $n \geq n_2$.

Let $m \geq \max\{n_1, n_2\}$. Then x_m is in $U \cap V = \emptyset$, which is a contradiction. Therefore, $x = y$. ∎

Remark The proof above illustrates a technique commonly used with sequences. Namely, in order to have two conditions occurring simultaneously, we need to go out far enough in the sequence to guarantee that each condition holds. As in the above proof we accomplish this by taking a maximum.

Limits Do Not Always Exist

The fact that limits do not always exist should come as no surprise to the reader. The sequence $(x_n)_{n \in \mathbb{N}}$ in $\mathbb{R}$ does not converge if and only if $\lim_{n \to \infty} x_n \neq x$ for all x in $\mathbb{R}$.

Let $(x_n)_{n \in \mathbb{N}}$ be a sequence in $\mathbb{R}$ and let x be in $\mathbb{R}$. Using the basic rule for negation as given on page 3, we negate both sides of the "if and only if" in Definition 3.4. Using Definition 3.2, we have that $\lim_{n \to \infty} x_n \neq x \iff$ there exists a neighborhood U of x such that the sequence $(x_n)_{n \in \mathbb{N}}$ is not eventually in $U \iff$ there exists a neighborhood U of x such that for all n_0 in $\mathbb{N}$, there exists an $n \geq n_0$ such that x_n is not in U. The last phrase is sometimes expressed as "the sequence $(x_n)_{n \in \mathbb{N}}$ is *frequently* outside of U."

Example 3.4 The sequences $(0, 1, 0, 1, 0, 1, \ldots)$, $(1, -1, 1, -1, 1, -1, \ldots)$, and $(1, 2, 1, 3, 1, 4, \ldots)$ do not converge. For instance, the first sequence is frequently outside the neighborhood $\left(-\frac{1}{2}, \frac{1}{2}\right)$ of 0 and is frequently outside the neighborhood $\left(\frac{1}{2}, \frac{3}{2}\right)$ of 1.

Let $(x_n)_{n \in \mathbb{N}}$ be a sequence in $\mathbb{R}$ and let x be in $\mathbb{R}$. Negating the statements in Proposition 3.1, we have that $\lim_{n \to \infty} x_n \neq x \iff$ there exists an $\varepsilon > 0$ such that for all n_0 in $\mathbb{N}$, there exists an $n \geq n_0$ with $|x_n - x| \geq \varepsilon \iff \exists \, \varepsilon > 0$ such that $\forall n_0 \in \mathbb{N}, \exists \, n \geq n_0$ with $|x_n - x| \geq \varepsilon$.

The reader should apply this to the sequences in Example 3.4.

Other Techniques for Convergence

We end this section with two examples that utilize different techniques. In the first example we use Bernoulli's inequality, given in Exercise 10 in Section 1.4, to derive an important result.

Example 3.5 Let $0 < r < 1$. Then $\lim_{n \to \infty} r^n = 0$. To see this, let $x = (1/r) - 1$. Then $x > 0$ and $r = 1/(1 + x)$. By Bernoulli's inequality, $(1 + x)^n \geq 1 + nx$ for all n in $\mathbb{N}$. Therefore,

$$0 < r^n = \frac{1}{(1+x)^n} \leq \frac{1}{1+nx} < \frac{1}{nx}$$

for all n in $\mathbb{N}$. Given $\varepsilon > 0$, choose n_0 in $\mathbb{N}$ such that $1/n_0 < x\varepsilon$. Then $n \geq n_0$ implies that

$$0 < r^n < \frac{1}{nx} \leq \frac{1}{n_0 x} < \frac{1}{x}(x\varepsilon) = \varepsilon,$$

and so $r^n \to 0$.

In the next example, we use results concerning the exponential function e^x and the natural logarithm function $\ln x$ from calculus. This example also requires L'Hôpital's rule.

Example 3.6 $\lim\limits_{n\to\infty} [1 + (1/n)]^n = e$.

Write

$$\left(1 + \frac{1}{n}\right)^n = e^{\ln[1+(1/n)]^n} = e^{n \ln[1+(1/n)]}.$$

Then $\lim\limits_{n\to\infty} [1 + (1/n)]^n = \lim\limits_{n\to\infty} e^{n \ln[1+(1/n)]} \underset{(*)}{=} e^{\lim\limits_{n\to\infty} n \ln[1+(1/n)]}$.

Since

$$\lim_{n\to\infty} n \ln\left(1 + \frac{1}{n}\right) = \lim_{n\to\infty} \frac{\ln[1 + (1/n)]}{1/n}$$

$$= \lim_{n\to\infty} \frac{(d/dn)[1 + (1/n)]}{\dfrac{1 + (1/n)}{\dfrac{d}{dn}\left(\dfrac{1}{n}\right)}} \qquad \text{(by L'Hôpital)}$$

$$= \lim_{n\to\infty} \frac{1}{1 + (1/n)}$$

$$= 1,$$

$\lim\limits_{n\to\infty} [1 + (1/n)]^n = e^1 = e$.

The reason for the equality at $(*)$ is because e^x is continuous on $\mathbb{R}$ (see Chapter 4).

Remark   A good exercise for the reader is to do Example 3.5 by the technique used in Example 3.6. This technique involves a sequence whose limit is negative infinity, which we consider in Section 3.7.

≡ Exercises ≡

1. Use Proposition 3.1 to establish the following limits.

(a) $\lim\limits_{n\to\infty} \dfrac{1}{n^2} = 0$

(d) $\lim\limits_{n\to\infty} \dfrac{n^2 + 1}{2n^2 + 5} = \dfrac{1}{2}$

(b) $\lim\limits_{n\to\infty} \dfrac{3n}{n + 2} = 3$

(e) $\lim\limits_{n\to\infty} \dfrac{\sin n}{n} = 0$

(c) $\lim\limits_{n\to\infty} \dfrac{3n + 7}{5n + 2} = \dfrac{3}{5}$

(f) $\lim\limits_{n\to\infty} \dfrac{-n}{n^2 + 1} = 0$

2. Let $x_n = \begin{cases} 1/n & \text{if } n \text{ is odd} \\ 0 & \text{if } n \text{ is even.} \end{cases}$ Does $\lim\limits_{n\to\infty} x_n$ exist?

3. Let $x_n = \begin{cases} 1/n & \text{if } n \text{ is odd} \\ 1 & \text{if } n \text{ is even.} \end{cases}$ Does $\lim\limits_{n\to\infty} x_n$ exist?

4. Let $(x_n)_{n\in\mathbb{N}}$ be a sequence in $\mathbb{R}$ and let x be in $\mathbb{R}$.

(a) Show that if $x_n \to x$, then $|x_n| \to |x|$. [*Hint:* Use Corollary 2.1.]

(b) Show that if $|x_n| \to 0$, then $x_n \to 0$.

(c) Show, by example, that $(|x_n|)_{n\in\mathbb{N}}$ may converge and $(x_n)_{n\in\mathbb{N}}$ may not converge.

5. Let $x_n \geq 0$ for each n in $\mathbb{N}$; let x be in $\mathbb{R}$ with $x_n \to x$. Note that $x \geq 0$. Show that $\sqrt{x_n} \to \sqrt{x}$. [*Hint:* Make two cases: $x = 0$ and $x > 0$. In the latter case, rationalize.]

6. Show that a convergent sequence in $\mathbb{R}$ is bounded. That is, if $(x_n)_{n\in\mathbb{N}}$ converges, show that there is a $B > 0$ such that $|x_n| \leq B$ for all n in $\mathbb{N}$. [*Hint:* Use Proposition 3.1 with $\varepsilon = 1$.]

7. Show that the sequences $(n^2)_{n\in\mathbb{N}} = (1, 4, 9, \ldots)$ and $(-n)_{n\in\mathbb{N}} = (-1, -2, -3, \ldots)$ do not converge.

8. Show that if $|r| < 1$, then $\lim\limits_{n\to\infty} r^n = 0$.

9. Establish the following limits.

(a) $\lim\limits_{n\to\infty} \dfrac{e^n}{\pi^n} = 0$

(b) $\lim\limits_{n\to\infty} c^{1/n} = 1 \quad (c > 0)$

(c) $\lim\limits_{n\to\infty} n^{1/n} = 1$

(d) $\lim\limits_{n\to\infty} (1 + \dfrac{1}{n})^{2n} = e^2$

(e) $\lim\limits_{n\to\infty} \dfrac{n^2}{n!} = 0$

(f) $\lim\limits_{n\to\infty} \dfrac{n!}{n^n} = 0$

3.2 Limit Theorems

In this section, we obtain many of the standard results concerning convergent sequences. Since a sequence is a function, the following definition is a special case of Definition 2.8.

Definition 3.5   A sequence $(x_n)_{n\in\mathbb{N}}$ of real numbers is *bounded* if there exists a $B > 0$ such that $|x_n| \leq B$ for all n in $\mathbb{N}$.

Proposition 3.2   A convergent sequence is bounded.

Proof   Let $(x_n)_{n\in\mathbb{N}}$ be a sequence in $\mathbb{R}$ and let x be in $\mathbb{R}$ with $x_n \to x$. By Proposition 3.1 with $\varepsilon = 1$, there exists an n_0 in $\mathbb{N}$ such that $|x_n - x| < 1$ for all $n \geq n_0$. By Corollary 2.1, $|x_n| - |x| \leq |x_n - x| < 1$, which implies that $|x_n| < 1 + |x|$ for all $n \geq n_0$. Let $B = \max\{|x_1|, |x_2|, \ldots, |x_{n_0-1}|, 1 + |x|\}$. Then $B > 0$ and $|x_n| \leq B$ for all n in $\mathbb{N}$.   ∎

Note that $(0, 1, 0, 1, 0, 1, \ldots)$ is a bounded sequence that does not converge. Hence, the converse of Proposition 3.2 is false.

In Theorem 3.2 below, we obtain results for the usual combinations of convergent sequences. For the quotient of convergent sequences, we need the following lemma.

Lemma 3.2 Let $(x_n)_{n \in \mathbb{N}}$ be a sequence in $\mathbb{R}$. Let x be in $\mathbb{R}$ with $x \neq 0$ and $x_n \to x$. Then there exist an $\varepsilon > 0$ and an n_0 in $\mathbb{N}$ such that $|x_n| \geq \varepsilon$ for all $n \geq n_0$. In other words, the sequence is eventually bounded away from 0.

Proof   Let $\varepsilon = |x|/2$. By Exercise 4 in Section 3.1, $|x_n| \to |x|$, and so the sequence $(|x_n|)_{n \in \mathbb{N}}$ is eventually in the neighborhood $(|x| - \varepsilon, |x| + \varepsilon)$ of $|x|$. Hence, there is an n_0 in $\mathbb{N}$ such that if $n \geq n_0$, then

$$|x_n| > |x| - \varepsilon = |x| - \frac{|x|}{2} = \frac{|x|}{2} = \varepsilon. \qquad \blacksquare$$

Theorem 3.2   Let $(x_n)_{n \in \mathbb{N}}$ and $(y_n)_{n \in \mathbb{N}}$ be sequences in $\mathbb{R}$; let x and y be in $\mathbb{R}$ with $x_n \to x$ and $y_n \to y$. Then

1. $\lim\limits_{n \to \infty} (x_n + y_n) = x + y = \lim\limits_{n \to \infty} x_n + \lim\limits_{n \to \infty} y_n$;

2. $\lim\limits_{n \to \infty} x_n y_n = xy = \lim\limits_{n \to \infty} x_n \cdot \lim\limits_{n \to \infty} y_n$;

3. $\lim\limits_{n \to \infty} c x_n = cx = c \lim\limits_{n \to \infty} x_n$ for all real numbers c;

4. $\lim\limits_{n \to \infty} \dfrac{x_n}{y_n} = \dfrac{x}{y} = \dfrac{\lim\limits_{n \to \infty} x_n}{\lim\limits_{n \to \infty} y_n}$, provided $y_n \neq 0$ for all n in $\mathbb{N}$ and $y \neq 0$.

Proof   For part 1, let $\varepsilon > 0$. By Proposition 3.1, there exist n_1 and n_2 in $\mathbb{N}$ such that

$$|x_n - x| < \frac{\varepsilon}{2} \text{ for all } n \geq n_1$$

and

$$|y_n - y| < \frac{\varepsilon}{2} \text{ for all } n \geq n_2.$$

Let $n_0 = \max\{n_1, n_2\}$. Then $n \geq n_0$ implies that

$$
\begin{aligned}
|(x_n + y_n) - (x + y)| &= |(x_n - x) + (y_n - y)| \\
&\leq |x_n - x| + |y_n - y| \\
&< \varepsilon.
\end{aligned}
$$

For part 2, first note that

$$
\begin{aligned}
|x_n y_n - xy| &= |(x_n y_n - x_n y) + (x_n y - xy)| \\
&\leq |x_n| \, |y_n - y| + |y| \, |x_n - x|.
\end{aligned}
$$

Let $\varepsilon > 0$. [*Note:* We wish to make each part less than $\varepsilon/2$. $|y|$, being a single number, is not a problem. However, $|x_n|$, not being a constant, is a slight problem.] First, choose n_1 in $\mathbb{N}$ such that if $n \geq n_1$, then $|x_n - x| < \varepsilon/[2(|y| + 1)]$. ($|y|$ could be 0, so we added the 1.) By Proposition 3.2, $(x_n)_{n \in \mathbb{N}}$ is a bounded sequence, and so there is a $B > 0$ such that $|x_n| \leq B$ for all n in $\mathbb{N}$. Now choose n_2 in $\mathbb{N}$ such that if $n \geq n_2$, then $|y_n - y| < \varepsilon/2B$. Let

$n_0 = \max\{n_1, n_2\}$. Then $n \geq n_0$ implies that

$$|x_n y_n - xy| \leq B |y_n - y| + |y| |x_n - x|$$

$$< B \left(\frac{\varepsilon}{2B}\right) + |y| \left[\frac{\varepsilon}{2(|y| + 1)}\right]$$

$$< \varepsilon.$$

The proof of part 3 follows from part 2 by letting $(y_n)_{n \in \mathbb{N}}$ be the constant sequence $(c, c, c, \ldots)$.

For part 4, because of part 2, it suffices to show that $1/y_n \to 1/y$, where $y_n \neq 0$ for all n in $\mathbb{N}$ and $y \neq 0$. First note that

$$\left|\frac{1}{y_n} - \frac{1}{y}\right| = \frac{|y_n - y|}{|y_n| |y|}.$$

By the proof of Lemma 3.2, there is an n_1 in $\mathbb{N}$ such that $|y_n| \geq |y|/2$ for all $n \geq n_1$. Hence,

$$\left|\frac{1}{y_n} - \frac{1}{y}\right| \leq \frac{2 |y_n - y|}{|y|^2} \text{ for all } n \geq n_1.$$

Let $\varepsilon > 0$. Choose n_2 in $\mathbb{N}$ such that if $n \geq n_2$, then $|y_n - y| < \varepsilon |y|^2 /2$. If $n_0 = \max\{n_1, n_2\}$, then $n \geq n_0$ implies that

$$\left|\frac{1}{y_n} - \frac{1}{y}\right| < \frac{2}{|y|^2} \frac{\varepsilon |y|^2}{2} = \varepsilon. \qquad \blacksquare$$

By mathematical induction, parts 1 and 2 of Theorem 3.2 can be extended to a finite number of convergent sequences. In particular, if $(x_n)_{n \in \mathbb{N}}$ converges to x and if k is in $\mathbb{N}$, then

$$\lim_{n \to \infty} (x_n^k) = \lim_{n \to \infty} \underbrace{(x_n \cdot x_n \cdot \cdots \cdot x_n)}_{k \text{ times}} = (\lim_{n \to \infty} x_n)^k = x^k.$$

Example 3.7   We redo Example 3.3 using Theorem 3.2:

$$\lim_{n \to \infty} \frac{2n + 3}{3n + 5} = \lim_{n \to \infty} \frac{2 + (3/n)}{3 + (5/n)} = \frac{2 + 0}{3 + 0} = \frac{2}{3}.$$

Example 3.8   Let $p(t) = a_k t^k + a_{k-1} t^{k-1} + \cdots + a_1 t + a_0$ be a polynomial with real coefficients (the a_i's are in $\mathbb{R}$ and k is in $\mathbb{N} \cup \{0\}$). Let $x_n \to x$, where x is in $\mathbb{R}$. By Theorem 3.2, $p(x_n) \to p(x)$.

Example 3.9   Part 1 of Theorem 3.2 states for sequences that the limit of the sum is the sum of the limits if the limits exist in $\mathbb{R}$. If we let $(x_n)_{n \in \mathbb{N}} = (1, -1, 1, -1, \ldots)$ and $(y_n)_{n \in \mathbb{N}} = (-1, 1, -1, 1, \ldots)$, then $(x_n + y_n)_{n \in \mathbb{N}} = (0, 0, 0, \ldots)$. Thus, $\lim_{n \to \infty} (x_n + y_n) = 0$, but it makes no sense to write $\lim_{n \to \infty} x_n + \lim_{n \to \infty} y_n$.

We end this section with three important theorems whose proofs are straightforward. Their importance will be demonstrated in the following sections. By a closed interval $[a, b]$ we mean that a and b are in $\mathbb{R}$ with $a \leq b$ and $[a, b] = \{y \in \mathbb{R} : a \leq y \leq b\}$.

Theorem 3.3   Let $(x_n)_{n \in \mathbb{N}}$ be a sequence in the closed interval $[a, b]$. Let x be in $\mathbb{R}$ with $x_n \to x$. Then x is in $[a, b]$.

Figure 3.2

Proof Suppose that $x < a$. Let U be a neighborhood of x lying entirely to the left of a (see Figure 3.2). Since $(x_n)_{n \in \mathbb{N}}$ is eventually in U, $x_n < a$ eventually, which is a contradiction. Similarly, if $x > b$, let V be a neighborhood of x lying entirely to the right of b. Then $(x_n)_{n \in \mathbb{N}}$ is eventually in V and hence $x_n > b$ eventually, which is a contradiction. Thus, $a \leq x \leq b$. ■

Note that the argument above implies that if $(x_n)_{n \in \mathbb{N}}$ is a convergent sequence with $x_n \geq 0$ for all n in $\mathbb{N}$, then $\lim_{n \to \infty} x_n \geq 0$. This is the reason why $x \geq 0$ in Exercise 5 in Section 3.1.

Theorem 3.4 (Squeeze Theorem) Let $(x_n)_{n \in \mathbb{N}}$, $(y_n)_{n \in \mathbb{N}}$, and $(z_n)_{n \in \mathbb{N}}$ be sequences in $\mathbb{R}$ with $x_n \leq y_n \leq z_n$ for all n in $\mathbb{N}$. Suppose that $\lim_{n \to \infty} x_n = \lim_{n \to \infty} z_n = x$, where x is in $\mathbb{R}$. Then $\lim_{n \to \infty} y_n = x$.

Proof Let $\varepsilon > 0$. By Proposition 3.1, there is an n_1 in $\mathbb{N}$ such that $|x_n - x| < \varepsilon$ for all $n \geq n_1$ and there is an n_2 in $\mathbb{N}$ such that $|z_n - x| < \varepsilon$ for all $n \geq n_2$. Then $n \geq \max\{n_1, n_2\}$ implies that

$$-\varepsilon < x_n - x \leq y_n - x \leq z_n - x < \varepsilon,$$

and so

$$|y_n - x| < \varepsilon$$

for all $n \geq \max\{n_1, n_2\}$. Thus, $y_n \to x$. ■

Example 3.10 Since $0 \leq |(\sin n)/n| \leq 1/n$ and $1/n \to 0$, it follows from the Squeeze Theorem that $\lim_{n \to \infty} |(\sin n)/n| = 0$. By Exercise 4(b) in Section 3.1, $\lim_{n \to \infty} (\sin n)/n = 0$.

Theorem 3.5 Let x be in $\mathbb{R}$. Then there exist (1) a sequence of rational numbers that converges to x and (2) a sequence of irrational numbers that converges to x.

Proof The proof of part 1 is based on the fact that the rational numbers are dense in $\mathbb{R}$ (Theorem 2.2), and the proof of part 2 is based on the fact that the irrational numbers are dense in $\mathbb{R}$ (Exercise 4 in Section 2.3). We prove part 1, leaving part 2 as an exercise.

Choose a rational x_1 in the open interval $(x - 1, x + 1)$. Next, choose a rational x_2 in the open interval $(x - \frac{1}{2}, x + \frac{1}{2})$. In general, choose a rational x_n in the open interval $(x - (1/n), x + (1/n))$. Such rational numbers exist because $\mathbb{Q}$ is dense in $\mathbb{R}$. Then $(x_n)_{n \in \mathbb{N}}$ is a sequence in $\mathbb{Q}$. We claim that $x_n \to x$. Let $\varepsilon > 0$. Choose n_0 in $\mathbb{N}$ such that $1/n_0 < \varepsilon$. Then $n \geq n_0$ implies that $|x_n - x| < 1/n \leq 1/n_0 < \varepsilon$, and so $x_n \to x$. ■

The technique illustrated in the proof of Theorem 3.5 is a standard way to construct a sequence converging to a given real number. By a slight modification, we could choose all the x_n's smaller than x or all the x_n's larger than x. Thus, we could construct a sequence of rational (or irrational) numbers converging to x from the left or from the right. For example, the sequence

$$\left(1, \frac{14}{10}, \frac{141}{100}, \frac{1414}{1000}, \frac{14,142}{10,000}, \frac{141,421}{100,000}, \cdots\right)$$

converges to $\sqrt{2}$ from the left.

≡ Exercises ≡≡≡≡≡≡≡≡≡≡≡≡≡≡≡≡≡≡≡≡≡≡≡≡≡≡

1. Find the limits of the following sequences.

(a) $\left(\dfrac{n}{n+2} \right)_{n\in\mathbb{N}}$

(d) $\left(\left(3 + \dfrac{1}{n}\right)^2 \right)_{n\in\mathbb{N}}$

(b) $\left(\dfrac{(-1)^n n}{n+2} \right)_{n\in\mathbb{N}}$

(e) $\left(\sqrt{n} - \sqrt{n+1} \right)_{n\in\mathbb{N}}$

(c) $\left(\dfrac{n^2 + 4n}{2n^2 + 5} \right)_{n\in\mathbb{N}}$

(f) $\left(n - \sqrt{n^2 + n} \right)_{n\in\mathbb{N}}$

2. Give examples of two sequences that do not converge but whose

(a) sum converges.
(b) product converges.
(c) quotient converges.

3. Let $(x_n)_{n\in\mathbb{N}}$ and $(y_n)_{n\in\mathbb{N}}$ be two sequences in $\mathbb{R}$ such that $(x_n + y_n)_{n\in\mathbb{N}}$ and $(x_n - y_n)_{n\in\mathbb{N}}$ both converge. Show that $(x_n)_{n\in\mathbb{N}}$ and $(y_n)_{n\in\mathbb{N}}$ both converge.

4. Find a convergent sequence in $(0, 1]$ that does not converge to a point in $(0, 1]$. Note Theorem 3.3.

5. Use the Squeeze Theorem to show that $\lim_{n\to\infty} (\cos n)/n = 0$.

6. Let $(x_n)_{n\in\mathbb{N}}$ be a bounded sequence in $\mathbb{R}$ (not necessarily convergent), and let $(y_n)_{n\in\mathbb{N}}$ be a sequence in $\mathbb{R}$ with $y_n \to 0$. Show that $(x_n y_n)_{n\in\mathbb{N}}$ converges to 0.

7. In reference to Exercise 6, give an example of sequences $(x_n)_{n\in\mathbb{N}}$ and $(y_n)_{n\in\mathbb{N}}$ where

(a) $(x_n)_{n\in\mathbb{N}}$ is bounded and $(y_n)_{n\in\mathbb{N}}$ converges, but $(x_n y_n)_{n\in\mathbb{N}}$ does not converge.
(b) $y_n \to 0$ but $(x_n y_n)_{n\in\mathbb{N}}$ does not converge.

8. Prove part 2 of Theorem 3.5.

3.3 Subsequences

Often, the easiest way to show that a sequence does not have a limit is to use subsequences. The main result of this section is Theorem 3.6.

Definition 3.6   A function h from $\mathbb{N}$ into $\mathbb{N}$ is *strictly increasing* if, whenever m and n are in $\mathbb{N}$ with $m < n$, then $h(m) < h(n)$.

Definition 3.7   Let $(x_n)_{n\in\mathbb{N}}$ and $(y_n)_{n\in\mathbb{N}}$ be sequences in $\mathbb{R}$. Then $(y_n)_{n\in\mathbb{N}}$ is a *subsequence* of $(x_n)_{n\in\mathbb{N}}$ if there is a strictly increasing function h from $\mathbb{N}$ into $\mathbb{N}$ such that $y_n = x_{h(n)}$ for all n in $\mathbb{N}$.

Equivalently, $y = x \circ h$, or the diagram

$$
\begin{array}{ccc}
\mathbb{N} & \xrightarrow{x} & \mathbb{R} \\
h \uparrow & \nearrow & \\
& y & \\
\mathbb{N} & &
\end{array}
$$

commutes.

Notation Given the sequence $(x_n)_{n \in \mathbb{N}} = (x_1, x_2, x_3, \ldots)$, think of the subsequence $(y_n)_{n \in \mathbb{N}}$ as follows:

$y_1 = x_{h(1)}$ where $h(1) \geq 1$,

$y_2 = x_{h(2)}$ where $h(2) > h(1)$,

$y_3 = x_{h(3)}$ where $h(3) > h(2)$, etc.

Thus, a subsequence has the same ordering as the original sequence. It is not just a subset of the original sequence. We usually write $h(k)$ as n_k so that the subsequence $(y_n)_{n \in \mathbb{N}}$ is $(x_{n_k})_{k=1}^{\infty}$ where $n_1 < n_2 < n_3 < \cdots$. Note that $n_k \geq k$ for each k in $\mathbb{N}$.

A little thought should indicate that a subsequence of a subsequence of a sequence is a subsequence of the original sequence.

Example 3.11 A sequence is a subsequence of itself (let h be the identity map on $\mathbb{N}$).

Example 3.12 The sequence $(0, 1, 0, 1, 0, 1, \ldots)$ has as subsequences the constant sequence $(0, 0, 0, \ldots)$ and the constant sequence $(1, 1, 1, \ldots)$. It also has many other subsequences.

Example 3.13 The sequence $(1, 3, 2, 4, 5, 6, \ldots)$ is not a subsequence of $(1, 2, 3, 4, 5, 6, \ldots)$ because the ordering is not the same.

Theorem 3.6 Let $(x_n)_{n \in \mathbb{N}}$ be a sequence in $\mathbb{R}$ that converges to x. Then every subsequence of $(x_n)_{n \in \mathbb{N}}$ converges to x.

Proof Let $(x_{n_k})_{k=1}^{\infty}$ be a subsequence of $(x_n)_{n \in \mathbb{N}}$. We want to show that $x_{n_k} \xrightarrow{k} x$ or, equivalently, that $\lim_{k \to \infty} x_{n_k} = x$. We use Definition 3.4. Let U be a neighborhood of x. Since $x_n \xrightarrow{n} x$, there is an n_0 in $\mathbb{N}$ such that if $n \geq n_0$, then x_n is in U. Let $k \geq n_0$. Since $n_k \geq k \geq n_0$, x_{n_k} is in U. Thus the subsequence $(x_{n_k})_{k=1}^{\infty}$ is eventually in U, and so $(x_{n_k})_{k=1}^{\infty}$ converges to x. ∎

Remark In order to keep straight whether it is the sequence or the subsequence that is converging, we suggest subscripting the arrow with the appropriate letter, as we did in the proof of Theorem 3.6.

Example 3.14 From Theorem 3.6, it follows that if a sequence has two subsequences each with a different limit, then the sequence itself has no limit. This is an easy method for verifying Example 3.4. For instance, the sequence $(0, 1, 0, 1, 0, 1, \ldots)$ has a subsequence with limit 0 and a subsequence with limit 1. Thus, $(0, 1, 0, 1, 0, 1, \ldots)$ cannot converge. Also, the sequence $(1, 2, 1, 3, 1, 4, 1, 5, \ldots)$ has an unbounded subsequence and so this sequence cannot converge.

The following characterization of nonconvergence in terms of subsequences will be used throughout the text. For now, the reader should apply this to the sequences in Example 3.14.

Proposition 3.3 Let $(x_n)_{n\in\mathbb{N}}$ be a sequence in $\mathbb{R}$ and let x be in $\mathbb{R}$. Then $\lim_{n\to\infty} x_n \neq x$ if and only if there exist an $\varepsilon > 0$ and a subsequence $(x_{n_k})_{k=1}^{\infty}$ of $(x_n)_{n\in\mathbb{N}}$ such that $|x_{n_k} - x| \geq \varepsilon$ for all k in $\mathbb{N}$. [The last part can be restated as: there exists a subsequence of $(x_n)_{n\in\mathbb{N}}$ that is bounded away from x.]

Proof Suppose that $\lim_{n\to\infty} x_n \neq x$. From the discussion following Example 3.4, there is an $\varepsilon > 0$ such that for all n_0 in $\mathbb{N}$, there exists an $n \geq n_0$ with $|x_n - x| \geq \varepsilon$. Let $n_0 = 1$. Then there exists an $n_1 \geq n_0 = 1$ with $|x_{n_1} - x| \geq \varepsilon$. Letting $n_0 = n_1 + 1$, there exists an $n_2 \geq n_1 + 1$ with $|x_{n_2} - x| \geq \varepsilon$. Letting $n_0 = n_2 + 1$, there exists an $n_3 \geq n_2 + 1$ with $|x_{n_3} - x| \geq \varepsilon$. Continuing, we obtain a subsequence $(x_{n_k})_{k=1}^{\infty}$ of $(x_n)_{n\in\mathbb{N}}$ with $|x_{n_k} - x| \geq \varepsilon$ for all k in $\mathbb{N}$.

To show the reverse implication, suppose that $(x_n)_{n\in\mathbb{N}}$ has a subsequence $(x_{n_k})_{k=1}^{\infty}$ that is bounded away from x. Then the subsequence $(x_{n_k})_{k=1}^{\infty}$ cannot converge to x. (Why?) So, by Theorem 3.6, $(x_n)_{n\in\mathbb{N}}$ cannot converge to x. ∎

Exercises

1. (a) Give an example of an unbounded sequence with a convergent subsequence.
   (b) Give an example of an unbounded sequence without a convergent subsequence.
   (c) Can you give an example of a bounded sequence that does not have a convergent subsequence?

2. Find the limit of $(x_n)_{n\in\mathbb{N}}$ where

   (a) $x_n = \left(1 + \dfrac{1}{n^2}\right)^{n^2}$

   (b) $x_n = \left(1 + \dfrac{1}{2n}\right)^{n}$. [*Hint:* Recall Example 3.6.]

3. Show that $(\sin n)_{n\in\mathbb{N}}$ does not converge. [*Hint:* Find a subsequence each of whose terms is in $[\frac{1}{2}, 1]$ and another subsequence each of whose terms is in $[-1, -\frac{1}{2}]$.]

4. Let $(x_n)_{n\in\mathbb{N}}$ and $(y_n)_{n\in\mathbb{N}}$ be two sequences in $\mathbb{R}$. Let $(z_n)_{n\in\mathbb{N}}$ be the sequence $(x_1, y_1, x_2, y_2, x_3, y_3, \ldots)$. Show that $(z_n)_{n\in\mathbb{N}}$ has a limit in $\mathbb{R}$ if and only if both $(x_n)_{n\in\mathbb{N}}$ and $(y_n)_{n\in\mathbb{N}}$ have the same limit in $\mathbb{R}$.

5. Let $(x_n)_{n\in\mathbb{N}}$ be an unbounded sequence in $\mathbb{R}$.

   (a) If $(x_n)_{n\in\mathbb{N}}$ is unbounded above, show that $(x_n)_{n\in\mathbb{N}}$ has a subsequence $(x_{n_k})_{k=1}^{\infty}$ with $x_{n_k} > k$ for all k in $\mathbb{N}$.
   (b) If $(x_n)_{n\in\mathbb{N}}$ is unbounded below, show that $(x_n)_{n\in\mathbb{N}}$ has a subsequence $(x_{n_k})_{k=1}^{\infty}$ with $x_{n_k} < -k$ for all k in $\mathbb{N}$.

3.4 Monotone Sequences

In this section we prove three important theorems: the Monotone Convergence Theorem, the Nested Intervals Theorem, and the Monotone Subsequence Theorem. The first of these theorems allows us to show that a certain type of sequence converges without knowing (or guessing) the limit.

The Monotone Convergence Theorem

> ***Definition 3.8*** A sequence $(x_n)_{n \in \mathbb{N}}$ in $\mathbb{R}$ is *monotone increasing* (respectively, *strictly increasing*) if $x_n \leq x_{n+1}$ (respectively, $x_n < x_{n+1}$) for all n in $\mathbb{N}$. A sequence $(x_n)_{n \in \mathbb{N}}$ in $\mathbb{R}$ is *monotone decreasing* (respectively, *strictly decreasing*) if $x_n \geq x_{n+1}$ (respectively, $x_n > x_{n+1}$) for all n in $\mathbb{N}$. A sequence is *monotone* (or *monotonic*) if it is either monotone increasing or monotone decreasing.

For example, $(1/n)_{n \in \mathbb{N}}$ is strictly decreasing, $(n)_{n \in \mathbb{N}}$ is strictly increasing, and a constant sequence is both monotone increasing and monotone decreasing, whereas $((-1)^n)_{n \in \mathbb{N}}$ is not monotone and $(0, 1, 2, 2, \frac{1}{4}, \frac{1}{5}, \frac{1}{6}, \ldots)$ is eventually monotone decreasing. Note that a monotone increasing sequence is bounded below by x_1, and a monotone decreasing sequence is bounded above by x_1.

Recall that a convergent sequence must be bounded (Proposition 3.2) but that a bounded sequence need not converge.

Theorem 3.7 (Monotone Convergence Theorem) A bounded monotone sequence converges.

Proof First, suppose that $(x_n)_{n \in \mathbb{N}}$ is a bounded monotone increasing sequence. By the Completeness Axiom for $\mathbb{R}$ (Section 2.2), $\alpha = \sup\{x_n : n \in \mathbb{N}\}$ is in $\mathbb{R}$. We claim that $(x_n)_{n \in \mathbb{N}}$ converges to α. Let $\varepsilon > 0$. By Proposition 2.5, there is an n_0 in $\mathbb{N}$ such that $\alpha - \varepsilon < x_{n_0}$. Let $n \geq n_0$. Since $(x_n)_{n \in \mathbb{N}}$ is monotone increasing, $\alpha - \varepsilon < x_{n_0} \leq x_n \leq \alpha$. Hence, $|x_n - \alpha| < \varepsilon$ for all $n \geq n_0$, and so $x_n \to \alpha$.

Similarly, if $(x_n)_{n \in \mathbb{N}}$ is a bounded monotone decreasing sequence, then $x_n \to \inf\{x_n : n \in \mathbb{N}\}$, which is a real number by Proposition 2.4. The details of this proof are asked for in Exercise 5. ∎

Prior to this section, to show that a sequence converges, our strategy has been basically to "guess" the limit and then prove that it is the limit, which is what we did in the proof of Theorem 3.7. Now, if we can show that a sequence is bounded and monotone, we know that the sequence converges, without knowing the limit. Of course, we know that the limit is a sup or an inf, but calculating this sup or inf may or may not be easy.

Example 3.15 Let $x_1 = 2$ and $x_{n+1} = \sqrt{6 + x_n}$ for all n in $\mathbb{N}$. The first three terms are $x_1 = 2$, $x_2 = \sqrt{8}$, and $x_3 = \sqrt{6 + \sqrt{8}}$. So it appears that $(x_n)_{n \in \mathbb{N}}$ is strictly increasing. We show this by induction. Clearly, $x_1 < x_2$. Assume that

$x_k < x_{k+1}$. Then

$$x_{k+1} = \sqrt{6 + x_k} < \sqrt{6 + x_{k+1}} = x_{k+2},$$

and so $(x_n)_{n \in \mathbb{N}}$ is strictly increasing.

We next show that $(x_n)_{n \in \mathbb{N}}$ is bounded above by 3, using induction. Clearly, $x_1 < 3$. Assume that $x_k < 3$. Then

$$x_{k+1} = \sqrt{6 + x_k} < \sqrt{6 + 3} = 3.$$

By Theorem 3.7, $(x_n)_{n \in \mathbb{N}}$ converges to a real number, say x.

We next find x using a new technique. Since $(x_n)_{n \in \mathbb{N}}$ converges to x, by the limit theorems in Section 3.2 and by Exercise 5 in Section 3.1,

$$x = \lim_{n \to \infty} x_{n+1} = \lim_{n \to \infty} \sqrt{6 + x_n} = \sqrt{6 + x}.$$

So, $x^2 = 6 + x$ or $x^2 - x - 6 = 0$ or $(x - 3)(x + 2) = 0$. Hence, $x = 3$ or $x = -2$. Since $x_n \geq 2$ for all n, $x = 3$.

Example 3.16   For each n in $\mathbb{N}$, let

$$x_n = \sum_{k=1}^{n} \frac{1}{k^2} = \frac{1}{1^2} + \frac{1}{2^2} + \cdots + \frac{1}{n^2}.$$

Since $x_{n+1} - x_n = 1/(n + 1)^2 > 0$, $(x_n)_{n \in \mathbb{N}}$ is strictly increasing. We next show that $(x_n)_{n \in \mathbb{N}}$ is bounded above by 2. For any n,

$$
\begin{aligned}
x_n &= 1 + \frac{1}{2 \cdot 2} + \frac{1}{3 \cdot 3} + \cdots + \frac{1}{n \cdot n} \\
&< 1 + \frac{1}{1 \cdot 2} + \frac{1}{2 \cdot 3} + \cdots + \frac{1}{(n - 1)n} \\
&= 1 + \left(\frac{1}{1} - \frac{1}{2} \right) + \left(\frac{1}{2} - \frac{1}{3} \right) + \cdots + \left(\frac{1}{n - 1} - \frac{1}{n} \right) \\
&= 1 + 1 - \frac{1}{n} \qquad \text{(by "telescoping")} \\
&= 2 - \frac{1}{n} < 2.
\end{aligned}
$$

Thus, by Theorem 3.7, $(x_n)_{n \in \mathbb{N}}$ converges to a real number x with $1 \leq x \leq 2$. At this point we are not in a position to find x. However, it follows from infinite series that the series $\sum_{k=1}^{\infty} (1/k^2)$ converges and that x is the sum of this series. From other techniques, it is known that $x = \pi^2/6$ (Simmons).

The Nested Intervals Theorem

Definition 3.9   Let A_n be a subset of $\mathbb{R}$ for each n in $\mathbb{N}$. Then $(A_n)_{n \in \mathbb{N}}$ is a *sequence of subsets* of $\mathbb{R}$, which is *nested upward* if $A_n \subset A_{n+1}$ for each n in $\mathbb{N}$ and is *nested downward* if $A_n \supset A_{n+1}$ for each n in $\mathbb{N}$.

For example, $((0, 1/n))_{n \in \mathbb{N}}$ is a nested downward sequence of open intervals whose intersection is empty.

Theorem 3.8   (Nested Intervals Theorem) If $(I_n)_{n \in \mathbb{N}}$ is a nested downward sequence of closed intervals, then

$$\bigcap_{n=1}^{\infty} I_n \neq \emptyset.$$

Proof   For each n in $\mathbb{N}$, let $I_n = [a_n, b_n]$, where a_n and b_n are real numbers with $a_n \leq b_n$ and $[a_1, b_1] \supset [a_2, b_2] \supset [a_3, b_3] \supset \cdots$. See Figure 3.3. Then $(a_n)_{n \in \mathbb{N}}$ is a monotone increasing sequence that is bounded above by b_1. By Theorem 3.7, there is an α in $\mathbb{R}$ such that $a_n \to \alpha$. We claim that α is in $\bigcap_{n=1}^{\infty} I_n$.

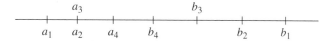

Figure 3.3

Since $\alpha = \sup\{a_n : n \in \mathbb{N}\}$, $a_n \leq \alpha$ for each n in $\mathbb{N}$. To show that α is in $\bigcap_{n=1}^{\infty} I_n$, it suffices to show that $\alpha \leq b_n$ for all n in $\mathbb{N}$. Fix an n in $\mathbb{N}$. We show that b_n is an upper bound of $\{a_k : k \in \mathbb{N}\}$. If $k \leq n$, then $a_k \leq a_n \leq b_n$, while if $k > n$, $a_k \leq b_k \leq b_n$ since $(b_k)_{k \in \mathbb{N}}$ is monotone decreasing. Therefore, b_n is an upper bound of $\{a_k : k \in \mathbb{N}\}$; and since α is the least upper bound of $\{a_k : k \in \mathbb{N}\}$, $\alpha \leq b_n$. ∎

Remark   Let $(a_n)_{n \in \mathbb{N}}$ and $(b_n)_{n \in \mathbb{N}}$ be sequences in $\mathbb{Q}$ with $a_n < \pi < b_n$ for all n in $\mathbb{N}$ and such that $a_n \to \pi$ and $b_n \to \pi$. (Such sequences exist by Theorem 3.5 and the paragraph following Theorem 3.5.) Then $\bigcap_{n=1}^{\infty}[a_n, b_n] = \{\pi\}$. Thus, $\bigcap_{n=1}^{\infty}([a_n, b_n] \cap \mathbb{Q}) = \emptyset$. This is another way to say that $\mathbb{Q}$ is not complete. (We are using the fact that π is irrational; any other irrational number may be substituted for π.)

The Monotone Subsequence Theorem

The following proof is taken from Bartle and Sherbert because the authors could not improve on their elegant argument.

Theorem 3.9   (Monotone Subsequence Theorem) Every sequence in $\mathbb{R}$ has a monotone subsequence.

Proof   Let $(x_n)_{n \in \mathbb{N}}$ be a sequence in $\mathbb{R}$. For the purpose of this proof, we call the mth term x_m a *peak* if $x_m \geq x_n$ for all $n \geq m$. That is, x_m is a peak if x_m is never exceeded by any term that follows it.

    Case 1 $(x_n)_{n \in \mathbb{N}}$ has infinitely many peaks. We pick off the peaks in order. Let m_1 be the smallest positive integer such that x_{m_1} is a peak. Let m_2 be the smallest positive integer larger than m_1 such that x_{m_2} is a peak. Continuing, we obtain the subsequence $(x_{m_k})_{k \in \mathbb{N}}$ of $(x_n)_{n \in \mathbb{N}}$. Since each x_{m_k} is a peak, we have $x_{m_1} \geq x_{m_2} \geq \cdots$ and hence $(x_{m_k})_{k \in \mathbb{N}}$ is monotone decreasing.

Case 2 $(x_n)_{n\in\mathbb{N}}$ has only a finite number of peaks. Let the peaks be (in order) $x_{m_1}, x_{m_2}, x_{m_3}, \ldots, x_{m_r}$. We go out beyond the last peak. Let $n_1 = m_r + 1$ (if the number of peaks is 0, let $n_1 = 1$.) Since x_{n_1} is not a peak, there is an $n_2 > n_1$ such that $x_{n_1} < x_{n_2}$. Since x_{n_2} is not a peak, there is an $n_3 > n_2$ such that $x_{n_2} < x_{n_3}$. Continuing, we obtain a strictly increasing subsequence $(x_{n_k})_{k\in\mathbb{N}}$ of $(x_n)_{n\in\mathbb{N}}$.   ∎

Exercises

1. Let $x_1 = 1$ and $x_{n+1} = \frac{1}{4}(2x_n + 3)$ for all n in $\mathbb{N}$. Show that $(x_n)_{n\in\mathbb{N}}$ converges, and find the limit.

2. Let $x_1 = 3$ and $x_{n+1} = 2 - (1/x_n)$ for all n in $\mathbb{N}$. Show that $(x_n)_{n\in\mathbb{N}}$ converges, and find the limit.

3. Let $x_1 = \sqrt{2}$ and $x_{n+1} = \sqrt{2 + x_n}$ for all n in $\mathbb{N}$. Show that $(x_n)_{n\in\mathbb{N}}$ converges, and find the limit.

4. For each n in $\mathbb{N}$, let

$$x_n = 1 + \frac{1}{2} + \frac{1}{3} + \cdots + \frac{1}{n} = \sum_{k=1}^{n} \frac{1}{k}.$$

   Show that $(x_n)_{n\in\mathbb{N}}$ is monotone but does not converge.

5. Complete the proof of Theorem 3.7.

6. Find a nested downward sequence of half-open intervals whose intersection is empty. [Half-open intervals are of the form $(a, b] = \{x \in \mathbb{R} : a < x \le b\}$ or of the form $[a, b) = \{x \in \mathbb{R} : a \le x < b\}$, where a and b are in $\mathbb{R}$.]

7. Find a nested downward sequence of open rays (respectively, closed rays) whose intersection is empty. [Open rays are of the form $(a, \infty) = \{x \in \mathbb{R} : x > a\}$ or $(-\infty, a) = \{x \in \mathbb{R} : x < a\}$, and closed rays are of the form $[a, \infty) = \{x \in \mathbb{R} : a \le x < \infty\}$ or $(-\infty, a] = \{x \in \mathbb{R} : -\infty < x \le a\}$, where a is in $\mathbb{R}$. Also, $(-\infty, \infty)$ is an open ray.]

8. In the notation of Theorem 3.8 and its proof, show that

   (a) $\beta = \lim_{n\to\infty} b_n$ is in $\bigcap_{n=1}^{\infty} I_n$;

   (b) $[\alpha, \beta] = \bigcap_{n=1}^{\infty} I_n$;

   (c) if $\lim_{n\to\infty} (b_n - a_n) = 0$, then $\bigcap_{n=1}^{\infty} I_n$ is a single point.

9. Let A be a nonempty bounded subset of $\mathbb{R}$ with $\alpha = \sup A$ and $\beta = \inf A$. Show that A contains a monotone increasing sequence with limit α and that A contains a monotone decreasing sequence with limit β. [*Hint:* By Theorem 3.9 it suffices to find a sequence in A with limit α. Consider two cases: α in A and α in $\mathbb{R} \setminus A$.]

3.5 Bolzano-Weierstrass Theorems

In this section we prove the Bolzano-Weierstrass Theorem for sequences and the Bolzano-Weierstrass Theorem for sets. The Bolzano-Weierstrass Theorem for sequences will be used in Sections 3.6 and 3.8 and also in Sections 4.4 and 4.5. The terminology set forth in our discussion of the Bolzano-Weierstrass Theorem for sets will be used in Section 4.3.

Bolzano-Weierstrass Theorem for Sequences

For Theorem 3.10 below, we give two proofs. The first proof is an easy application of the results obtained in the previous section, but this proof cannot be generalized to $\mathbb{R}^n$ because it depends on the order axioms for $\mathbb{R}$. The second proof, although more complicated, is extendable to $\mathbb{R}^n$.

Theorem 3.10   (Bolzano-Weierstrass Theorem for sequences) A bounded sequence in $\mathbb{R}$ has a convergent subsequence (that is, a subsequence that converges to a real number).

Proof 1   Let $(x_n)_{n \in \mathbb{N}}$ be a bounded sequence in $\mathbb{R}$. By the Monotone Subsequence Theorem, $(x_n)_{n \in \mathbb{N}}$ has a monotone subsequence. Since $(x_n)_{n \in \mathbb{N}}$ is bounded, this monotone subsequence is bounded. By the Monotone Convergence Theorem, this subsequence converges to a real number.

Proof 2   Let $(x_n)_{n \in \mathbb{N}}$ be a bounded sequence in $\mathbb{R}$.

Case 1 $\{x_n\}_{n \in \mathbb{N}}$ is finite. Then some value, say a, must be repeated an infinite number of times. Then $(x_n)_{n \in \mathbb{N}}$ has a constant subsequence (each term is a) that, of course, converges to a. (To construct this subsequence, let n_1 be the least positive integer such that $x_{n_1} = a$. Let n_2 be the least positive integer greater than n_1 such that $x_{n_2} = a$. Continue by induction.)

Case 2 $\{x_n\}_{n \in \mathbb{N}}$ is infinite. Since $(x_n)_{n \in \mathbb{N}}$ is bounded, there are real numbers a and b such that $a \le x_n \le b$ for all n in $\mathbb{N}$. Consider the intervals $[a, (a+b)/2]$ and $[(a+b)/2, b]$. An infinite number of the x_n's must be in at least one of these two subintervals, because otherwise $\{x_n\}_{n \in \mathbb{N}}$ would be finite. Let I_1 be one of these two subintervals containing an infinite number of the x_n's and choose an n_1 in $\mathbb{N}$ with x_{n_1} in I_1. Also note that the length of I_1 is $(b-a)/2$. See Figure 3.4.

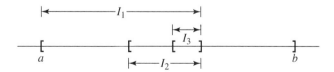

Figure 3.4

We next bisect I_1 into two closed subintervals of length $(b-a)/2^2$. By an argument analogous to obtaining I_1, let I_2 be one of these subintervals containing an infinite number of the x_n's. Choose a positive integer $n_2 > n_1$ with x_{n_2} in I_2.

Continuing by induction, we obtain a nested downward sequence $(I_n)_{n \in \mathbb{N}}$ of closed intervals with the length of $I_n = (b-a)/2^n$ for each n and a subsequence $(x_{n_k})_{k=1}^{\infty}$ of $(x_n)_{n \in \mathbb{N}}$ with x_{n_k} in I_k for each k in $\mathbb{N}$. By the Nested Intervals Theorem, $\bigcap_{n=1}^{\infty} I_n \neq \emptyset$. (By Exercise 8 in Section 3.4, $\bigcap_{n=1}^{\infty} I_n$ is a single point.)

Let x be in $\bigcap_{n=1}^{\infty} I_n$. We claim that $(x_{n_k})_{k=1}^{\infty}$ converges to x. Let $\varepsilon > 0$. Choose n_0 in $\mathbb{N}$ with $(b-a)/2^{n_0} < \varepsilon$. For $k \geq n_0$, $n_k \geq k \geq n_0$, and so $x_{n_k} \in I_k \subset I_{n_0}$. Thus,

$$\left| x_{n_k} - x \right| \leq \text{ length of } I_{n_0} = (b-a)/2^{n_0} < \varepsilon \text{ for all } k \geq n_0$$

and, therefore, $x_{n_k} \underset{k}{\to} x$. ∎

Remark   Let $(x_n)_{n \in \mathbb{N}}$ be a sequence in $\mathbb{Q}$ with $x_n \underset{n}{\to} \pi$. Then $(x_n)_{n \in \mathbb{N}}$ is a bounded sequence in $\mathbb{Q}$. By Theorem 3.6, no subsequence of $(x_n)_{n \in \mathbb{N}}$ can converge to a point of $\mathbb{Q}$. This is another way to say that $\mathbb{Q}$ is not complete.

Bolzano-Weierstrass Theorem for Sets

Recall that a neighborhood of a real number x is an open interval centered at x.

Definition 3.10   Let A be a subset of $\mathbb{R}$ and let x be in $\mathbb{R}$. Then x is an *accumulation point of A* if every neighborhood of x contains a point of A different from x. If x is in A and x is not an accumulation point of A, then x is an *isolated point of A*.

Thus, x in $\mathbb{R}$ is an accumulation point of A if and only if for every neighborhood U of x, we have $(U \setminus \{x\}) \cap A \neq \emptyset$.

Some books use the terms "limit point" and "cluster point" for accumulation point. Informally, accumulation points of a set are "close" to the set. Also note that x is an isolated point of A if and only if there exists a neighborhood V of x such that $V \cap A = \{x\}$.

Example 3.17   The set of accumulation points of $[0, 1] = $ the set of accumulation points of $(0, 1) = [0, 1]$.

Example 3.18   The set of accumulation points of $\mathbb{Q} = \mathbb{R}$ since $\mathbb{Q}$ is dense in $\mathbb{R}$.

Example 3.19   A finite set has no accumulation points.

Example 3.20 $\mathbb{N}$ has no accumulation points. Thus, every point of $\mathbb{N}$ is isolated.

Example 3.21   The set of accumulation points of $A = [0, 1] \cup \{2\}$ is $[0, 1]$, while 2 is an isolated point of A.

Example 3.22 The set of accumulation points of $\{1/n : n \in \mathbb{N}\} = \{0\}$ and each $1/n$ is an isolated point of $\{1/n : n \in \mathbb{N}\}$.

Proposition 3.4 Let A be a subset of $\mathbb{R}$ and let x be in $\mathbb{R}$. Then x is an accumulation point of A if and only if every neighborhood of x contains infinitely many points of A.

Proof Let x be an accumulation point of A and let U be a neighborhood of x. We want to show that $U \cap A$ is infinite. Suppose U contains only a finite number of points of A, say $(U \setminus \{x\}) \cap A = \{x_1, x_2, \ldots, x_n\}$. Let $\varepsilon = \min\{|x - x_1|, |x - x_2|, \ldots, |x - x_n|\}$. Then $\varepsilon > 0$ and the neighborhood $(x - (\varepsilon/2), x + (\varepsilon/2))$ of x contains no points of A distinct from x, which is a contradiction. Thus, $U \cap A$ is infinite.

The other implication is obvious. ∎

The following theorem is another way to say that $\mathbb{R}$ is complete. In Exercise 8, we ask the reader to show that the corresponding result in $\mathbb{Q}$ is false.

Theorem 3.11 (Bolzano-Weierstrass Theorem for sets) Every bounded infinite subset of $\mathbb{R}$ has an accumulation point in $\mathbb{R}$.

Proof Let A be a bounded infinite subset of $\mathbb{R}$. Since A is infinite, A contains a sequence $(x_n)_{n \in \mathbb{N}}$ of distinct points by Theorem 2.3. Since A is bounded, $(x_n)_{n \in \mathbb{N}}$ is a bounded sequence. By the Bolzano-Weierstrass Theorem for sequences, there exist a subsequence $(x_{n_k})_{k=1}^{\infty}$ of $(x_n)_{n \in \mathbb{N}}$ and an x in $\mathbb{R}$ such that $(x_{n_k})_{k=1}^{\infty}$ converges to x.

We claim that x is an accumulation point of A. Let U be any neighborhood of x. Since $x_{n_k} \xrightarrow{k} x$, the sequence $(x_{n_k})_{k=1}^{\infty}$ is eventually in U and thus U contains an infinite number of points of A. ∎

Exercises

1. Let $0 < x_n < 7$ for each n in $\mathbb{N}$. By Theorem 3.10, $(x_n)_{n \in \mathbb{N}}$ has a convergent subsequence. In what interval will the limit of this subsequence lie?

2. Let $(x_n)_{n \in \mathbb{N}}$ be a bounded sequence of integers. Show that $(x_n)_{n \in \mathbb{N}}$ has a subsequence that eventually is constant.

3. Show that a bounded sequence in $\mathbb{R}$ that does not converge has more than one subsequential limit. That is, show that a nonconvergent bounded sequence has two subsequences each with a different limit.

4. What is the set of accumulation points of the irrational numbers?

5. Give an example of a bounded set of real numbers with exactly three accumulation points.

6. Let $A \subset \mathbb{R}$ and let x be in $\mathbb{R}$. Show that x is an accumulation point of A if and only if there exists a sequence of distinct points in A that converges to x.

7. Let $A \subset \mathbb{R}$ and let x be an isolated point of A. What are the only types of sequences in A that can converge to x?

8. Give an example of a bounded infinite subset of $\mathbb{Q}$ that has no accumulation point in $\mathbb{Q}$. (This is another way to say that $\mathbb{Q}$ is not complete.)

3.6 Cauchy Sequences

We show that convergent sequences and Cauchy sequences are equivalent in $\mathbb{R}$ but not in $\mathbb{Q}$. This is another way to state that $\mathbb{R}$ is complete and $\mathbb{Q}$ is not complete.

> **Definition 3.11** A sequence $(x_n)_{n \in \mathbb{N}}$ in $\mathbb{R}$ is a *Cauchy sequence* (or simply *Cauchy*) if for every $\varepsilon > 0$ there exists an n_0 in $\mathbb{N}$ such that if $n \geq n_0$ and $m \geq n_0$, then $|x_n - x_m| < \varepsilon$.

Although the Cauchy sequences in $\mathbb{R}$ are precisely the convergent sequences in $\mathbb{R}$ (Theorem 3.12 below), the difference between Definition 3.11 and Proposition 3.1 is that you do not need to know the limit point of the sequence. This will be illustrated in Example 3.23 and the exercises.

Proposition 3.5 A convergent sequence is Cauchy.

Proof Let $(x_n)_{n \in \mathbb{N}}$ converge to x and let $\varepsilon > 0$. By Proposition 3.1, there is an n_0 in $\mathbb{N}$ such that if $n \geq n_0$, then $|x_n - x| < \varepsilon/2$. Let $n \geq n_0$ and $m \geq n_0$. Then

$$\begin{aligned} |x_n - x_m| &= |(x_n - x) + (x - x_m)| \\ &\leq |x_n - x| + |x - x_m| \\ &< \frac{\varepsilon}{2} + \frac{\varepsilon}{2} = \varepsilon. \end{aligned}$$

Therefore, $(x_n)_{n \in \mathbb{N}}$ is Cauchy. ∎

Lemma 3.3 A Cauchy sequence is bounded.

Proof Let $(x_n)_{n \in \mathbb{N}}$ be a Cauchy sequence and let $\varepsilon = 1$. By Definition 3.11, there exists an n_0 in $\mathbb{N}$ such that if $n \geq n_0$, then $|x_n - x_{n_0}| < 1$ (we are using $m = n_0$). From Corollary 2.1, $|x_n| < 1 + |x_{n_0}|$ for all $n \geq n_0$. Letting $B = \max\{|x_1|, |x_2|, \ldots, |x_{n_0 - 1}|, 1 + |x_{n_0}|\}$, we have $|x_n| \leq B$ for all n in $\mathbb{N}$. ∎

Theorem 3.12 A sequence in $\mathbb{R}$ is Cauchy if and only if the sequence converges.

Proof That convergence implies Cauchy is Proposition 3.5.

Let $(x_n)_{n \in \mathbb{N}}$ be a Cauchy sequence in $\mathbb{R}$. By Lemma 3.3, $(x_n)_{n \in \mathbb{N}}$ is bounded. By the Bolzano-Weierstrass Theorem for sequences, there exist a subsequence $(x_{n_k})_{k=1}^{\infty}$ of $(x_n)_{n \in \mathbb{N}}$ and an x in $\mathbb{R}$ with $x_{n_k} \underset{k}{\to} x$. We show that $x_n \underset{n}{\to} x$.

Let $\varepsilon > 0$. Since $(x_n)_{n \in \mathbb{N}}$ is Cauchy, there exists an N_1 in $\mathbb{N}$ such that if $n \geq N_1$ and $m \geq N_1$, then $|x_n - x_m| < \varepsilon/2$. Since $x_{n_k} \underset{k}{\to} x$, there exists an N_2 in $\mathbb{N}$ such that if $k \geq N_2$, then $\left| x_{n_k} - x \right| < \varepsilon/2$. Let $n_0 = \max\{N_1, N_2\}$. Fix a $k \geq n_0$ (so $n_k \geq k \geq n_0$). Then for $n \geq n_0$,

$$|x_n - x| \leq \left| x_n - x_{n_k} \right| + \left| x_{n_k} - x \right|$$
$$< \frac{\varepsilon}{2} + \frac{\varepsilon}{2} = \varepsilon.$$

Therefore, $x_n \underset{n}{\to} x$.   ∎

Remark If each x_n is in $\mathbb{Q}$ and $(x_n)_{n \in \mathbb{N}}$ is Cauchy, then $(x_n)_{n \in \mathbb{N}}$ need not converge to a point of $\mathbb{Q}$. This is another way to say that $\mathbb{Q}$ is not complete. For example, let $(x_n)_{n \in \mathbb{N}}$ be a sequence in $\mathbb{Q}$ with $x_n \to \pi$. Then $(x_n)_{n \in \mathbb{N}}$ is a Cauchy sequence in $\mathbb{Q}$ that has no limit in $\mathbb{Q}$.

Remark In Definition 3.11, we consider the nth and mth terms of the sequence. It is not sufficient to consider only the nth and $(n+1)$st terms. For example, let $x_n = \sqrt{n}$ for each n in $\mathbb{N}$. Then $(x_n)_{n \in \mathbb{N}}$ is not bounded and hence not Cauchy. However,

$$|x_{n+1} - x_n| = \frac{1}{\sqrt{n+1} + \sqrt{n}} < \frac{1}{2\sqrt{n}} \underset{n}{\to} 0.$$

So, given $\varepsilon > 0$, there exists an n_0 in $\mathbb{N}$ with $|x_{n+1} - x_n| < \varepsilon$ for all $n \geq n_0$.

As with monotone sequences, showing that a sequence is Cauchy provides us with another way of showing that a sequence converges without knowing the limit. Given $(x_n)_{n \in \mathbb{N}}$, if we can show that $(x_n)_{n \in \mathbb{N}}$ is Cauchy, then we know that $(x_n)_{n \in \mathbb{N}}$ converges. We may or may not be able to find the limit.

Example 3.23 Let $(x_n)_{n \in \mathbb{N}}$ be a sequence in $\mathbb{R}$. Let $0 < r < 1$ and suppose that $|x_{n+1} - x_n| < r^n$ for all n in $\mathbb{N}$. Then $(x_n)_{n \in \mathbb{N}}$ converges.

We show that $(x_n)_{n \in \mathbb{N}}$ is Cauchy. First suppose that $m > n$. Then

$$|x_m - x_n| \leq |x_m - x_{m-1}| + |x_{m-1} - x_{m-2}| + \cdots + |x_{n+1} - x_n|$$
$$< r^{m-1} + r^{m-2} + \cdots + r^n$$
$$= r^n(1 + r + r^2 + \cdots + r^{m-1-n})$$
$$= r^n \cdot \frac{1 - r^{m-n}}{1 - r}$$
$$< \frac{r^n}{1 - r} \underset{n}{\to} 0 \qquad \text{(by Example 3.5).}$$

Let $\varepsilon > 0$. Then there is an n_0 in $\mathbb{N}$ such that $r^n/(1 - r) < \varepsilon$ for all $n \geq n_0$. So, if $n \geq n_0$ and $m \geq n_0$, then $|x_n - x_m| < \varepsilon$; hence $(x_n)_{n \in \mathbb{N}}$ is Cauchy.

In the example above, we could have proceeded as follows:

$$r^n(1 + r + r^2 + \cdots + r^{m-1-n}) < r^n(1 + r + r^2 + r^3 + \cdots)$$
$$= r^n \cdot \frac{1}{1 - r} \qquad \text{(sum of a geometric series).}$$

≡ Exercises ≡

1. (a) Find a Cauchy sequence in $(0, 1)$ that does not converge to a point of $(0, 1)$.

   (b) Show that a Cauchy sequence in $[0, 1]$ must converge to a point of $[0, 1]$. (See Theorem 3.3.)

2. For each n in $\mathbb{N}$, let

   $$x_n = 1 + \frac{1}{2} + \frac{1}{3} + \cdots + \frac{1}{n} = \sum_{k=1}^{n} \frac{1}{k}.$$

   (a) Show that $(x_n)_{n \in \mathbb{N}}$ is not Cauchy.

   (b) Show that $\lim_{n \to \infty} |x_{n+1} - x_n| = 0$.

3. Let x_n be in $\mathbb{Z}$ (the integers) for each n in $\mathbb{N}$. Show that if $(x_n)_{n \in \mathbb{N}}$ is Cauchy, then $(x_n)_{n \in \mathbb{N}}$ is eventually constant. Conclude that a sequence in $\mathbb{Z}$ converges if and only if it is eventually constant.

4. Let $a < b$. Let $x_1 = a$, $x_2 = b$, and

   $$x_{n+2} = \frac{x_{n+1} + x_n}{2} \text{ for } n \geq 1.$$

   Follow these steps to show that $(x_n)_{n \in \mathbb{N}}$ is Cauchy.

   (a) Draw a picture and let $L = b - a$.

   (b) Use induction to show that $|x_{n+1} - x_n| = L/2^{n-1}$ for each n.

   (c) Proceed as in Example 3.23 to show for $m > n$ that

   $$|x_m - x_n| < \frac{L}{2^{n-2}} \xrightarrow[n]{} 0.$$

   (d) Conclude that $(x_n)_{n \in \mathbb{N}}$ is Cauchy.

5. Refer to Exercise 4. By Theorem 3.12, we have that $(x_n)_{n \in \mathbb{N}}$ converges to some x in $\mathbb{R}$. Follow these steps to find x.

   (a) From your picture (or by induction) and from Exercise 4(b), show that

   $$x_{2k+2} - x_{2k+1} = \frac{L}{2^{2k}}.$$

   (b) Use induction to show that

   $$x_{2n+1} = a + \frac{L}{2} + \frac{L}{2^3} + \cdots + \frac{L}{2^{2n-1}} \text{ for } n \geq 1.$$

   (c) $x = \lim_{n \to \infty} x_{2n+1}$ and

   $$x_{2n+1} = a + \frac{L}{2} \left[\frac{1 - \left(\frac{1}{4}\right)^n}{1 - \frac{1}{4}} \right].$$

6. Let $(x_n)_{n \in \mathbb{N}}$ be a sequence in $\mathbb{R}$. Let $0 < r < 1$ and suppose that

   $$|x_{n+2} - x_{n+1}| \leq r |x_{n+1} - x_n| \text{ for } n \geq 1.$$

   Show that $(x_n)_{n \in \mathbb{N}}$ is Cauchy. [*Hint:* First show that $|x_{n+2} - x_{n+1}| \leq r^n |x_2 - x_1|$; then proceed as in Example 3.23.]

7. Let $x_1 > 0$ and let $x_{n+1} = 1/(2 + x_n)$ for $n \geq 1$.

   (a) Use Exercise 6 to show that $(x_n)_{n \in \mathbb{N}}$ is Cauchy.

   (b) Find $\lim_{n \to \infty} x_n$.

3.7 Limits at Infinity

Recall Definition 2.2 for the ordering on $\mathbb{R}^{\#} = \mathbb{R} \cup \{\pm\infty\}$. In this section we extend the notion of the limit of a sequence to points in $\mathbb{R}^{\#}$, thus allowing a sequence to have limit $\pm\infty$. However, the term "convergent" is still reserved for those sequences whose limits are real numbers. In this section we examine which theorems of Sections 3.1 through 3.4 have analogous results in $\mathbb{R}^{\#}$.

Basic Results

Definition 3.12   Let α and β be in $\mathbb{R}$. The open ray

$$(\alpha, \infty) = \{x \in \mathbb{R} : x > \alpha\}$$

is a *neighborhood of* ∞, while the open ray

$$(-\infty, \beta) = \{x \in \mathbb{R} : x < \beta\}$$

is a *neighborhood of* $-\infty$.

The following definition extends Definition 3.4 to $\mathbb{R}^{\#}$.

Definition 3.13   Let $(x_n)_{n\in\mathbb{N}}$ be a sequence in $\mathbb{R}$ and let x be in $\mathbb{R}^{\#}$. Then $(x_n)_{n\in\mathbb{N}}$ has *limit* x, denoted by $\lim_{n\to\infty} x_n = x$ or $\lim_n x_n = x$ or $x_n \to x$ as $n \to \infty$, if for every neighborhood U of x, the sequence $(x_n)_{n\in\mathbb{N}}$ is eventually in U.

Note that if $(x_n)_{n\in\mathbb{N}}$ has limit $\pm\infty$, then $(x_n)_{n\in\mathbb{N}}$ is a divergent sequence.

Paraphrasing Definitions 3.12 and 3.13, for $(x_n)_{n\in\mathbb{N}}$ a sequence in $\mathbb{R}$, we have that

$x_n \to \infty$ if and only if for all $\alpha > 0$, there exists an n_0 in $\mathbb{N}$ such that if $n \geq n_0$, then $x_n > \alpha$;

and

$x_n \to -\infty$ if and only if for all $\beta < 0$, there exists an n_0 in $\mathbb{N}$ such that if $n \geq n_0$, then $x_n < \beta$.

Example 3.24 $\lim_{n\to\infty} (n^2 - 3n + 2) = \infty$. Let $\alpha > 0$. Choose $n_0 = ?$ Let $n \geq n_0$. For $n \geq 3$,

$$n^2 - 3n + 2 > n^2 - 3n = n(n - 3) \geq n - 3 \geq n_0 - 3.$$

So choose n_0 in $\mathbb{N}$ such that $n_0 > \alpha + 3$. Then $n^2 - 3n + 2 > n_0 - 3 > \alpha$.

Example 3.25 $\lim_{n\to\infty} (-n) = -\infty$. Let $\beta < 0$. Choose n_0 in $\mathbb{N}$ such that $n_0 > -\beta$. Let $n \geq n_0$. Then $-n \leq -n_0 < \beta$.

Theorem 3.13 Limits of sequences are unique.

Proof If x is in $\mathbb{R}$, then $U = (x - 1, x + 1)$ and $V = (x + 1, \infty)$ are disjoint neighborhoods of x and ∞ while U and $W = (-\infty, x - 1)$ are disjoint neighborhoods of x and $-\infty$. Clearly, $(-\infty, 0)$ and $(0, \infty)$ are disjoint neighborhoods of $-\infty$ and ∞. Thus, distinct points in $\mathbb{R}^{\#}$ can be separated by disjoint neighborhoods. The remainder of the proof is completely similar to the proof of Theorem 3.1. ■

When the limits are $\pm\infty$, an exact analogue of Theorem 3.2 cannot be obtained since indeterminate forms can arise: for example, $\infty - \infty$, $0 \cdot \infty$, or ∞/∞. However, we have the following.

Theorem 3.14 Let $(x_n)_{n\in\mathbb{N}}$, $(y_n)_{n\in\mathbb{N}}$, and $(z_n)_{n\in\mathbb{N}}$ be sequences in $\mathbb{R}$. Let x be in $\mathbb{R}^{\#}$ and suppose that $x_n \to x$, $y_n \to \infty$, and $z_n \to -\infty$.

1. If $-\infty < x \le \infty$, then $x_n + y_n \to \infty$.

2. If $-\infty \le x < \infty$, then $x_n + z_n \to -\infty$.

3. If $0 < x \le \infty$, then $x_n y_n \to \infty$ and $x_n z_n \to -\infty$.

4. If $-\infty \le x < 0$, then $x_n y_n \to -\infty$ and $x_n z_n \to \infty$.

5. If x is in $\mathbb{R}$, then $\dfrac{x_n}{y_n} \to 0$ and $\dfrac{x_n}{z_n} \to 0$.

Proof Note that the conditions in the different parts of the theorem do not allow the limits to be indeterminate forms such as $\infty - \infty$, $0 \cdot \infty$, etc.

For part 3, first suppose that $0 < x < \infty$. Then, as in Lemma 3.2, there is an n_1 in $\mathbb{N}$ such that $x_n > x/2$ for all $n \ge n_1$. Let $\alpha > 0$. Since $y_n \to \infty$, there exists an n_2 in $\mathbb{N}$ with $y_n > 2\alpha/x$ for all $n \ge n_2$. Let $n_0 = \max\{n_1, n_2\}$. Then $n \ge n_0$ implies that $x_n y_n > (x/2)y_n > (x/2)(2\alpha/x) = \alpha$. Thus, $x_n y_n \to \infty$. Let $\beta < 0$. Since $z_n \to -\infty$, there exists an n_3 in $\mathbb{N}$ with $z_n < 2\beta/x$ for all $n \ge n_3$. Then, since $z_n < 0$, $n \ge \max\{n_1, n_3\}$ implies that $x_n z_n < (x/2)z_n < (x/2)(2\beta/x) = \beta$. Thus, $x_n z_n \to -\infty$.

Next, let $x = \infty$. Then there is an N_1 in $\mathbb{N}$ such that $x_n > 1$ for all $n \ge N_1$. Since eventually $y_n > 0$ and $z_n < 0$, we have that eventually $x_n y_n > y_n$ and $x_n z_n < z_n$. It follows that $x_n y_n \to \infty$ and $x_n z_n \to -\infty$.

For part 5, because $y_n \to \infty$ and $z_n \to -\infty$, we assume that no y_n or z_n is zero. Since x is in $\mathbb{R}$, by Proposition 3.2, there is a $B > 0$ such that $|x_n| \le B$ for all n in $\mathbb{N}$. Let $\varepsilon > 0$. Since $y_n \to \infty$, choose n_0 in $\mathbb{N}$ such that $y_n > B/\varepsilon$ for all $n \ge n_0$. Then $n \ge n_0$ implies that $|x_n/y_n| \le B/y_n < B(\varepsilon/B) = \varepsilon$. Since $z_n \to -\infty$, choose n_1 in $\mathbb{N}$ such that $z_n < -(B/\varepsilon)$ for all $n \ge n_1$. Then $n \ge n_1$ implies that $|x_n/z_n| \le B/|z_n| < B(\varepsilon/B) = \varepsilon$.

The rest of the proof is left as an exercise. ■

Subsequences

We first obtain the analogue of Theorem 3.6.

Theorem 3.15 Let $(x_n)_{n\in\mathbb{N}}$ be a sequence in $\mathbb{R}$; let x be in $\mathbb{R}^{\#}$ with $x_n \to x$. Then every subsequence of $(x_n)_{n\in\mathbb{N}}$ has limit x.

Proof In the proof of Theorem 3.6, replace Definition 3.4 with Definition 3.13. ■

To obtain the analogue of Proposition 3.3, we first negate the statements preceding Example 3.24. For $(x_n)_{n \in \mathbb{N}}$ a sequence in $\mathbb{R}$,

$$\lim_{n \to \infty} x_n \neq \infty \Longleftrightarrow \text{ there exists an } \alpha > 0 \text{ such that for all } n_0 \text{ in } \mathbb{N}, \text{ there}$$
$$\text{exists an } n \geq n_0 \text{ with } x_n \leq \alpha \Longleftrightarrow \exists\, \alpha > 0 \text{ such that } \forall\, n_0 \in \mathbb{N}, \exists\, n \geq$$
$$n_0 \text{ with } x_n \leq \alpha$$

and

$$\lim_{n \to \infty} x_n \neq -\infty \Longleftrightarrow \text{ there exists a } \beta < 0 \text{ such that for all } n_0 \text{ in } \mathbb{N}, \text{ there}$$
$$\text{exists an } n \geq n_0 \text{ with } x_n \geq \beta \Longleftrightarrow \exists\, \beta < 0 \text{ such that } \forall\, n_0 \in \mathbb{N}, \exists\, n \geq$$
$$n_0 \text{ with } x_n \geq \beta.$$

Proposition 3.6   Let $(x_n)_{n \in \mathbb{N}}$ be a sequence in $\mathbb{R}$. Then

1. $\displaystyle\lim_{n \to \infty} x_n \neq \infty$ if and only if $(x_n)_{n \in \mathbb{N}}$ has a subsequence that is bounded above;

2. $\displaystyle\lim_{n \to \infty} x_n \neq -\infty$ if and only if $(x_n)_{n \in \mathbb{N}}$ has a subsequence that is bounded below.

Proof   We prove part 1, leaving the proof of part 2 to the reader.

Suppose that $\displaystyle\lim_{n \to \infty} x_n \neq \infty$. Then there exists an $\alpha > 0$ such that for all n_0 in $\mathbb{N}$, there exists an $n \geq n_0$ with $x_n \leq \alpha$.

Let $n_0 = 1$. Then there exists an $n_1 \geq n_0 = 1$ with $x_{n_1} \leq \alpha$.

Letting $n_0 = n_1 + 1$, there exists an $n_2 \geq n_1 + 1$ with $x_{n_2} \leq \alpha$. Continuing, we obtain a subsequence $(x_{n_k})_{k=1}^{\infty}$ of $(x_n)_{n \in \mathbb{N}}$ with $x_{n_k} \leq \alpha$ for all k in $\mathbb{N}$.

For the reverse implication, if $(x_n)_{n \in \mathbb{N}}$ has a subsequence $(x_{n_k})_{k=1}^{\infty}$ that is bounded above, say by α, then $(x_{n_k})_{k=1}^{\infty}$ is never in (α, ∞) and hence $(x_{n_k})_{k=1}^{\infty}$ cannot have limit ∞. So, by Theorem 3.15, $(x_n)_{n \in \mathbb{N}}$ cannot have limit ∞.   ∎

As an example, the sequence $(1, 2, 1, 3, 1, 4, 1, 5, \ldots)$ has no limit in $\mathbb{R}^{\#}$ but has the constant sequence $(1, 1, 1, 1, \ldots)$ as a subsequence. Also, the sequence $(1, -2, 3, -4, 5, -6, \ldots)$ has no limit in $\mathbb{R}^{\#}$ but has the subsequence $(-2, -4, -6, \ldots)$, which is bounded above, and the subsequence $(1, 3, 5, \ldots)$, which is bounded below.

Monotone Sequences

The following theorem is the analogue of Theorem 3.7.

Theorem 3.16   If $(x_n)_{n \in \mathbb{N}}$ is a monotone sequence in $\mathbb{R}$, then $(x_n)_{n \in \mathbb{N}}$ has a limit in $\mathbb{R}^{\#}$.

Proof   If $(x_n)_{n \in \mathbb{N}}$ is bounded, then $(x_n)_{n \in \mathbb{N}}$ has a limit in $\mathbb{R}$ by Theorem 3.7. Suppose $(x_n)_{n \in \mathbb{N}}$ is monotone increasing but unbounded. Then $(x_n)_{n \in \mathbb{N}}$ is unbounded above. We claim that $(x_n)_{n \in \mathbb{N}}$ has limit ∞, which is also $\sup\{x_n : n \in \mathbb{N}\}$. Let $\alpha > 0$. Since $(x_n)_{n \in \mathbb{N}}$ is unbounded above, there is an n_0 in $\mathbb{N}$ such that $x_{n_0} > \alpha$. Since $(x_n)_{n \in \mathbb{N}}$ is monotone increasing, $\alpha < x_{n_0} \leq x_n$ for all $n \geq n_0$. Hence, $x_n \to \infty$.

Similarly, if $(x_n)_{n \in \mathbb{N}}$ is monotone decreasing but unbounded, then $x_n \to \inf\{x_n : n \in \mathbb{N}\} = -\infty$.   ∎

The reader should observe that the sequence in Exercise 4 in Section 3.4 has limit ∞.

Remark   If $(x_n)_{n \in \mathbb{N}}$ is a sequence in $\mathbb{R}$, then $(x_n)_{n \in \mathbb{N}}$ has a monotone subsequence by Theorem 3.9. By Theorem 3.16, this monotone subsequence has a limit in $\mathbb{R}^{\#}$; and by Theorem 3.7, this limit is in $\mathbb{R}$ if $(x_n)_{n \in \mathbb{N}}$ is bounded. These observations are deduced at the beginning of the next section from a different point of view.

Exercises

1. Use the method of Examples 3.24 and 3.25 to establish the following limits.

   (a) $\lim\limits_{n \to \infty} \sqrt{n} = \infty$

   (b) $\lim\limits_{n \to \infty} \dfrac{\sqrt{n^2 + 1}}{\sqrt{n}} = \infty$

   (c) $\lim\limits_{n \to \infty} (n^2 - 6n + 1) = \infty$

   (d) $\lim\limits_{n \to \infty} (n - 6\sqrt{n}) = \infty$

   (e) $\lim\limits_{n \to \infty} -\sqrt{n - 7} = -\infty$

   (f) $\lim\limits_{n \to \infty} (-n + \sin n) = -\infty$

2. Let $(x_n)_{n \in \mathbb{N}}$ be a sequence in $\mathbb{R}$.

   (a) Show that $x_n \to \infty$ if and only if $-x_n \to -\infty$.

   (b) If $x_n > 0$ for all n in $\mathbb{N}$, show that

   $$\lim\limits_{n \to \infty} x_n = 0 \text{ if and only if } \lim\limits_{n \to \infty} \frac{1}{x_n} = \infty.$$

3. Let $(x_n)_{n \in \mathbb{N}}$ be a sequence in $\mathbb{R}$.

   (a) If $x_n < 0$ for all n in $\mathbb{N}$, show that $x_n \to -\infty$ if and only if $|x_n| \to \infty$.

   (b) Show, by example, that if $|x_n| \to \infty$, then $(x_n)_{n \in \mathbb{N}}$ need not have a limit in $\mathbb{R}^{\#}$.

4. Finish the proof of Theorem 3.14.

5. Let $(x_n)_{n \in \mathbb{N}}$ and $(y_n)_{n \in \mathbb{N}}$ be sequences in $\mathbb{R}$ such that $x_n \to \pm\infty$ and $(x_n y_n)_{n \in \mathbb{N}}$ converges. Show that $y_n \to 0$.

6. Let $(x_n)_{n \in \mathbb{N}}$ and $(y_n)_{n \in \mathbb{N}}$ be sequences of positive real numbers and suppose that $x_n/y_n \to L$ where $0 < L < \infty$. Show that $x_n \to \infty$ if and only if $y_n \to \infty$. [*Hint:* Note that, eventually, $\frac{1}{2}L < x_n/y_n < \frac{3}{2}L$.]

7. Let $(x_n)_{n \in \mathbb{N}}$ and $(y_n)_{n \in \mathbb{N}}$ be sequences of positive real numbers and suppose that $x_n/y_n \to 0$. Show that

   (a) if $x_n \to \infty$, then $y_n \to \infty$;

   (b) if $(y_n)_{n \in \mathbb{N}}$ is bounded, then $x_n \to 0$.

8. Let $(x_n)_{n \in \mathbb{N}}$ be a sequence in $\mathbb{R}$. Show that

   (a) if $(x_n)_{n \in \mathbb{N}}$ is unbounded above, then $(x_n)_{n \in \mathbb{N}}$ has a subsequence with limit ∞;

   (b) if $(x_n)_{n \in \mathbb{N}}$ is unbounded below, then $(x_n)_{n \in \mathbb{N}}$ has a subsequence with limit $-\infty$. [*Hint:* See Exercise 5 in Section 3.3.]

9. Let $x_n > 0$ for each n in $\mathbb{N}$. Show that if $(x_n)_{n \in \mathbb{N}}$ does not have limit ∞, then $(x_n)_{n \in \mathbb{N}}$ has a convergent subsequence.

10. Let A be a nonempty subset of $\mathbb{R}$ with $\alpha = \sup A$ and $\beta = \inf A$. Show that A contains a monotone increasing sequence with limit α and a monotone decreasing sequence with limit β. [*Hint:* By Exercise 9 in Section 3.4, you need only consider the cases $\alpha = \infty$ and $\beta = -\infty$.]

11. For the purpose of this exercise, we allow our sequence to have values in $\mathbb{R}^{\#}$; that is, $\pm\infty$ are permissible terms of the sequence. The term "monotone" is defined as in Definition 3.8, using the ordering on $\mathbb{R}^{\#}$.

    Let A be a nonempty subset of $\mathbb{R}^{\#}$ with $\alpha = \sup A$ and $\beta = \inf A$. Show that A contains a monotone increasing sequence with limit α and that A contains a monotone decreasing sequence with limit β. [*Note:* If $\alpha = -\infty$, then $A = \{-\infty\}$, so that the constant sequence with each term $-\infty$ has limit α.]

3.8 Limit Superior and Limit Inferior

In this section we utilize results from Sections 3.5 and 3.7. First observe that if $(x_n)_{n \in \mathbb{N}}$ is a sequence in $\mathbb{R}$, then $(x_n)_{n \in \mathbb{N}}$ has a subsequence that has a limit in $\mathbb{R}^{\#} = \mathbb{R} \cup \{\pm\infty\}$. If $(x_n)_{n \in \mathbb{N}}$ is bounded, then $(x_n)_{n \in \mathbb{N}}$ has a subsequence that converges to a real number by the Bolzano-Weierstrass Theorem for sequences, whereas if $(x_n)_{n \in \mathbb{N}}$ is unbounded, then $(x_n)_{n \in \mathbb{N}}$ has a subsequence with limit $\pm\infty$ by Exercise 8 in Section 3.7.

> **Definition 3.14**   Let $(x_n)_{n \in \mathbb{N}}$ be a sequence in $\mathbb{R}$. Let
>
> $$E = \{x \in \mathbb{R}^{\#} : x_{n_k} \underset{k}{\to} x \text{ for some subsequence } (x_{n_k})_{k=1}^{\infty} \text{ of } (x_n)_{n \in \mathbb{N}}\}.$$
>
> Thus, E consists of all subsequential limits of $(x_n)_{n \in \mathbb{N}}$ in the extended reals and E is never empty. The *limit superior* of $(x_n)_{n \in \mathbb{N}}$, denoted by $\limsup_{n \to \infty} x_n$, and the *limit inferior* of $(x_n)_{n \in \mathbb{N}}$, denoted by $\liminf_{n \to \infty} x_n$, are given by
>
> $$\limsup_{n \to \infty} x_n = \sup E$$
>
> and
>
> $$\liminf_{n \to \infty} x_n = \inf E.$$

We will usually omit $n \to \infty$ in the notation above. Clearly, $\liminf x_n \leq \limsup x_n$. For the remainder of this section, E will denote the set given in Definition 3.14. If $E = \{-\infty\}$, then $\sup E = -\infty$. If $E \neq \{-\infty\}$, then $\sup E$ is a real number if E is bounded above and $+\infty$ if E is not bounded above. If

$E = \{\infty\}$, then $\inf E = \infty$. If $E \neq \{\infty\}$, then $\inf E$ is a real number if E is bounded below and $-\infty$ if E is not bounded below.

Example 3.26   Let $(x_n)_{n \in \mathbb{N}} = (1, \frac{1}{2}, 2, \frac{1}{3}, 3, \frac{1}{4}, 4, \frac{1}{5}, \ldots)$. Then $\liminf x_n = 0$ and $\limsup x_n = \infty$.

Example 3.27   Let $(x_n)_{n \in \mathbb{N}} = (1, -2, 3, -4, 5, -6, \ldots)$. Then $\liminf x_n = -\infty$ and $\limsup x_n = \infty$.

Example 3.28   Let $(a_n)_{n \in \mathbb{N}} = (0, 1, 0, 1, 0, 1, \ldots)$ and $(b_n)_{n \in \mathbb{N}} = (1, 0, 1, 0, 1, 0, \ldots)$. Then $\liminf a_n = \liminf b_n = 0$ and $\limsup a_n = \limsup b_n = 1$. The sequence $(a_n + b_n)_{n \in \mathbb{N}} = (1, 1, 1, 1, \ldots)$. Thus,

$$\liminf a_n + \liminf b_n = 0 < 1 = \liminf(a_n + b_n)$$

and

$$\limsup(a_n + b_n) = 1 < 2 = \limsup a_n + \limsup b_n.$$

Proposition 3.7   Let $(x_n)_{n \in \mathbb{N}}$ be a sequence in $\mathbb{R}$ and let x be in $\mathbb{R}^{\#}$. Then $x_n \to x$ if and only if $\limsup x_n = \liminf x_n = x$.

Proof   If $x_n \to x$, then all subsequences of $(x_n)_{n \in \mathbb{N}}$ have limit x by Theorem 3.15. Thus, $E = \{x\}$ and $\sup E = \inf E = x$.

Suppose that $(x_n)_{n \in \mathbb{N}}$ does not have limit x. We will show that $E \neq \{x\}$ and so $\sup E$ and $\inf E$ cannot both be x.

If x is in $\mathbb{R}$, then, by Proposition 3.3, there exist an $\varepsilon > 0$ and a subsequence $(x_{n_k})_{k=1}^{\infty}$ of $(x_n)_{n \in \mathbb{N}}$ such that $|x_{n_k} - x| \geq \varepsilon$ for all k in $\mathbb{N}$. By the first paragraph of this section, the sequence $(x_{n_k})_{k=1}^{\infty}$ will have a subsequence $(y_n)_{n \in \mathbb{N}}$ that has a limit y in $\mathbb{R}^{\#}$. Then $y \neq x$ since $|y_n - x| \geq \varepsilon$ for all n in $\mathbb{N}$, and y is in E since $(y_n)_{n \in \mathbb{N}}$ is a subsequence of $(x_n)_{n \in \mathbb{N}}$.

If $x = \infty$, then, by Proposition 3.6, $(x_n)_{n \in \mathbb{N}}$ has a subsequence that is bounded above. This subsequence is either unbounded below or bounded below. Hence this subsequence has a subsequence with limit $-\infty$ (Exercise 8 in Section 3.7) or with limit a real number (Theorem 3.10). In either case, $E \neq \{\infty\}$.

The case $x = -\infty$ is handled similarly.   ■

It is intuitively clear that $\limsup x_n$ is the largest subsequential limit of $(x_n)_{n \in \mathbb{N}}$, and that $\liminf x_n$ is the smallest subsequential limit of $(x_n)_{n \in \mathbb{N}}$. For the sake of clarity, we will postpone justifying this statement until Proposition 3.9 at the end of this section; for now we will use this as a given fact.

Remark   The advantage of the limit superior and limit inferior is that, given a sequence $(x_n)_{n \in \mathbb{N}}$ in $\mathbb{R}$, $\limsup x_n$ and $\liminf x_n$ always exist in $\mathbb{R}^{\#}$, whereas $(x_n)_{n \in \mathbb{N}}$ may or may not have a limit in $\mathbb{R}^{\#}$, and in the latter case it makes no sense to write $\lim x_n$. A new way to show that $(x_n)_{n \in \mathbb{N}}$ has a limit is to show that $\limsup x_n = \liminf x_n$. This is illustrated in Exercise 5.

For the purpose of the next proposition and some of the exercises, we extend addition to $\mathbb{R}^{\#}$ by:

$$x + \infty = \infty + x = \infty \quad \text{for} \ -\infty < x \leq \infty$$

and

$$x + (-\infty) = (-\infty) + x = -\infty \quad \text{for} \ -\infty \leq x < \infty$$

and we do not define

$$-\infty + \infty \text{ or } \infty + (-\infty).$$

Proposition 3.8   Let $(a_n)_{n\in\mathbb{N}}$ and $(b_n)_{n\in\mathbb{N}}$ be sequences in $\mathbb{R}$. Then

$$\limsup(a_n + b_n) \le \limsup a_n + \limsup b_n$$

whenever the right side is defined.

Proof   Let $\alpha = \limsup a_n$, $\beta = \limsup b_n$, and $\gamma = \limsup(a_n + b_n)$. By the right side being defined, we mean that $\alpha + \beta$ is not of the form $\infty + (-\infty)$ or $-\infty + \infty$. We want to show that $\gamma \le \alpha + \beta$. We may assume that $\gamma \ne -\infty$ and $\alpha + \beta \ne \infty$.

Since γ is a subsequential limit of $(a_n + b_n)_{n\in\mathbb{N}}$, there is a subsequence $(a_{n_k} + b_{n_k})_{k=1}^{\infty}$ of $(a_n + b_n)_{n\in\mathbb{N}}$ with $a_{n_k} + b_{n_k} \underset{k}{\to} \gamma$. The sequence $(a_{n_k})_{k=1}^{\infty}$ has a subsequence $(a_{n_{k_j}})_{j=1}^{\infty}$ with limit u in $\mathbb{R}^{\#}$ where $u < \infty$ since $\alpha \ne \infty$, and the sequence $(b_{n_k})_{k=1}^{\infty}$ has a subsequence $(b_{n_{k_j}})_{j=1}^{\infty}$ with limit v in $\mathbb{R}^{\#}$ where $v < \infty$ since $\beta \ne \infty$. Then $(a_{n_{k_j}} + b_{n_{k_j}}) \underset{j}{\to} \gamma$ and $(a_{n_{k_j}} + b_{n_{k_j}}) \underset{j}{\to} u + v$. By the uniqueness of limits, $\gamma = u + v$. Since α and β are the largest subsequential limits of $(a_n)_{n\in\mathbb{N}}$ and $(b_n)_{n\in\mathbb{N}}$, respectively, $u \le \alpha$ and $v \le \beta$. Therefore, $\gamma = u + v \le \alpha + \beta$.   ∎

Proposition 3.9   Let $(x_n)_{n\in\mathbb{N}}$ be a sequence in $\mathbb{R}$. Then $\limsup x_n$ and $\liminf x_n$ are both in E.

Proof   We show that $\limsup x_n$ is in E, the proof for $\liminf x_n$ being similar. Let $\alpha = \limsup x_n$. If $\alpha = -\infty$, then $x_n \underset{n}{\to} -\infty$ by Proposition 3.7 and so $E = \{-\infty\}$. Assume that $\alpha > -\infty$.

By Exercise 11 in Section 3.7, there exists a monotone increasing sequence $(y_n)_{n\in\mathbb{N}}$ in E with $y_n \to \alpha$. Thus, each y_i is a limit of a subsequence of $(x_n)_{n\in\mathbb{N}}$. For each i in $\mathbb{N}$, let $(x_{i,n_k})_{k=1}^{\infty}$ be a subsequence of $(x_n)_{n\in\mathbb{N}}$ with limit y_i. To construct a subsequence of $(x_n)_{n\in\mathbb{N}}$ with limit α, we basically use a diagonalization argument on the following array:

$$
\begin{array}{ccccc}
x_{1,n_1} & x_{1,n_2} & x_{1,n_3} & \cdots & \to & y_1 \\
x_{2,n_1} & x_{2,n_2} & x_{2,n_3} & \cdots & \to & y_2 \\
\vdots & & & & & \vdots \\
x_{i,n_1} & x_{i,n_2} & x_{i,n_3} & \cdots & \to & y_i \\
\vdots & & & & & \vdots \\
& & & & & \downarrow \\
& & & & & \alpha
\end{array}
$$

Note that the n_k's in each row are not necessarily the same. So we cannot go directly down the diagonal.

If each y_i is real, we can choose $x_{1,n_{k_1}}$ from row 1, $x_{2,n_{k_2}}$ from row 2, $x_{3,n_{k_3}}$ from row 3, etc., with $n_{k_i} < n_{k_j}$ for $i < j$ (we move to the right as we go down the rows) and $\left| x_{i,n_{k_i}} - y_i \right| < 1/i$ for each i in $\mathbb{N}$. Then $(x_{i,n_{k_i}})_{i=1}^{\infty}$ is a subsequence of $(x_n)_{n\in\mathbb{N}}$ with limit α.

Suppose $y_m = \infty$ for some m in $\mathbb{N}$. Then $y_i = \infty$ for all $i \ge m$. Choose $x_{i,n_{k_i}}$ from each row i such that $n_{k_i} < n_{k_j}$ for $i < j$ with $x_{i,n_{k_i}} > i$ for all $i \ge m$. Then $(x_{i,n_{k_i}})_{i=1}^{\infty}$ is a subsequence of $(x_n)_{n\in\mathbb{N}}$ with limit $\alpha = \infty$.   ∎

Exercises

1. Find $\limsup x_n$ and $\liminf x_n$ if x_n is given by
   (a) $(-1)^n$,
   (b) $(-1)^n n$,
   (c) $(-1)^n \left(1 + \dfrac{1}{n}\right)$,
   (d) $\cos(n\pi)$,
   (e) $\left(1 + \dfrac{1}{n}\right) \sin \dfrac{n\pi}{2}$,
   (f) e^{-n}.

2. Give an example of a sequence $(x_n)_{n\in\mathbb{N}}$ in $\mathbb{R}$ with $\liminf x_n = \infty$.

3. Let $x_n \le y_n$ for each n in $\mathbb{N}$. Show that $\liminf x_n \le \liminf y_n$ and $\limsup x_n \le \limsup y_n$.

4. Let $(a_n)_{n\in\mathbb{N}}$ and $(b_n)_{n\in\mathbb{N}}$ be sequences in $\mathbb{R}$. Show that
   $$\liminf a_n + \liminf b_n \le \liminf(a_n + b_n)$$
   whenever the left side is defined.

5. Let $(x_n)_{n\in\mathbb{N}}$ be a sequence in $\mathbb{R}$. For each n in $\mathbb{N}$, let
   $$y_n = \frac{x_1 + x_2 + \cdots + x_n}{n}.$$
   Show that if $(x_n)_{n\in\mathbb{N}}$ converges to x, then $(y_n)_{n\in\mathbb{N}}$ converges to x. [*Hint:* Write
   $$y_n - x = \frac{x_1 + x_2 + \cdots + x_n}{n} - \frac{nx}{n}$$
   $$= \frac{(x_1 - x) + \cdots + (x_{n_0} - x)}{n} + \frac{(x_{n_0+1} - x) + \cdots + (x_n - x)}{n}$$
   and, given $\varepsilon > 0$ and suitably choosing n_0,
   $$|y_n - x| \le \frac{|x_1 - x| + \cdots + |x_{n_0} - x|}{n} + \left(\frac{n - n_0}{n}\right)\varepsilon.$$
   Now take the limit superior of both sides of this inequality.]

6. Refer to Exercise 5. Show that there are nonconvergent sequences $(x_n)_{n\in\mathbb{N}}$ for which $(y_n)_{n\in\mathbb{N}}$ converges. [*Hint:* Consider $(0, 1, 0, 1, 0, 1, \ldots).$]

4 Continuity

One of the concepts essential to the study of functions is that of continuity. In this chapter we explore the concept of continuity and derive some of its consequences. We begin with basic definitions and examples in Section 4.1. Section 4.2 ties together continuity and sequences. Section 4.3 introduces the formal notion of the "limit" of a function at a point. Section 4.4 deals with the consequences of continuity and Section 4.5 explores "uniform continuity." In the last section we consider "monotone functions" and types of discontinuities.

4.1 Continuous Functions

Basic Results and Examples

Definition 4.1 Let $D \subset \mathbb{R}$, let $f : D \to \mathbb{R}$ be a function, and let c be in D. Then f is *continuous at c* if for every neighborhood V of $f(c)$ there is a neighborhood U of c such that if x is in $U \cap D$, then $f(x)$ is in V. If f is not continuous at c, then f is *discontinuous at c.*

Using the definition of neighborhood (Definition 3.3), the definition of continuity at a point can be restated as follows: $f : D \to \mathbb{R}$ is continuous at a point c in D if and only if for every $\varepsilon > 0$ there is a $\delta > 0$ such that if x is in D and $|x - c| < \delta$, then $|f(x) - f(c)| < \varepsilon$. See Figure 4.1. By examining graphically $f(x) = 1/x$ for $x > 0$, the reader should easily see that δ usually depends on both c and ε.

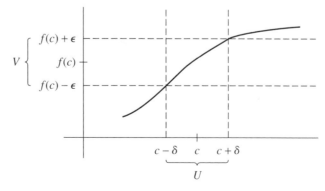

Figure 4.1

Example 4.1   Let $f : \mathbb{R} \to \mathbb{R}$ be given by $f(x) = x^2$. We show that f is continuous at 3. Let $\varepsilon > 0$. We want a $\delta > 0$ such that if $|x - 3| < \delta$, then $|f(x) - f(3)| < \varepsilon$. First note that

$$|f(x) - f(3)| = \left|x^2 - 9\right| = |x - 3|\,|x + 3|.$$

If $|x - 3| < 1$, then $2 < x < 4$ and so $|x + 3| < 7$. Thus, $|x - 3| < 1$ implies that $|f(x) - f(3)| < 7\,|x - 3|$. Let $\delta = \min\{1, \varepsilon/7\}$. If $|x - 3| < \delta$, then

$$|f(x) - f(3)| < 7\delta \le 7\frac{\varepsilon}{7} = \varepsilon$$

and so f is continuous at 3.

The same argument can be modified to show that $f(x) = x^2$ is continuous at any real number.

Example 4.2   Let f be a function with domain $(0, 1) \cup \{2\}$. We claim that f is continuous at 2. To see this, note that we can always produce a neighborhood around 2 containing no other element of the domain of f except 2 (for example, choose $\delta = \frac{1}{2}$). In this case, for all $\varepsilon > 0$, $|f(x) - f(2)| = |f(2) - f(2)| = 0 < \varepsilon$ since 2 is the only element in the domain of f and in this δ-neighborhood of 2.

A similar proof shows that a function is always continuous at an isolated point of its domain.

Definition 4.2   Let $f : D \to \mathbb{R}$ and let $A \subset D$. If f is continuous at each point of A, then f is *continuous on A*.

When we simply say that f is continuous, we mean that f is continuous on its domain.

Example 4.3   Define $f : (0, 1) \cup (3, 4) \to \mathbb{R}$ by

$$f(x) = \begin{cases} x & \text{if } 0 < x < 1 \\ 2 & \text{if } 3 < x < 4. \end{cases}$$

This function is continuous on its domain. Let $0 < c < 1$. Given $0 < \varepsilon < 1$, choose $\delta = \varepsilon$. Then, if x is in the domain of f and $|x - c| < \delta$, $|f(x) - f(c)| = |x - c| < \delta = \varepsilon$ and so f is continuous on $(0, 1)$.

If $3 < c < 4$ and $\varepsilon > 0$, choose $\delta = 1$. Then, if x is in the domain of f and in this δ-neighborhood of c, $|f(x) - f(c)| = |2 - 2| = 0 < \varepsilon$. So f is continuous on $(3, 4)$.

Example 4.4   Define a function f by

$$f(x) = \begin{cases} x & \text{if } 0 < x < 1 \\ 5 & \text{if } x = 1 \\ 2 & \text{if } 1 < x < 4. \end{cases}$$

We claim that f is not continuous at $x = 1$. To see this, take $\varepsilon = 1$. We show that there is no $\delta > 0$ with the property that if x is in $(0, 4)$ and x is in a δ-neighborhood of 1, then $f(x)$ is within 1 $(= \varepsilon)$ of 5 $[= f(c)]$. Now,

$|f(x) - 5| < 1$ implies $4 < f(x) < 6$. If $0 < x < 1$, then $f(x) = x$ and so $0 < f(x) < 1$. If $1 < x < 4$, then $f(x) = 2$. So there is no neighborhood of 1 on which $f(x)$ is within 1 of 5. Hence, f is not continuous at $x = 1$.

The reader should notice the similarity between the following proof and the solution to Exercise 5 in Section 3.1.

Proposition 4.1   Let $f : [0, \infty) \to \mathbb{R}$ be defined by $f(x) = \sqrt{x}$. Then f is continuous on $[0, \infty)$.

Proof   Let c be a nonnegative real number. We must show that for any ε-neighborhood of $f(c)$, there is a δ-neighborhood of c such that if $x \geq 0$ and x is in that δ-neighborhood, then $f(x)$ is in the ε-neighborhood of $f(c)$. Let $\varepsilon > 0$.

If $c = 0$, choose $\delta = \varepsilon^2$. If $0 \leq x < \delta$, then $|f(x) - f(0)| = \sqrt{x} < \sqrt{\delta} = \varepsilon$.

Assume now that $c > 0$. Then $\sqrt{x} + \sqrt{c} \geq \sqrt{c}$. Choose $\delta = \varepsilon \sqrt{c}$. If $x \geq 0$ and $|x - c| < \delta$, then

$$|f(x) - f(c)| = \left|\sqrt{x} - \sqrt{c}\right| = \frac{|x - c|}{\sqrt{x} + \sqrt{c}} < \frac{1}{\sqrt{c}} \delta = \frac{1}{\sqrt{c}} \cdot \varepsilon \sqrt{c} = \varepsilon.$$

So f is continuous at c. Since c is an arbitrary member of the domain, f is continuous on $[0, \infty)$.   ∎

Combinations of Continuous Functions

We show below that the usual combinations of continuous functions are again continuous. For the quotient part of Proposition 4.2, we need the following lemma.

Lemma 4.1   Let $f : D \to \mathbb{R}$ be continuous at a point c in D. If $f(c) \neq 0$, then there exist an $\varepsilon > 0$ and a neighborhood U of c such that $|f(x)| > \varepsilon$ for all x in $U \cap D$.

Proof   Let $\varepsilon = |f(c)|/2 > 0$. Since f is continuous at c, corresponding to this value of ε there is a $\delta > 0$ such that if x is in D and $|x - c| < \delta$, then $|f(x) - f(c)| < \varepsilon$, and so $f(c) - \varepsilon < f(x) < f(c) + \varepsilon$. If $f(c) > 0$, then $f(x) > f(c) - \varepsilon = f(c)/2$. Similarly, if $f(c) < 0$, then $f(x) < f(c)/2$. Hence, $|f(x)| > |f(c)|/2 = \varepsilon$ for all x in $U \cap D$ where $U = (c - \delta, c + \delta)$.   ∎

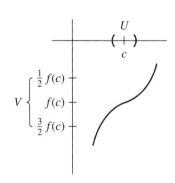

Figure 4.2

The hypothesis of Lemma 4.1 implies that f is bounded away from 0 on $U \cap D$ for some neighborhood U of c. In particular, if $f(c) > 0$, then f is positive on $U \cap D$, and if $f(c) < 0$, then f is negative on $U \cap D$. Figure 4.2 pictorially describes the situation in Lemma 4.1 for $f(c) < 0$. Letting $V = (\frac{3}{2} f(c), \frac{1}{2} f(c))$ be our neighborhood of $f(c)$, by Definition 4.1, there must be a neighborhood U of c such that $f(x)$ is in V for all x in $U \cap D$. In particular, $f(x) < \frac{1}{2} f(c)$ for all x in $U \cap D$.

Proposition 4.2   Let $f : D \to \mathbb{R}$ and $g : D \to \mathbb{R}$ be functions. Let c be an element of D and assume that f and g are continuous at c. Then the functions

$f + g, f - g, \; f \cdot g$, and kf (for any k in $\mathbb{R}$) are continuous at c. Also, if $g(c) \neq 0$, then f/g is continuous at c.

Proof   We first show that $f + g$ is continuous at c. Let $\varepsilon > 0$. Since f is continuous at c, there is a $\delta_1 > 0$ such that if x is in D and $|x - c| < \delta_1$, then $|f(x) - f(c)| < \varepsilon/2$. Since g is continuous at c, there is a $\delta_2 > 0$ such that if x is in D and $|x - c| < \delta_2$, then $|g(x) - g(c)| < \varepsilon/2$. Let $\delta = \min\{\delta_1, \delta_2\} > 0$. (This is a standard way to force both conditions to occur simultaneously.) Let x be in D with $|x - c| < \delta$. Then

$$|(f + g)(x) - (f + g)(c)| \leq |f(x) - f(c)| + |g(x) - g(c)|$$
$$< \frac{\varepsilon}{2} + \frac{\varepsilon}{2} = \varepsilon.$$

Hence, $f + g$ is continuous at c.

We next show that if $g(c) \neq 0$, then $1/g$ is continuous at c. This result, combined with the continuity of the product of two continuous functions, will show that f/g is continuous at c.

Let $\varepsilon > 0$. By Lemma 4.1, there exist a $\delta_1 > 0$ and a $B > 0$ such that if x is in D and $|x - c| < \delta_1$, then $|g(x)| \geq B$. Also, by the continuity of g at c, there is a $\delta_2 > 0$ such that if x is in D and $|x - c| < \delta_2$, then $|g(x) - g(c)| < B^2\varepsilon$.

Let $\delta = \min\{\delta_1, \delta_2\} > 0$. Then for x in D with $|x - c| < \delta$, we have

$$\left| \frac{1}{g(x)} - \frac{1}{g(c)} \right| = \left| \frac{g(c) - g(x)}{g(x) \cdot g(c)} \right|$$
$$= \frac{|g(x) - g(c)|}{|g(x)| \, |g(c)|}$$
$$\leq \frac{|g(x) - g(c)|}{B^2}$$
$$< \frac{1}{B^2} B^2 \varepsilon$$
$$= \varepsilon.$$

Hence, $1/g$ is continuous at c.

The rest of the proof is left as an exercise.   ■

We next show that the composition of continuous functions is continuous.

Proposition 4.3   Let A and B be subsets of $\mathbb{R}$. Let $f : A \to \mathbb{R}$ and $g : B \to \mathbb{R}$ be functions with $f(A) \subset B$. Assume that f is continuous at a point c in A and that g is continuous at $f(c)$ in B. Then the composite function $g \circ f : A \to \mathbb{R}$ is continuous at c.

Proof   Let $\varepsilon > 0$. Since g is continuous at $f(c)$, there is a $\delta_1 > 0$ such that if y is in B and $|y - f(c)| < \delta_1$, then $|g(y) - g(f(c))| < \varepsilon$. Since f is continuous at c, there is a $\delta > 0$ such that if x is in A and $|x - c| < \delta$, then $|f(x) - f(c)| < \delta_1$.

Thus, if x is in A and $|x - c| < \delta$, then

$$|(g \circ f)(x) - (g \circ f)(c)| = |g(f(x)) - g(f(c))| < \varepsilon,$$

and so $g \circ f$ is continuous at c.   ■

≡ Exercises ≡

1. Prove that $f(x) = |x|$ is continuous on $\mathbb{R}$.

2. Show that $f(x) = 1/x$ is continuous at any $c \neq 0$. [*Hint:* Choose your δ so that you stay away from 0.]

3. Show that
$$f(x) = \begin{cases} x \sin \dfrac{1}{x} & \text{if } x \neq 0 \\ 0 & \text{if } x = 0 \end{cases}$$
is continuous on $\mathbb{R}$.

4. Let $f : D \to \mathbb{R}$ be continuous at c in D. Let h be in $\mathbb{R}$ with $f(c) < h$ [respectively, $f(c) > h$]. Show that there is a neighborhood U of c such that if x is in $U \cap D$, then $f(x) < h$ [respectively, $f(x) > h$].

5. Let $f : D \to \mathbb{R}$ be a function. Assume that f is continuous at a point c in D. Show that there is a neighborhood U of c such that f is bounded on $U \cap D$.

6. Let $f : D \to \mathbb{R}$ be a continuous function whose range is a subset of the nonnegative real numbers. Prove that $g(x) = \sqrt{f(x)}$ is continuous on D.

7. Show that any polynomial is continuous on $\mathbb{R}$.

8. Recall that a rational function is the quotient of two polynomials. Show that any rational function is continuous except at a finite set of real numbers (those where the denominator is 0).

9. Finish the proof of Proposition 4.2.

10. Let $f, g : D \to \mathbb{R}$ be continuous functions. Define $h(x) = \max\{f(x), g(x)\}$ and $k(x) = \min\{f(x), g(x)\}$ for x in D. Prove that both h and k are continuous functions. [*Hint:* Observe that for any real numbers a and b, $\max\{a, b\} = \frac{1}{2}(a + b + |a - b|)$ and $\min\{a, b\} = \frac{1}{2}(a + b - |a - b|)$.]

4.2 Continuity and Sequences

We first establish a sequential criterion for continuity in Theorem 4.1. We then use this criterion to examine continuity or discontinuity of various functions.

Theorem 4.1   Let $f : D \to \mathbb{R}$ and let c be in D. Then f is continuous at c if and only if whenever $(x_n)_{n \in \mathbb{N}}$ is a sequence in D that converges to c, then $(f(x_n))_{n \in \mathbb{N}}$ converges to $f(c)$.

Proof   Suppose that f is continuous at c and let $(x_n)_{n \in \mathbb{N}}$ be a sequence in D with $x_n \to c$. We want to show that $f(x_n) \to f(c)$.

Argument 1   Let $\varepsilon > 0$. Since f is continuous at c, there is a $\delta > 0$ such that if x is in D and $|x - c| < \delta$, then $|f(x) - f(c)| < \varepsilon$. Since $x_n \to c$, there is an n_0 in $\mathbb{N}$ such that if $n \geq n_0$, then $|x_n - c| < \delta$. Therefore, $n \geq n_0$ implies that $|f(x_n) - f(c)| < \varepsilon$ and so $f(x_n) \to f(c)$.

Argument 2 Let V be a neighborhood of $f(c)$. Since f is continuous at c, there is a neighborhood U of c such that if x is in $U \cap D$, then $f(x)$ is in V. Since $x_n \to c$, $(x_n)_{n \in \mathbb{N}}$ is eventually in U and hence $(f(x_n))_{n \in \mathbb{N}}$ is eventually in V. Thus, $f(x_n) \to f(c)$.

To show the reverse implication, suppose that f is not continuous at c. We will construct a sequence $(x_n)_{n \in \mathbb{N}}$ in D with $(x_n)_{n \in \mathbb{N}}$ converging to c but $(f(x_n))_{n \in \mathbb{N}}$ not converging to $f(c)$.

Since f is not continuous at c, there exists an $\varepsilon > 0$ such that for all $\delta > 0$, there is an x in D with $|x - c| < \delta$ but $|f(x) - f(c)| \geq \varepsilon$. So for each n in $\mathbb{N}$, using $\delta = 1/n$, there is an x_n in D with $|x_n - c| < 1/n$ but $|f(x_n) - f(c)| \geq \varepsilon$. Therefore, $(x_n)_{n \in \mathbb{N}}$ is a sequence in D, $x_n \to c$, but $(f(x_n))_{n \in \mathbb{N}}$ does not converge to $f(c)$ [because $(f(x_n))_{n \in \mathbb{N}}$ is never in the ε-neighborhood of $f(c)$]. ∎

Remark We point out that we could now prove that the usual combinations of continuous functions are continuous (Proposition 4.2) by combining Theorem 4.1 with Theorem 3.2. For example, if f and g are continuous at c and $(x_n)_{n \in \mathbb{N}}$ is a sequence in their common domain with $x_n \to c$, then

$$(f + g)(x_n) = f(x_n) + g(x_n) \to f(c) + g(c) = (f + g)(c).$$

Thus, $f + g$ is continuous at c.

Remark In the last part of the proof of Theorem 4.1, we negated both sides of the if and only if statement following Definition 4.1. We now do the same for the statement of Theorem 4.1. Given $f : D \to \mathbb{R}$ with c in D, then f is not continuous at c if and only if there is a sequence $(x_n)_{n \in \mathbb{N}}$ in D converging to c but $(f(x_n))_{n \in \mathbb{N}}$ does not converge to $f(c)$.

Example 4.5 Let

$$f(x) = \begin{cases} 1 & \text{if} \quad x \text{ is rational} \\ -1 & \text{if} \quad x \text{ is irrational.} \end{cases}$$

We show that f is discontinuous at every real number, thus providing us with an example showing that $|f|$ continuous does not imply that f is continuous.

Let c be irrational. Let $(x_n)_{n \in \mathbb{N}}$ be a sequence of rational numbers with $x_n \to c$. Since $f(c) = -1$ and $f(x_n) = 1$ for all n in $\mathbb{N}$, $(f(x_n))_{n \in \mathbb{N}}$ does not converge to $f(c)$.

Let c be rational. Let $(y_n)_{n \in \mathbb{N}}$ be a sequence of irrational numbers with $y_n \to c$. Since $f(c) = 1$ and $f(y_n) = -1$ for all n in $\mathbb{N}$, $(f(y_n))_{n \in \mathbb{N}}$ does not converge to $f(c)$.

The sequences above exist by Theorem 3.5.

Example 4.6 Let $f(x) = \sin(1/x)$ for $x \neq 0$. We claim that f cannot be continuously extended to 0; that is, no matter how we define $f(0)$, f would not be continuous at 0. To see this, let $x_n = 1/n\pi$ and $y_n = 1/[(\pi/2) + 2n\pi]$

for each n in $\mathbb{N}$. Then $x_n \to 0$ and $y_n \to 0$, but $f(x_n) = 0$ and $f(y_n) = 1$ for all n in $\mathbb{N}$. Since $\lim\limits_{n\to\infty} f(x_n) \ne \lim\limits_{n\to\infty} f(y_n)$, $f(0)$ could not equal both of these limits and so Theorem 4.1 could not be satisfied.

Proposition 4.4   Let $f : \mathbb{R} \to \mathbb{R}$ be continuous on $\mathbb{R}$ with $f(x) = 0$ for all x in $\mathbb{Q}$. Then $f(x) = 0$ for all x in $\mathbb{R}$.

Proof   Let c be an irrational number and let $(x_n)_{n\in\mathbb{N}}$ be a sequence of rational numbers with $x_n \to c$. Then $f(c) = \lim\limits_{n\to\infty} f(x_n) = \lim\limits_{n\to\infty} 0 = 0$.   ■

Intuitively, in reference to Proposition 4.4, a function continuous on $\mathbb{R}$ is determined by its action on a dense subset of $\mathbb{R}$. The next proposition gives a solution to Exercise 5 in Section 4.1 using sequences.

Proposition 4.5   Let $f : D \to \mathbb{R}$ with c in D. If f is continuous at c, then there is a neighborhood U of c such that f is bounded on $U \cap D$.

Proof   Suppose that f is unbounded on $U \cap D$ for all neighborhoods U of c. Then, for each n in $\mathbb{N}$, f is unbounded on $(c - (1/n), c + (1/n)) \cap D$. So for each n in $\mathbb{N}$, there exists an x_n in $(c - (1/n), c + (1/n)) \cap D$ with $|f(x_n)| > n$. Then $(x_n)_{n\in\mathbb{N}}$ is a sequence in D and $x_n \to c$, but $(f(x_n))_{n\in\mathbb{N}}$ cannot converge since it is unbounded (Proposition 3.2). Therefore, f is not continuous at c.   ■

Example 4.7   Let $f(x) = 1/x$ for $x \ne 0$. Since f is unbounded on every neighborhood of 0 (intersected with the nonzero reals), f cannot be continuously extended to 0.

We end this section with a very nonintuitive function given in Example 4.8 below. First, we need a lemma.

Lemma 4.2   Let $f : D \to \mathbb{R}$ with c in D. Make the following suppositions:

1. If $(x_n)_{n\in\mathbb{N}}$ is a sequence in $\mathbb{Q} \cap D$ converging to c, then $(f(x_n))_{n\in\mathbb{N}}$ converges to $f(c)$;

and

2. If $(x_n)_{n\in\mathbb{N}}$ is a sequence in $(\mathbb{R} \setminus \mathbb{Q}) \cap D$ converging to c, then $(f(x_n))_{n\in\mathbb{N}}$ converges to $f(c)$.

Then f is continuous at c.

Proof   Suppose that f is not continuous at c. From the proof of Theorem 4.1, there exist an $\varepsilon > 0$ and a sequence $(x_n)_{n\in\mathbb{N}}$ in D with $x_n \to c$ but $|f(x_n) - f(c)| \ge \varepsilon$ for all n in $\mathbb{N}$. Each x_n is either rational or irrational. Therefore, either $A = \{n \in \mathbb{N} : x_n \in \mathbb{Q}\}$ or $B = \{n \in \mathbb{N} : x_n \in \mathbb{R} \setminus \mathbb{Q}\}$ is infinite.

If A is infinite, then there exists a subsequence $(x_{n_k})_{k=1}^{\infty}$ of $(x_n)_{n\in\mathbb{N}}$ with each x_{n_k} in $\mathbb{Q}$ (see the Remark below). By Theorem 3.6, $x_{n_k} \to c$. Since $|f(x_{n_k}) - f(c)| \ge \varepsilon$ for all k in $\mathbb{N}$, $(f(x_{n_k}))_{k=1}^{\infty}$ does not converge to $f(c)$, which is a contradiction to supposition 1. If B is infinite, then there exists a subsequence of $(x_n)_{n \in \mathbb{N}}$ consisting entirely of irrational numbers which gives a contradiction to supposition 2.   ∎

Remark   For A infinite, we construct the subsequence as follows. Let n_1 be the smallest positive integer such that x_{n_1} is in $\mathbb{Q}$. Let n_2 be the smallest positive integer greater than n_1 such that x_{n_2} is in $\mathbb{Q}$. Continue by induction.

Example 4.8   Define $f : (0, \infty) \to \mathbb{R}$ by

$$f(x) = \begin{cases} 0 & \text{if } x \text{ is irrational} \\ \dfrac{1}{n} & \text{if } x \text{ is rational and } x = \dfrac{m}{n} \\ & \text{in lowest terms with } m \text{ and } n \text{ in } \mathbb{N}. \end{cases}$$

We show that f is continuous at every irrational number and discontinuous at every rational number in $(0, \infty)$.

Let c be a rational number in $(0, \infty)$. Then $f(c) > 0$. Let $(x_n)_{n \in \mathbb{N}}$ be a sequence of irrational numbers in $(0, \infty)$ with $x_n \to c$. Since $f(x_n) = 0$ for each n in $\mathbb{N}$, $f(x_n) \to 0 \neq f(c)$. Therefore, f is not continuous at c.

Let c be an irrational number in $(0, \infty)$. We use Lemma 4.2 to show that f is continuous at c. If $(x_n)_{n \in \mathbb{N}}$ is a sequence of irrational numbers in $(0, \infty)$ with $x_n \to c$, then $f(x_n) = 0 \to 0 = f(c)$.

Let $(p_n/q_n)_{n \in \mathbb{N}}$ be a sequence of rational numbers in $(0, \infty)$ where every p_n and every q_n are in $\mathbb{N}$, each p_n/q_n is in lowest terms, and $p_n/q_n \to c$. Since $f(p_n/q_n) = 1/q_n$ and $f(c) = 0$, we want to show that $1/q_n \to 0$ or, equivalently, $q_n \to \infty$ (Exercise 2 in Section 3.7). Suppose that $(q_n)_{n \in \mathbb{N}}$ does not have limit ∞. By Exercise 9 in Section 3.7 and Theorem 3.3, there exist a subsequence $(q_{n_k})_{k=1}^{\infty}$ of $(q_n)_{n \in \mathbb{N}}$ and an x in $[1, \infty)$ such that $q_{n_k} \underset{k}{\to} x$. Since a convergent sequence of integers eventually must be constant (Exercise 3 in Section 3.6), the subsequence $(q_{n_k})_{k \in \mathbb{N}}$ eventually is constant. So there is a k_0 in $\mathbb{N}$ such that $q_{n_k} = x$ for all $k \geq k_0$. In particular, x is in $\mathbb{N}$. Since $(p_{n_k}/q_{n_k})_{k \in \mathbb{N}}$ is a subsequence of $(p_n/q_n)_{n \in \mathbb{N}}$, $p_{n_k}/q_{n_k} \underset{k}{\to} c$. Therefore, $p_{n_k}/x \underset{k}{\to} c$ or $p_{n_k} \underset{k}{\to} cx$.

Again, by Exercise 3 in Section 3.6, $(p_{n_k})_{k \in \mathbb{N}}$ eventually is the constant cx and so cx is in $\mathbb{N}$. Thus c is in $\mathbb{Q}$, which is a contradiction to c being irrational. Therefore, $q_n \to \infty$.

≡ Exercises ≡

1. Use sequences to show that the composition of two continuous functions is continuous.

2. Prove that

$$f(x) = \begin{cases} 1 & \text{if } x \in \mathbb{Q} \\ 0 & \text{if } x \in \mathbb{R} \setminus \mathbb{Q} \end{cases}$$

is discontinuous everywhere.

3. Let
$$f(x) = \begin{cases} x & \text{if} \quad x \in \mathbb{Q} \\ 0 & \text{if} \quad x \in \mathbb{R} \setminus \mathbb{Q}. \end{cases}$$

Show that f is continuous only at $x = 0$.

4. Let
$$f(x) = \begin{cases} x & \text{if} \quad x \in \mathbb{Q} \\ 1 - x & \text{if} \quad x \in \mathbb{R} \setminus \mathbb{Q}. \end{cases}$$

Show that f is continuous only at $x = \frac{1}{2}$.

5. Show that $\cos(1/x)$ cannot be continuously extended to 0.

6. Let f be continuous on $\mathbb{R}$. Let c be in $\mathbb{R}$ with $f(x) = c$ for all x in $\mathbb{Q}$. Show that $f(x) = c$ for all x in $\mathbb{R}$.

7. Give an example of a continuous function on $(0, 1)$ and a Cauchy sequence $(x_n)_{n \in \mathbb{N}}$ in $(0, 1)$ such that $(f(x_n))_{n \in \mathbb{N}}$ is not Cauchy.

8. In Example 4.8, extend the domain of f to include 0 by defining $f(0) = 1$. Show that f is discontinuous at 0.

9. Define $f : (0, \infty) \to \mathbb{R}$ by
$$f(x) = \begin{cases} 0 & \text{if } x \text{ is irrational} \\ n & \text{if } x \text{ is rational and } x = \dfrac{m}{n} \\ & \text{in lowest terms with } m \text{ and } n \text{ in } \mathbb{N}. \end{cases}$$

Show that f is discontinuous at every point of $(0, \infty)$. [*Hint:* See Example 4.8.]

4.3 Limits of Functions

The reader has seen the notion of a limit of a function in calculus. Given $f : D \to \mathbb{R}$ and c in $\mathbb{R}^{\#}$, we are going to define the limit of $f(x)$ as x approaches c. For this to make sense, there must be points in D "close to" c. "Close to" will be handled by the notion of neighborhood. In this section we will first consider c in $\mathbb{R}$ and then allow c to be $\pm\infty$.

We now recall Definition 3.10, Proposition 3.4, and Exercise 6 in Section 3.5. For $D \subset \mathbb{R}$ and c in $\mathbb{R}$,

c is an accumulation point of D

if and only if every neighborhood of c contains a point of D different from c

if and only if every neighborhood of c contains infinitely many points of D

if and only if there exists a sequence of distinct points in D converging to c.

Thus, if c is an accumulation point of D, then D contains an infinite number of points arbitrarily "close to" c.

Definition 4.3 Let $D \subset \mathbb{R}$, let $f : D \to \mathbb{R}$, let c be an accumulation point of D, and let L be in $\mathbb{R}^{\#}$. Then the *limit of f at c is L*, denoted by

$$\lim_{x \to c} f(x) = L \text{ or } \lim_{x \to c} f = L \text{ or } f(x) \to L \text{ as } x \to c,$$

if for every neighborhood V of L there is a neighborhood U of c such that if x is in $U \cap D$ and $x \neq c$, then $f(x)$ is in V.

Since neighborhoods of real numbers are open intervals and neighborhoods of $\pm\infty$ are open rays, we can rewrite Definition 4.3 as follows.

For L in $\mathbb{R}$, $\lim_{x \to c} f(x) = L$ if and only if for all $\varepsilon > 0$ there is a $\delta > 0$ such that if x is in D and $0 < |x - c| < \delta$, then $|f(x) - L| < \varepsilon$. Here, V is $(L - \varepsilon, L + \varepsilon)$ and U is $(c - \delta, c + \delta)$. If we replace $f(c)$ by L in Figure 4.1 and ignore the value of f at c, then Figure 4.1 pictorially describes $\lim_{x \to c} f(x) = L$.

Also, $\lim_{x \to c} f(x) = \infty$ (respectively, $-\infty$) if and only if for all $\alpha > 0$ (respectively, $\beta < 0$) there is a $\delta > 0$ such that if x is in D and $0 < |x - c| < \delta$, then $f(x) > \alpha$ [respectively, $f(x) < \beta$]. Here, V is (α, ∞) [respectively, V is $(-\infty, \beta)$] and U is $(c - \delta, c + \delta)$. See Figure 4.3.

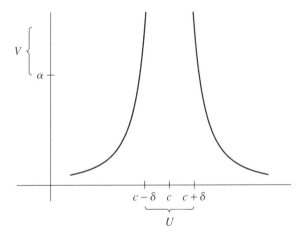

Figure 4.3

Remark In Definition 4.3 it should be noted that c need not be in D. Moreover, if c is in D, it may turn out that $f(c) \neq \lim_{x \to c} f(x)$ (see Example 4.9 below).

Remark The reader should note (for L real) the similarity between Definition 4.3 and Definition 4.1. By comparing these definitions it should be clear that for c in D and c an accumulation point of D, we have

$$f \text{ is continuous at } c \text{ if and only if } f(c) = \lim_{x \to c} f(x) = f(\lim_{x \to c} x).$$

We used this in Example 3.6.

Proposition 4.6   Let D, f, c, and L be as in Definition 4.3. Then $\lim\limits_{x \to c} f(x) = L$ if and only if $\lim\limits_{n \to \infty} f(x_n) = L$ for all sequences $(x_n)_{n \in \mathbb{N}}$ in D such that $x_n \neq c$ for all n and $\lim\limits_{n \to \infty} x_n = c$.

Proof   For L in $\mathbb{R}$ the proof is almost identical (replace $f(c)$ by L) to the proof of Theorem 4.1. By using neighborhoods we give one proof for all three cases, that is, L real or $L = \pm\infty$.

Suppose that $\lim\limits_{x \to c} f(x) = L$ and let $(x_n)_{n \in \mathbb{N}}$ be a sequence in $D \setminus \{c\}$ with $x_n \to c$. We want to show that $f(x_n) \to L$. Let V be a neighborhood of L. Since $\lim\limits_{x \to c} f(x) = L$, there is a neighborhood U of c such that if x is in $U \cap D$ and $x \neq c$, then $f(x)$ is in V. Since $x_n \to c$, $(x_n)_{n \in \mathbb{N}}$ eventually is in U. Thus, $(f(x_n))_{n \in \mathbb{N}}$ eventually is in V and so $f(x_n) \to L$.

To show the reverse implication, suppose that $\lim\limits_{x \to c} f(x) \neq L$. Then there exists a neighborhood V of L such that for all neighborhoods U of c, there is an x in $U \cap D$ with $x \neq c$ and $f(x)$ not in V. So for each n in $\mathbb{N}$, using $U = (c - (1/n), c + (1/n))$, there is an x_n in $U \cap D$ with $x_n \neq c$ and $f(x_n)$ not in V. Therefore, $(x_n)_{n \in \mathbb{N}}$ is a sequence in $D \setminus \{c\}$, $x_n \to c$, but $\lim\limits_{n \to \infty} f(x_n) \neq L$ since the sequence $(f(x_n))_{n \in \mathbb{N}}$ is never in V.   ∎

Corollary 4.1   If f has a limit at c, then this limit is unique.

Proof   This follows from Proposition 4.6 and Theorem 3.13 (limits of sequences are unique).   ∎

Proposition 4.7   For $D \subset \mathbb{R}$, c an accumulation point of D, f and g two real valued functions on D, L_1 and L_2 in $\mathbb{R}^{\#}$, suppose that

$$\lim_{x \to c} f(x) = L_1 \text{ and } \lim_{x \to c} g(x) = L_2.$$

Then

1. $\lim\limits_{x \to c} (f(x) + g(x)) = L_1 + L_2$;

2. $\lim\limits_{x \to c} f(x)g(x) = L_1 L_2$;

3. $\lim\limits_{x \to c} \dfrac{f(x)}{g(x)} = \dfrac{L_1}{L_2}$, if $L_2 \neq 0$

provided the right members of parts 1, 2, and 3 are defined. Note that $\infty - \infty$, $0 \cdot \infty$, ∞/∞, and $L_1/0$ are not defined.

Proof   For L_1 and L_2 real, the proof is almost identical to the proof of Proposition 4.2. Moreover, in view of Proposition 4.6, this follows from the analogous properties of sequences (Theorems 3.2 and 3.14).   ∎

For the following examples we suggest drawing pictures.

Example 4.9   Let $f : \mathbb{R} \to \mathbb{R}$ with

$$f(x) = \begin{cases} 7 & \text{if } x \neq 3 \\ 9 & \text{if } x = 3. \end{cases}$$

Then $\lim\limits_{x \to 3} f(x) = 7 \neq f(3)$.

Example 4.10   Let $f : \mathbb{R} \to \mathbb{R}$ with

$$f(x) = \begin{cases} 1 & \text{if} \quad x \geq 0 \\ -1 & \text{if} \quad x < 0. \end{cases}$$

Then $\lim_{x \to 0} f(x)$ does not exist.

Example 4.11   Let $f : (0, \infty) \to \mathbb{R}$ with $f(x) = 1/x$. Then $\lim_{x \to 0} f(x) = +\infty$.

Example 4.12   Let $g : \mathbb{R} \setminus \{0\} \to \mathbb{R}$ with $g(x) = 1/x$. Then $\lim_{x \to 0} g(x)$ does not exist.

Example 4.13   Let $f(x) = \sin(1/x)$ for $x \neq 0$. Then $\lim_{x \to 0} f(x)$ does not exist. This follows from Proposition 4.6 as follows. If $x_n = 1/n\pi$ for each n in $\mathbb{N}$, then $x_n \to 0$ and $f(x_n) = 0 \to 0$; but if $y_n = 1/[(\pi/2) + 2n\pi]$ for each n in $\mathbb{N}$, then $y_n \to 0$ and $f(y_n) = 1 \to 1$.

Example 4.14   Let $f(x) = x \sin(1/x)$ for $x \neq 0$. Then $\lim_{x \to 0} f(x) = 0$. To see this, let $\varepsilon > 0$. Let $\delta = \varepsilon$. Let x be in $\mathbb{R}$ with $0 < |x - 0| < \delta$. Then

$$|f(x) - 0| = \left| x \sin \frac{1}{x} \right| \leq |x| < \delta = \varepsilon.$$

Remark   Let $D \subset \mathbb{R}$ and let $f : D \to \mathbb{R}$ be continuous on D. Suppose that c is an accumulation point of D and c is not in D. The question is whether or not we can continuously extend f to c. That is, does there exist a continuous function $F : D \cup \{c\} \to \mathbb{R}$ such that $F(x) = f(x)$ for all x in D? Such a continuous extension F of f exists if and only if $\lim_{x \to c} F(x) = F(c)$ if and only if $\lim_{x \to c} f(x) = F(c)$ (since $x \neq c$). Thus, f can be continuously extended to $D \cup \{c\}$ if and only if $\lim_{x \to c} f(x)$ is a real number.

From Examples 4.13 and 4.14 it follows that $\sin(1/x)$ cannot be continuously extended to 0, but $x \sin(1/x)$ can be continuously extended to 0 by defining

$$F(x) = \begin{cases} x \sin \dfrac{1}{x} & \text{if} \quad x \neq 0 \\ 0 & \text{if} \quad x = 0. \end{cases}$$

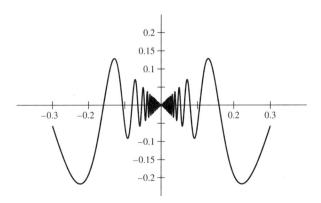

Given $D \subset \mathbb{R}$ and $f : D \to \mathbb{R}$, we now consider the limit of $f(x)$ as x approaches ∞ or $-\infty$. The only difference between this and what we have done previously is that now there must be points in D "close to" ∞ or $-\infty$. If D is unbounded above (respectively, below), then every neighborhood of ∞ (respectively, $-\infty$) contains points of D and hence D contains a sequence of points with limit ∞ (respectively, $-\infty$). This follows because if D is unbounded above (respectively, below), then there exists an x_n in D with $x_n > n$ (respectively, $x_n < -n$) for each n in $\mathbb{N}$. The point we make is that if D is bounded above (respectively, below), then it makes no sense to approach ∞ (respectively, $-\infty$) through points of D.

Definition 4.4 Let D be an unbounded subset of $\mathbb{R}$, let $f : D \to \mathbb{R}$, and let L be in $\mathbb{R}^{\#}$. If D is unbounded above (respectively, below), then the *limit of f at ∞ (respectively, $-\infty$) is L* if for every neighborhood V of L there is a neighborhood U of ∞ (respectively, $-\infty$) such that if x is in $U \cap D$, then $f(x)$ is in V.

Notation   The limit of f at ∞ is L is denoted by $\lim\limits_{x \to \infty} f(x) = L$ or $\lim\limits_{x \to \infty} f = L$ or $f(x) \to L$ as $x \to \infty$. The limit of f at $-\infty$ is L is denoted by $\lim\limits_{x \to -\infty} f(x) = L$ or $\lim\limits_{x \to -\infty} f = L$ or $f(x) \to L$ as $x \to -\infty$.

The reader should see the similarity between Definitions 4.3 and 4.4. For the sake of completeness, we rewrite parts of Definition 4.4 using the definition of a neighborhood. The reader should rewrite the other parts.

For L in $\mathbb{R}$, $\lim\limits_{x \to \infty} f(x) = L$ if and only if for all $\varepsilon > 0$ there is an $\alpha > 0$ such that if x is in D and $x > \alpha$, then $|f(x) - L| < \varepsilon$. Here, V is $(L - \varepsilon, L + \varepsilon)$ and U is (α, ∞). See Figure 4.4, where $\lim\limits_{x \to \infty} f(x) = 0$ is illustrated.

Also, $\lim\limits_{x \to -\infty} f(x) = \infty$ if and only if for all $\alpha > 0$ there is a $\beta < 0$ such that if x is in D and $x < \beta$, then $f(x) > \alpha$. Here, V is (α, ∞) and U is $(-\infty, \beta)$.

We point out that Proposition 4.6, Corollary 4.1, and Proposition 4.7 have corresponding analogues for limits at $\pm \infty$. For example, $\lim\limits_{x \to \infty} f(x) = L$ if and only if $\lim\limits_{n \to \infty} f(x_n) = L$ for all sequences $(x_n)_{n \in \mathbb{N}}$ in D with $x_n \to \infty$. Also, limits are unique, and the limit of a sum, product, and quotient is the sum, product, and quotient of the limits if everything is defined. Since the proofs offer nothing new, we omit them.

Example 4.15   The reader is probably familiar with the statement that a polynomial behaves as its leading term behaves for $|x|$ large. To see this, let

$$f(x) = a_n x^n + a_{n-1} x^{n-1} + \cdots + a_1 x + a_0,$$

where n is in $\mathbb{N}$, each a_i is real, and $a_n \neq 0$. For $x \neq 0$,

$$f(x) = x^n \left(a_n + \frac{a_{n-1}}{x} + \cdots + \frac{a_1}{x^{n-1}} + \frac{a_0}{x^n} \right).$$

Since

$$\lim_{x \to \pm\infty} \left(\frac{a_{n-1}}{x} + \cdots + \frac{a_1}{x^{n-1}} + \frac{a_0}{x^n} \right) = 0,$$

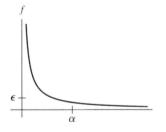

Figure 4.4

$f(x)$ behaves as $a_n x^n$ behaves for $|x|$ large. So if $a_n > 0$, then

$$\lim_{x \to \infty} f(x) = \infty \text{ and } \lim_{x \to -\infty} f(x) = \begin{cases} \infty & \text{if } n \text{ is even} \\ -\infty & \text{if } n \text{ is odd,} \end{cases}$$

while if $a_n < 0$,

$$\lim_{x \to \infty} f(x) = -\infty \text{ and } \lim_{x \to -\infty} f(x) = \begin{cases} \infty & \text{if } n \text{ is odd} \\ -\infty & \text{if } n \text{ is even.} \end{cases}$$

Exercises

In the exercises where c occurs, either c is an accumulation point of D or $c = \pm\infty$.

1. Find two functions f and g such that f and g do not have limits at the real number c, but both $f + g$ and fg have limits at c.

2. Let
$$f(x) = \begin{cases} 1 & \text{if } x \text{ is rational} \\ -1 & \text{if } x \text{ is irrational.} \end{cases}$$
Show that f does not have a limit at any point. Note that $|f|$ has a limit everywhere.

3. Show that $\cos(1/x)$ ($x \neq 0$) cannot be continuously extended to 0. Show that $x \cos(1/x)$ ($x \neq 0$) can be continuously extended to 0.

4. Let $f : D \to \mathbb{R}$ with $a \leq f(x) \leq b$ for all x in $D \setminus \{c\}$. Show that if $\lim_{x \to c} f(x)$ exists (in $\mathbb{R}^{\#}$), then $a \leq \lim_{x \to c} f(x) \leq b$.

5. (The Squeeze Theorem) Let f, g, and $h : D \to \mathbb{R}$ with $f(x) \leq g(x) \leq h(x)$ for all x in $D \setminus \{c\}$. Show that if $\lim_{x \to c} f(x) = \lim_{x \to c} h(x) = L$, then $\lim_{x \to c} g(x) = L$.

6. Let $f : D \to \mathbb{R}$ such that $\lim_{x \to c} f(x)$ is a real number. Show that there is a neighborhood U of c such that f is bounded on $U \cap D$.

7. Let $f : D \to \mathbb{R}$ such that $\lim_{x \to c} f(x)$ is a positive real number. Show that there is a neighborhood U of c such that $f(x) > 0$ for all x in $U \cap D$, $x \neq c$.

8. Let $f : D \to \mathbb{R}$ such that $\lim_{x \to c} f(x)$ is a nonzero real number. Show that there is a neighborhood U of c such that f is bounded away from 0 on $(U \setminus \{c\}) \cap D$. That is, show that there is a neighborhood U of c and an $\varepsilon > 0$ such that $|f(x)| \geq \varepsilon$ for all x in $(U \setminus \{c\}) \cap D$.

9. Let $f : D \to \mathbb{R}$ with $f(x) > 0$ for all x in D. Show that $\lim_{x \to c} f(x) = \infty$ if and only if $\lim_{x \to c} 1/f(x) = 0$.

10. Let $f : \mathbb{R} \to \mathbb{R}$ such that $f(x + y) = f(x) + f(y)$ for all x and y in $\mathbb{R}$. Such a function is called *additive*.

    (a) Show that $f(n) = nf(1)$ for all integers n.

    (b) Show that $f(x) = xf(1)$ for all x in $\mathbb{Q}$.

    (c) Show that if f is continuous at some c in $\mathbb{R}$, then f is continuous on $\mathbb{R}$. [*Hint:* Use the second Remark on page 78.]

    (d) Show that if f is continuous on $\mathbb{R}$, then $f(x) = xf(1)$ for all x in $\mathbb{R}$. [*Hint:* Use sequences to pass from the rationals to the irrationals.]

4.4 Consequences of Continuity

In this section we derive some important properties of functions that are continuous on an interval, usually a closed interval. These properties will be used often throughout the text.

Proposition 4.8 If $f : [a, b] \to \mathbb{R}$ is continuous on $[a, b]$, then f is bounded on $[a, b]$.

Proof   Suppose that f is not bounded on $[a, b]$. Then for each n in $\mathbb{N}$ there is an element x_n in $[a, b]$ with $|f(x_n)| \geq n$. The sequence $(x_n)_{n \in \mathbb{N}}$ is bounded since each x_n is in $[a, b]$. By Theorem 3.10 (Bolzano-Weierstrass Theorem for sequences), $(x_n)_{n \in \mathbb{N}}$ has a convergent subsequence, say $(x_{n_k})_{k=1}^{\infty}$. Let $\lim_{k \to \infty} x_{n_k} = x$. Since each x_{n_k} is in $[a, b]$, x is in $[a, b]$ by Theorem 3.3. By the continuity of f on $[a, b]$, $f(x_{n_k}) \to_k f(x)$. Proposition 3.2 then implies that the sequence $(f(x_{n_k}))_{k \in \mathbb{N}}$ is bounded. This is a contradiction since $\left| f(x_{n_k}) \right| \geq n_k \geq k$ for each k in $\mathbb{N}$. Hence, f must be bounded on $[a, b]$. ∎

By constructing examples, the reader can show that Proposition 4.8 need not hold if $[a, b]$ is replaced by an open or half-open interval, or if f is not continuous.

Definition 4.5   Let $f : D \to \mathbb{R}$. Then f has an *absolute maximum* (respectively, *absolute minimum*) on D if there is an element x_1 in D (respectively, x_2 in D) such that $f(x) \leq f(x_1)$ [respectively, $f(x) \geq f(x_2)$] for all x in D.

In general, a function may not have an absolute maximum or an absolute minimum on its domain. For example, consider $f(x) = x$ defined on $(0, 1)$.

Theorem 4.2   If $f : [a, b] \to \mathbb{R}$ is continuous on $[a, b]$, then f has an absolute maximum and an absolute minimum on $[a, b]$.

Proof   We must show that there is an element M in $[a, b]$ such that $f(x) \leq f(M)$ for all x in $[a, b]$. Similarly, we must show that there is an element m in

$[a, b]$ such that $f(x) \geq f(m)$ for all x in $[a, b]$. We prove the second part and leave the first part as an exercise.

By Proposition 4.8, f is bounded on $[a, b]$. By Proposition 2.4, $\alpha = \inf\{f(x) : x \in [a, b]\}$ exists in $\mathbb{R}$. We want to find an m in $[a, b]$ such that $f(m) = \alpha$.

By Proposition 2.5, for each n in $\mathbb{N}$ there is an element x_n in $[a, b]$ such that $\alpha \leq f(x_n) < \alpha + 1/n$. The sequence $(x_n)_{n\in\mathbb{N}}$ is bounded; so by Theorems 3.10 and 3.3 there exist a subsequence $(x_{n_k})_{k=1}^{\infty}$ of $(x_n)_{n\in\mathbb{N}}$ and an m in $[a, b]$ such that $x_{n_k} \underset{k}{\to} m$. By the continuity of f, $f(x_{n_k}) \underset{k}{\to} f(m)$. Since $\alpha \leq f(x_n) < \alpha + 1/n$ for all n in $\mathbb{N}$, $f(x_{n_k}) \underset{k}{\to} \alpha$. Since the limit of a sequence is unique, $f(m) = \alpha$. ∎

Example 4.16 Let $f : [a, b] \to \mathbb{R}$ with $f(x) > 0$ for all x in $[a, b]$. If $\alpha = \inf\{f(x) : x \in [a, b]\}$, we can conclude that $\alpha \geq 0$. However, if f is also continuous on $[a, b]$, then Theorem 4.2 implies that α is in the range of f, and so $\alpha > 0$. Thus, f is bounded away from 0 on $[a, b]$.

Theorem 4.3 (Intermediate Value Theorem) Let $f : [a, b] \to \mathbb{R}$ be continuous on $[a, b]$. Assume that $f(a) \neq f(b)$. Then, for any k between $f(a)$ and $f(b)$, there is a c in $[a, b]$ such that $f(c) = k$.

Proof For definiteness, assume that $f(a) < k < f(b)$. Let $S = \{x \in [a, b] : f(x) < k\}$. Since a is in S, $S \neq \emptyset$. Also, S is bounded above by b. By the Completeness Axiom for $\mathbb{R}$, S has a supremum in $\mathbb{R}$. Let $c = \sup S$. By Exercise 4 in Section 4.1, $a < c < b$. We claim that $f(c) = k$.

Suppose that $f(c) < k$. Since f is continuous at c, there is a $\delta > 0$ such that if x is in $[a, b]$ with $|x - c| < \delta$, then $f(x) < k$. Thus there exists a d in $[a, b]$ with $c < d$ and $f(d) < k$, which is a contradiction to c being an upper bound of S.

Suppose that $f(c) > k$. Since f is continuous at c, there is a $\delta > 0$ such that if x is in $[a, b]$ with $|x - c| < \delta$, then $f(x) > k$. Thus, every x in $[a, b]$ with $c - \delta < x < c$ satisfies $f(x) > k$, which is a contradiction to c being the least upper bound of S. Therefore, $f(c) = k$. ∎

The conclusion of Theorem 4.3 may hold even if f is not continuous (see Exercise 10). In Chapter 5 we show that a derivative, which may not be continuous, always satisfies the conclusion of Theorem 4.3.

Remark Note that the Intermediate Value Theorem applies to a continuous function on any interval or ray I. If a and b are in I with $f(a) < k < f(b)$, then, by applying Theorem 4.3 to $f|_{[a,b]}$, there is a c in I such that $f(c) = k$.

Corollary 4.2 Let f be continuous on $[a, b]$. Assume that $f(a) \cdot f(b) < 0$. Then there is an element c in (a, b) with $f(c) = 0$. (That is, f has a root between a and b.)

Proof Apply Theorem 4.3 with $k = 0$. ∎

As an application of Corollary 4.2, we have the following elegant result.

Proposition 4.9   Let $f : [0, 1] \to [0, 1]$ be continuous on $[0, 1]$. Then there is a point c in $[0, 1]$ such that $f(c) = c$. (Such a point c is called a *fixed point* of f. The graph of f and the line $y = x$ must intersect at $x = c$.)

Proof   If $f(0) = 0$ or $f(1) = 1$, we are done. So assume that $f(0) \neq 0$ and $f(1) \neq 1$. Let $g(x) = f(x) - x$. Since f is continuous on $[0, 1]$, g is continuous on $[0, 1]$. Observe that $g(0) = f(0) - 0 > 0$ and that $g(1) = f(1) - 1 < 0$. By Corollary 4.2, g has a root in $(0, 1)$. That is, there is some point c in $(0, 1)$ with $g(c) = 0$. Since $g(c) = f(c) - c$, $f(c) = c$.   ∎

Corollary 4.3   The continuous image of a closed interval is again a closed interval (where we allow $\{c\}$ to be the closed interval $[c, c]$).

Proof   Let $f : [a, b] \to \mathbb{R}$ be continuous on $[a, b]$. By Theorem 4.2, there exist m and M in $[a, b]$ such that $f(m) \leq f(x) \leq f(M)$ for all x in $[a, b]$. By the Intermediate Value Theorem, $f([a, b]) = [f(m), f(M)]$.   ∎

Exercises

1. (a) Give an example of a continuous function on an interval (necessarily not closed) that is not bounded.
   (b) Give an example of a continuous function on an open interval that does not have an absolute maximum or an absolute minimum on that interval.

2. Give an example to show that the continuous image of an open interval need not be an open interval.

3. Complete the proof of Theorem 4.2.

4. Show that $f(x) = x^4 - 3x^2 + 1$ has two roots in $[0, 2]$.

5. Show that a polynomial of odd degree has at least one real root.

6. Show that a polynomial of even degree has either an absolute maximum or an absolute minimum.

7. Let $f : (a, b) \to \mathbb{Z}$ be continuous on (a, b) where $\mathbb{Z}$ is the integers. Show that f must be a constant function.

8. Let $f : [0, 1] \to \mathbb{R}$ be continuous on $[0, 1]$ with $f(0) = f(1)$. Show that there exists a c in $[0, \frac{1}{2}]$ with $f(c) = f(c + \frac{1}{2})$.

9. Show that the equation $x = \cos x$ has a solution in $(0, \pi/2)$. Does $x = \sin x$ have a solution in $(0, \pi/2)$?

10. A function f is said to satisfy the *intermediate value property* if whenever a and b are in the domain of f with $f(a) < f(b)$, then $[f(a), f(b)] \subset$ range of f. Give an example of a discontinuous function that satisfies the intermediate value property.

4.5 Uniform Continuity

The main results of this section are Theorem 4.4 and, as discussed in Theorems 4.5 and 4.6, the interplay between uniform continuity and Cauchy sequences.

Recall that for $D \subset \mathbb{R}$ and $f : D \to \mathbb{R}$, f is continuous on D if and only if f is continuous at each point of D; and f is continuous at a point c in D if and only if for all $\varepsilon > 0$ there is a $\delta > 0$ such that if x is in D and $|x - c| < \delta$, then $|f(x) - f(c)| < \varepsilon$. The δ here depends on both ε and c. For example, consider $f(x) = 1/x$ on $(0, \infty)$. Fix $c > 0$ and $\varepsilon > 0$ and draw a picture. First note that for $x > 0$,

$$\left| \frac{1}{x} - \frac{1}{c} \right| = \frac{|x - c|}{cx}.$$

If $x > c/2$, then $|(1/x) - (1/c)| < (2/c^2)|x - c|$. Let $\delta = \min\{c/2, \varepsilon c^2/2\}$. Then $|x - c| < \delta$ implies that

$$\left| \frac{1}{x} - \frac{1}{c} \right| < \frac{2}{c^2}\delta \leq \varepsilon.$$

The point here is that δ depends on both ε and c.

Definition 4.6 Let $D \subset \mathbb{R}$ and $f : D \to \mathbb{R}$. Then f is *uniformly continuous on D* if for all $\varepsilon > 0$ there is a $\delta > 0$ such that if x and y are in D with $|x - y| < \delta$, then $|f(x) - f(y)| < \varepsilon$.

Observe that the δ for uniform continuity depends only on ε. It should be clear that if f is uniformly continuous on D, then f is continuous on D.

Example 4.17 Let $f : [1, \infty) \to \mathbb{R}$ be defined by $f(x) = 1/x$. Then f is uniformly continuous on $[1, \infty)$. To see this, let $\varepsilon > 0$. Let $\delta = \varepsilon$. Let x and y be in $[1, \infty)$ with $|x - y| < \delta$. Since $xy \geq 1$,

$$|f(x) - f(y)| = \frac{|x - y|}{xy} \leq |x - y| < \delta = \varepsilon.$$

Example 4.18 Let $f : (0, 1] \to \mathbb{R}$ be defined by $f(x) = 1/x$. Then f is not uniformly continuous on $(0, 1]$. To see this, first note that f is not uniformly continuous on D if and only if there is an $\varepsilon > 0$ such that for all $\delta > 0$ there exist x and y in D with $|x - y| < \delta$ but $|f(x) - f(y)| \geq \varepsilon$.

Let $\varepsilon = 1$. First observe that if a certain δ works in Definition 4.6, then any smaller δ must also work. So suppose $0 < \delta < 1$. Let $x = \delta$ and $y = \delta/2$. Then $|x - y| = \frac{\delta}{2} < \delta$, but

$$|f(x) - f(y)| = \left| \frac{1}{\delta} - \frac{2}{\delta} \right| = \frac{1}{\delta} > 1.$$

Thus f is not uniformly continuous on $(0, 1]$.

Example 4.19 $f(x) = x^2$ is not uniformly continuous on $\mathbb{R}$. Let $\varepsilon = 1$ and let $\delta > 0$. Let $x = 1/\delta$ and $y = 1/\delta + \delta/2$. Then $|x - y| = \delta/2 < \delta$, but

$$
\begin{aligned}
|f(x) - f(y)| &= \left| x^2 - y^2 \right| \\
&= |x - y| \, |x + y| \\
&= \frac{\delta}{2} \left(\frac{2}{\delta} + \frac{\delta}{2} \right) \\
&= 1 + \frac{\delta^2}{4} > 1.
\end{aligned}
$$

The following very important result is needed in Chapter 6.

Theorem 4.4   A continuous function on a closed interval is uniformly continuous there.

Proof   Assume that $f : [a, b] \to \mathbb{R}$ is not uniformly continuous on $[a, b]$. Then there is an $\varepsilon > 0$ such that for all $\delta > 0$ there exist x and y in $[a, b]$ with $|x - y| < \delta$ but $|f(x) - f(y)| \geq \varepsilon$. For each n in $\mathbb{N}$, using $\delta = 1/n$, there exist x_n and y_n in $[a, b]$ with $|x_n - y_n| < 1/n$ but $|f(x_n) - f(y_n)| \geq \varepsilon$. Since each x_n is in $[a, b]$, by the Bolzano-Weierstrass Theorem for sequences (Theorem 3.10) and Theorem 3.3, there exist a subsequence $(x_{n_k})_{k=1}^{\infty}$ of $(x_n)_{n \in \mathbb{N}}$ and an x in $[a, b]$ with $x_{n_k} \to x$. Also, the subsequence $(y_{n_k})_{k=1}^{\infty}$ of $(y_n)_{n \in \mathbb{N}}$ converges to x for

$$
\lim_{k \to \infty} y_{n_k} = \lim_{k \to \infty} [x_{n_k} + (y_{n_k} - x_{n_k})] = x + 0 = x.
$$

Since $\left| f(x_{n_k}) - f(y_{n_k}) \right| \geq \varepsilon$ for all k in $\mathbb{N}$, the two sequences $(f(x_{n_k}))_{k \in \mathbb{N}}$ and $(f(y_{n_k}))_{k \in \mathbb{N}}$ cannot both converge to $f(x)$. By Theorem 4.1, f is not continuous at x.   ∎

We now consider uniform continuity and sequences. A continuous function need not take a Cauchy sequence to a Cauchy sequence. For example, let $f(x) = 1/x$ for $x > 0$. Then $(1/n)_{n \in \mathbb{N}}$ is Cauchy, but $(f(1/n))_{n \in \mathbb{N}} = (n)_{n \in \mathbb{N}}$ is unbounded and hence not Cauchy.

Theorem 4.5   Let $f : D \to \mathbb{R}$ be uniformly continuous on D. If $(x_n)_{n \in \mathbb{N}}$ is a Cauchy sequence in D, then $(f(x_n))_{n \in \mathbb{N}}$ is a Cauchy sequence in $\mathbb{R}$.

Proof   Let $\varepsilon > 0$. By Definition 3.11, we want an n_0 in $\mathbb{N}$ such that if $n \geq n_0$ and $m \geq n_0$, then $|f(x_n) - f(x_m)| < \varepsilon$.

Since f is uniformly continuous on D, there is a $\delta > 0$ such that if x and y are in D with $|x - y| < \delta$, then $|f(x) - f(y)| < \varepsilon$. Since $(x_n)_{n \in \mathbb{N}}$ is Cauchy, there is an n_0 in $\mathbb{N}$ such that if $n \geq n_0$ and $m \geq n_0$, then $|x_n - x_m| < \delta$. Thus, for $n \geq n_0$ and $m \geq n_0$, $|f(x_n) - f(x_m)| < \varepsilon$.   ∎

Example 4.20   Let $f : (1, \infty) \to \mathbb{R}$ be defined by $f(x) = 1/(x - 1)$. If $x_n = 1 + 1/n$ for each n in $\mathbb{N}$, then $x_n \to 1$, but $f(x_n) = n$ for each n. Hence, $(f(x_n))_{n \in \mathbb{N}}$ is not Cauchy. By Theorem 4.5, f is not uniformly continuous on $(1, \infty)$. [This sequence approach always works when the function has a vertical asymptote. The reader should redo Example 4.18 using this approach.]

Remark   The result of Theorem 4.5 may hold even if f is not uniformly continuous. By Example 4.19, $f(x) = x^2$ is not uniformly continuous on $\mathbb{R}$. However, let $(x_n)_{n \in \mathbb{N}}$ be a Cauchy sequence in $\mathbb{R}$. By Theorem 3.12, there is an x in $\mathbb{R}$ such that $x_n \to x$. Since f is continuous, $f(x_n) \to f(x)$. Thus, $(f(x_n))_{n \in \mathbb{N}}$ is Cauchy.

Theorem 4.6   Let D be a bounded subset of $\mathbb{R}$ and let $f : D \to \mathbb{R}$. Assume that whenever $(x_n)_{n \in \mathbb{N}}$ is a Cauchy sequence in D, $(f(x_n))_{n \in \mathbb{N}}$ is a Cauchy sequence. Then f is uniformly continuous on D.

Proof   Suppose f is not uniformly continuous on D. From the proof of Theorem 4.4 (with $[a, b]$ replaced by D), there exist sequences $(x_n)_{n \in \mathbb{N}}$ and $(y_n)_{n \in \mathbb{N}}$, both in D, and an $\varepsilon > 0$ such that $|x_n - y_n| < 1/n$ and $|f(x_n) - f(y_n)| \geq \varepsilon$ for each n in $\mathbb{N}$.

Since D is bounded, by Theorem 3.10, there is a subsequence $(x_{n_k})_{k=1}^{\infty}$ of $(x_n)_{n \in \mathbb{N}}$ and an L in $\mathbb{R}$ such that $x_{n_k} \underset{k}{\to} L$. Note that L may or may not be in D.

Also, as in the proof of Theorem 4.4, $y_{n_k} \underset{k}{\to} L$.

Consider the sequence $(z_n)_{n \in \mathbb{N}} = (x_{n_1}, y_{n_1}, x_{n_2}, y_{n_2}, x_{n_3}, y_{n_3}, \ldots)$. Then $z_n \underset{n}{\to} L$ (Exercise 4 in Section 3.3) and hence $(z_n)_{n \in \mathbb{N}}$ is Cauchy. For each k in $\mathbb{N}$,

$$|f(z_{2k-1}) - f(z_{2k})| = |f(x_{n_k}) - f(y_{n_k})| \geq \varepsilon,$$

and so $(f(z_n))_{n \in \mathbb{N}}$ is not Cauchy.   ∎

Remark   We give another proof of Theorem 4.4. Let $f : [a, b] \to \mathbb{R}$ be continuous on $[a, b]$. Let $(x_n)_{n \in \mathbb{N}}$ be a Cauchy sequence in $[a, b]$. Then there is an x in $[a, b]$ with $x_n \to x$. Since f is continuous, $f(x_n) \to f(x)$. Thus, $(f(x_n))_{n \in \mathbb{N}}$ is Cauchy. By Theorem 4.6, f is uniformly continuous on $[a, b]$.

Recall that a continuous function on an open interval (a, b) may not be continuously extendable to either a or b. For example, $f(x) = 1/x$ on $(0, 1)$ cannot be continuously extended to 0. In the following theorem, we see that uniform continuity is sufficient to guarantee a continuous extension.

Theorem 4.7   If f is uniformly continuous on (a, b), then f has a continuous extension to $[a, b]$.

Proof   We show that f can be continuously extended to a, leaving the continuous extension to b as an exercise. By the Remark on page 80, we must show that $\lim_{x \to a} f(x)$ is a real number. Let $(x_n)_{n \in \mathbb{N}}$ be a sequence in (a, b) with $x_n \to a$. By Theorem 4.5, $(f(x_n))_{n \in \mathbb{N}}$ is a Cauchy sequence and so there is an L in $\mathbb{R}$ such that $\lim_{n \to \infty} f(x_n) = L$. We claim that $\lim_{x \to a} f(x) = L$. Let $(y_n)_{n \in \mathbb{N}}$ be a sequence in (a, b) with $y_n \to a$. As above, there is an L_1 in $\mathbb{R}$ such that $f(y_n) \to L_1$. By Proposition 4.6, it suffices to show that $L_1 = L$.

Consider the sequence $(z_n)_{n \in \mathbb{N}} = (x_1, y_1, x_2, y_2, x_3, y_3, \ldots)$ in (a, b). Since $z_n \to a$ by Exercise 4 in Section 3.3, $(f(z_n))_{n \in \mathbb{N}}$ converges to some M in $\mathbb{R}$. Since $(f(x_n))_{n \in \mathbb{N}}$ and $(f(y_n))_{n \in \mathbb{N}}$ are subsequences of $(f(z_n))_{n \in \mathbb{N}}$, $f(x_n) \to M$ and $f(y_n) \to M$. Since limits are unique, $L = M = L_1$.   ∎

Exercises

1. Show that $f(x) = x^3$ is not uniformly continuous on $\mathbb{R}$.

2. Show that $f(x) = 1/(x - 2)$ is not uniformly continuous on $(2, \infty)$.

3. Show that $f(x) = \sin(1/x)$ is not uniformly continuous on $(0, \pi/2]$.

4. Show that $f(x) = mx + b$ is uniformly continuous on $\mathbb{R}$.

5. Show that $f(x) = 1/x^2$ is uniformly continuous on $[1, \infty)$, but not on $(0, 1]$.

6. Show that if f is uniformly continuous on $[a, b]$ and uniformly continuous on D (where D is either $[b, c]$ or $[b, \infty)$), then f is uniformly continuous on $[a, b] \cup D$.

7. Show that $f(x) = \sqrt{x}$ is uniformly continuous on $[1, \infty)$. Use Exercise 6 to conclude that f is uniformly continuous on $[0, \infty)$.

8. Show that if D is bounded and f is uniformly continuous on D, then f is bounded on D.

9. Let f and g be uniformly continuous on D. Show that $f + g$ is uniformly continuous on D. Show, by example, that fg need not be uniformly continuous on D.

10. Complete the proof of Theorem 4.7.

11. Give an example of a continuous function on $\mathbb{Q}$ that cannot be continuously extended to $\mathbb{R}$.

12. Let $f : \mathbb{Q} \to \mathbb{R}$ be uniformly continuous on $\mathbb{Q}$. Show that f has a unique continuous extension to $\mathbb{R}$. [*Hint:* Define $g : \mathbb{R} \to \mathbb{R}$ by

$$g(x) = \begin{cases} f(x) & \text{if } x \in \mathbb{Q} \\ \lim_{n \to \infty} f(x_n) & \text{if } x \in \mathbb{R} \setminus \mathbb{Q} \text{ and } (x_n)_{n \in \mathbb{N}} \text{ is a} \\ & \text{sequence in } \mathbb{Q} \text{ with } x_n \to x. \end{cases}$$

First show that g is well-defined; that is, $\lim_{n \to \infty} f(x_n)$ is a real number and if $(y_n)_{n \in \mathbb{N}}$ is another sequence in $\mathbb{Q}$ with $y_n \to x$, then $\lim_{n \to \infty} f(x_n) = \lim_{n \to \infty} f(y_n)$. Then show directly ($\varepsilon - \delta$ argument) that g is uniformly continuous on $\mathbb{R}$.]

4.6 Discontinuities and Monotone Functions

In this section we first define one-sided limits, then types of discontinuities, then monotone functions. Our main results are Corollary 4.4 and Theorem 4.9. Much of this material is taken from Rudin.

Our setting in this section is as follows. I is an interval or ray in $\mathbb{R}$—that is, $I = (a, b), [a, b], (a, b], [a, b), (a, \infty), [a, \infty), (-\infty, b), (-\infty, b]$ where a and b are real or $I = (-\infty, \infty)$; f is a function from I into $\mathbb{R}$; and c is a real number (because limits at $\pm\infty$ are automatically one-sided). Recall that for $D \subset I$, $f|_D$ denotes the restriction of f to D; that is, $f|_D(x) = f(x)$ for all x in D.

One-Sided Limits

Definition 4.7 If there is a b in $\mathbb{R}$ such that $(c, b) \subset I$, the *right-hand limit of f at c*, denoted by

$$f(c+) \text{ or } \lim_{\substack{x \to c \\ x > c}} f(x),$$

is $\lim_{x \to c} f|_{I \cap (c, \infty)}$; and if there is an a in $\mathbb{R}$ such that $(a, c) \subset I$, the *left-hand limit of f at c*, denoted by

$$f(c-) \text{ or } \lim_{\substack{x \to c \\ x < c}} f(x),$$

is $\lim_{x \to c} f|_{I \cap (-\infty, c)}$.

Example 4.21 Define $f : \mathbb{R} \to \mathbb{R}$ by

$$f(x) = \begin{cases} 1 & \text{if } x \geq 0 \\ -1 & \text{if } x < 0. \end{cases}$$

Then $f(0+) = 1$ and $f(0-) = -1$.

Example 4.22 Define $f : \mathbb{R} \setminus \{0\} \to \mathbb{R}$ by $f(x) = 1/x$. Then $f(0+) = +\infty$ and $f(0-) = -\infty$.

From Section 4.3 or by proofs completely similar to those of Section 4.3, it follows that

$f(c+) = L$ (in $\mathbb{R}^\#$)

if and only if for all neighborhoods V of L, there is a $\delta > 0$ such that if x is in I and $c < x < c + \delta$, then $f(x)$ is in V

if and only if $f(x_n) \to L$ for all sequences $(x_n)_{n \in \mathbb{N}}$ in I with $x_n > c$ for all n and $x_n \to c$;

and

$f(c-) = L$ (in $\mathbb{R}^\#$)

if and only if for all neighborhoods V of L, there is a $\delta > 0$ such that if x is in I and $c - \delta < x < c$, then $f(x)$ is in V

if and only if $f(x_n) \to L$ for all sequences $(x_n)_{n \in \mathbb{N}}$ in I with $x_n < c$ for all n and $x_n \to c$.

Remark If c is an interior point of I (that is, c is not an endpoint), then it makes sense to consider both right-hand and left-hand limits of f at c. If c is an endpoint of I, then it makes sense to consider only one of the one-sided limits.

Also note that if c is an endpoint of I, then the appropriate one-sided limit of f at c and the limit of f at c as defined in Section 4.3 are the same.

It should be clear that for c an interior point of I, $\lim\limits_{x \to c} f(x)$ exists if and only if $f(c+) = f(c-) = \lim\limits_{x \to c} f(x)$.

Types of Discontinuities

Definition 4.8 Let c be in I and suppose that f is discontinuous at c. If c is an interior point of I, then f has a discontinuity of the *first kind*, or a *simple discontinuity*, at c if both $f(c+)$ and $f(c-)$ exist in $\mathbb{R}$; otherwise, f has a discontinuity of the *second kind* at c. If c is an endpoint of I, then f has a discontinuity of the *first kind* at c if the appropriate one-sided limit of f at c exists in $\mathbb{R}$; otherwise, f has a discontinuity of the *second kind* at c.

Example 4.23 The function in Example 4.21 has a discontinuity of the first kind at 0.

Example 4.24 Let
$$f(x) = \begin{cases} 1 & \text{if } x \text{ is rational} \\ 0 & \text{if } x \text{ is irrational.} \end{cases}$$
Since neither $f(c+)$ nor $f(c-)$ exists at any point c, f has a discontinuity of the second kind at every real number.

Example 4.25 Let
$$f(x) = \begin{cases} \sin \dfrac{1}{x} & \text{if } x \neq 0 \\ 0 & \text{if } x = 0. \end{cases}$$
Then f is continuous at every $x \neq 0$ and f has a discontinuity of the second kind at 0. That $f(0+)$ does not exist follows from Example 4.13.

Remark When f has a discontinuity of the first kind at a point c in I, there is a jump in the graph of f at c. Consequently, some authors call this a *jump discontinuity*. For c an interior point of I, this can occur if (1) $f(c-) \neq f(c+)$ or if (2) $f(c-) = f(c+) \neq f(c)$. In (2), the discontinuity is removable by redefining $f(c) = f(c+)$, whereas in (1) the discontinuity is not removable.

Monotone Functions

Definition 4.9 The function f is *monotone increasing* (respectively, *strictly increasing*) on I if whenever x and y are in I with $x < y$, then $f(x) \leq f(y)$ [respectively, $f(x) < f(y)$]. Also, f is *monotone decreasing* (respectively, *strictly decreasing*) on I if whenever x and y are in I with $x < y$, then $f(x) \geq f(y)$ [respectively, $f(x) > f(y)$]. f is *monotone* on I if f is either monotone increasing or monotone decreasing on I.

Remark The words "monotone" and "monotonic" are used interchangeably. Observe that a constant function is both monotone increasing and monotone decreasing. The function in Example 4.21 is monotone increasing, and keeping this example in mind with $c = 0$ will aid the reader in following the next proof.

Theorem 4.8 Let f be monotone on I and let c be in I. For c an interior point of I, both $f(c+)$ and $f(c-)$ exist in $\mathbb{R}$. If I contains a left endpoint a, then $f(a+)$ exists in $\mathbb{R}$; while if I contains a right endpoint b, then $f(b-)$ exists in $\mathbb{R}$.

Proof Assume that f is monotone increasing on I. Let c be an interior point of I or a right endpoint of I. We will show that

$$f(c-) = \sup_{x \in I \cap (-\infty, c)} f(x).$$

Since f is monotone increasing, $\{f(x) : x \in I \cap (-\infty, c)\}$ is bounded above by $f(c)$ and therefore has a least upper bound in $\mathbb{R}$, which we denote by α. Thus, $\alpha \le f(c)$. We want to show that $\alpha = f(c-)$.

Let $\varepsilon > 0$. Since $\alpha = \sup\{f(x) : x \in I \cap (-\infty, c)\}$, there is an x_0 in $I \cap (-\infty, c)$ such that $\alpha - \varepsilon < f(x_0) \le \alpha$. Let $\delta = c - x_0$. Let x be in I with $x_0 = c - \delta < x < c$. Since f is monotone increasing, $\alpha - \varepsilon < f(x_0) \le f(x) \le \alpha$, and so $|f(x) - \alpha| < \varepsilon$. Hence, $f(c-) = \alpha$.

Next, let c be an interior point of I or a left endpoint of I. We will show that

$$f(c+) = \inf_{x \in I \cap (c, \infty)} f(x).$$

Since f is monotone increasing, $\{f(x) : x \in I \cap (c, \infty)\}$ is bounded below by $f(c)$ and therefore has a greatest lower bound in $\mathbb{R}$, which we denote by β. Thus, $f(c) \le \beta$. We want to show that $\beta = f(c+)$.

Let $\varepsilon > 0$. Since $\beta = \inf\{f(x) : x \in I \cap (c, \infty)\}$, there is an x_0 in $I \cap (c, \infty)$ such that $\beta \le f(x_0) < \beta + \varepsilon$. Let $\delta = x_0 - c$. Let x be in I with $c < x < c + \delta = x_0$. Since f is monotone increasing, $\beta \le f(x) \le f(x_0) < \beta + \varepsilon$, and so $|f(x) - \beta| < \varepsilon$. Hence, $f(c+) = \beta$.

In summary, for f monotone increasing and c an interior point of I, we have shown that

$$\sup_{x \in I \cap (-\infty, c)} f(x) = f(c-) \le f(c) \le f(c+) = \inf_{x \in I \cap (c, \infty)} f(x).$$

The analogous results for f monotone decreasing are asked for in Exercise 6. ■

Our main results, given below, are now an easy consequence of Theorem 4.8.

Corollary 4.4 Monotone functions have no discontinuities of the second kind.

Proof This follows from Definition 4.8 and Theorem 4.8. ■

Lemma 4.3   Let c and d be in I with $c < d$. If f is monotone increasing (respectively, decreasing) on I, then $f(c+) \leq f(d-)$ [respectively, $f(c+) \geq f(d-)$].

Proof   Assume that f is monotone increasing on I. From the proof of Theorem 4.8,

$$f(c+) = \inf_{x \in I \cap (c,\infty)} f(x) = \inf_{x \in I \cap (c,d)} f(x),$$

where the last equality follows from f being monotone increasing. Similarly,

$$f(d-) = \sup_{x \in I \cap (-\infty,d)} f(x) = \sup_{x \in I \cap (c,d)} f(x).$$

Thus, $f(c+) \leq f(d-)$.

The proof for f monotone decreasing is asked for in Exercise 7.   ∎

Theorem 4.9   The set of discontinuities of a monotone function is countable.

Proof   Assume f is monotone increasing on I and let E be the set of interior points of I at which f is discontinuous. For each c in E, $f(c-) < f(c+)$ by the proof of Theorem 4.8, and so there exists a rational number $r(c)$ with

$$f(c-) < r(c) < f(c+).$$

Given c and d in E with $c < d$, by Lemma 4.3, $f(c+) \leq f(d-)$, and so $r(c) \neq r(d)$.

Thus, there exists a bijection from E onto a subset of $\mathbb{Q}$, namely $c \rightarrow r(c)$, and so E is countable. Since I can have at most two endpoints, the set of discontinuities of f is countable.

The proof for f monotone decreasing is asked for in Exercise 8.   ∎

Example 4.26   Let E be any countably infinite subset of $\mathbb{R}$ (for instance, E could be $\mathbb{Q}$). We construct a monotone function on $\mathbb{R}$ whose set of discontinuities is E. First write E as a sequence of distinct points $(x_n)_{n \in \mathbb{N}}$.

We need two facts from infinite series:

1. The geometric series $\sum\limits_{n=1}^{\infty} \left(\frac{1}{2}\right)^n$ converges;

2. The order in which the terms of this series are arranged is immaterial since $\sum\limits_{n=1}^{\infty} \left(\frac{1}{2}\right)^n$ converges absolutely (we will prove this in Chapter 7).

Define $f : \mathbb{R} \rightarrow \mathbb{R}$ by

$$f(x) = \sum_{\{n : x_n < x\}} \left(\frac{1}{2}\right)^n$$

for each x in $\mathbb{R}$. This summation means to sum over those indices n for which $x_n < x$. If there are no indices n such that $x_n < x$, then $f(x) = 0$. Then

(a) f is monotone increasing on $\mathbb{R}$;

(b) f is discontinuous at every point of E, in fact, $f(x_n+) - f(x_n-) = \left(\frac{1}{2}\right)^n$;

(c) f is continuous at every point of $\mathbb{R} \setminus E$.

The verification of parts (a), (b), and (c) is asked for in Exercise 11.

Exercises

1. For x in $\mathbb{R}$, let $[x]$ denote the largest integer less than or equal to x. What kind of discontinuities does $[x]$ have?

2. Let $f(x) = x - [x]$. What kind of discontinuities does f have?

3. Verify the statements in Example 4.24. [*Hint:* Use sequences.]

4. What kind of discontinuities does the function in Exercise 3 in Section 4.2 have?

5. Show that f is monotone increasing if and only if $-f$ is monotone decreasing.

6. Complete the proof of Theorem 4.8. In doing so, for f monotone decreasing on I and c an interior point of I, you should obtain

$$\inf_{x \in I \cap (-\infty, c)} f(x) \;=\; f(c-) \geq f(c) \geq f(c+) = \sup_{x \in I \cap (c, \infty)} f(x).$$

7. Complete the proof of Lemma 4.3 for f monotone decreasing.

8. Complete the proof of Theorem 4.9 for f monotone decreasing.

9. Give an example of a monotone function defined on an open interval (a, b) such that $f(a+)$ and $f(b-)$ exist in $\mathbb{R}^{\#}$ but not in $\mathbb{R}$.

10. Let f be a monotone function on I that satisfies the intermediate value property. Show that f is continuous on I.

11. Verify parts (a), (b), and (c) of Example 4.26.

5 Differentiation

In Section 5.1 we show that, although a derivative need not be continuous, a derivative satisfies the intermediate value property. Section 5.2 deals with mean value theorems and their applications. We use Taylor's Theorem in Section 5.3 to show that e is irrational.

5.1 The Derivative

Our setting in this section is similar to the setting in Section 4.6: I is an interval or ray in $\mathbb{R}$, $f : I \to \mathbb{R}$, and c is in I.

Definition 5.1 The *derivative of f at c*, denoted by $f'(c)$, is defined by

$$f'(c) = \lim_{x \to c} \frac{f(x) - f(c)}{x - c}$$

provided this limit exists in $\mathbb{R}^{\#}$ (as in Definition 4.3). The function f is *differentiable at c* provided $f'(c)$ is a real number. For $D \subset I$, f is *differentiable on D* if f is differentiable at each point of D.

Remark The reader is probably familiar with the notations $f'(x) = df(x)/dx = dy/dx$, where $y = f(x)$, and with the fact that the derivative $f'(c)$ is the slope of the tangent line to f at c. When f is continuous at c and $f'(c) = \pm\infty$, the tangent line to f at c is vertical.

Remark By replacing $x - c$ with h in Definition 5.1, we obtain the equivalent expression

$$f'(c) = \lim_{h \to 0} \frac{f(c + h) - f(c)}{h}.$$

Remark Using the definition of limit in Section 4.3, we have that f is differentiable at c if and only if there is a real number L such that for all $\varepsilon > 0$ there is a $\delta > 0$ such that if x is in I and $0 < |x - c| < \delta$, then

$$\left| \frac{f(x) - f(c)}{x - c} - L \right| < \varepsilon.$$

In this case, $L = f'(c)$.

Also, $f'(c) = \infty$ (respectively, $-\infty$) if and only if for all $\alpha > 0$ (respectively, $\beta < 0$) there is a $\delta > 0$ such that if x is in I and $0 < |x - c| < \delta$, then $[f(x) - f(c)]/(x - c) > \alpha$ (respectively, $< \beta$).

Theorem 5.1   If f is differentiable at c, then f is continuous at c.

Proof   Observe that

$$\lim_{x \to c} [f(x) - f(c)] = \lim_{x \to c} \left[\frac{f(x) - f(c)}{x - c} \cdot (x - c) \right]$$

$$= \lim_{x \to c} \frac{f(x) - f(c)}{x - c} \cdot \lim_{x \to c} (x - c) \qquad \text{(since both limits exist)}$$

$$= f'(c) \cdot 0 = 0.$$

Hence $\lim_{x \to c} f(x) = f(c)$, and so f is continuous at c.   ∎

Example 5.1   The converse of Theorem 5.1 is false. The standard example is $f(x) = |x|$, which is continuous at 0 but not differentiable at 0. This is seen by noting that

$$f'(0) = \lim_{x \to 0} \frac{f(x) - f(0)}{x - 0} = \lim_{x \to 0} \frac{|x|}{x}$$

and that

$$\frac{|x|}{x} = \begin{cases} 1 & \text{if} \quad x > 0 \\ -1 & \text{if} \quad x < 0. \end{cases}$$

Hence, $f'(0)$ does not exist. The reader is probably familiar with the statement that if f is differentiable, then f has no "sharp" corners.

The following proposition and theorem give us the usual rules for differentiation normally encountered in calculus.

Proposition 5.1   Let $f, g : I \to \mathbb{R}$ both be differentiable at c in I. Let a be in $\mathbb{R}$. Then $f \pm g$, fg, af, and f/g [if $g(c) \neq 0$] are differentiable at c and

1. $(f \pm g)'(c) = f'(c) \pm g'(c)$,
2. $(fg)'(c) = f(c)g'(c) + g(c)f'(c)$,
3. $(af)'(c) = af'(c)$,
4. $\left(\dfrac{f}{g} \right)'(c) = \dfrac{g(c)f'(c) - f(c)g'(c)}{g^2(c)}.$

Proof   We prove part 4. First note that g is continuous at c by Theorem 5.1. Since $g(c) \neq 0$, by Lemma 4.1, there is a neighborhood U of c such that $g(x) \neq 0$ for all x in $U \cap I$. Let $h = f/g$. Then, for x in $U \cap I$, $x \neq c$,

$$\frac{h(x) - h(c)}{x - c} = \frac{[f(x)/g(x)] - [f(c)/g(c)]}{x - c}$$

$$= \frac{f(x)g(c) - g(x)f(c)}{(x - c)g(x)g(c)}$$

$$= \frac{[f(x)g(c) - f(c)g(c)] + [f(c)g(c) - f(c)g(x)]}{(x - c)g(x)g(c)}$$

$$= \frac{1}{g(x)g(c)} \left[\frac{f(x) - f(c)}{x - c} \cdot g(c) - f(c) \cdot \frac{g(x) - g(c)}{x - c} \right].$$

Letting $x \to c$ and using the continuity of g at c and Proposition 4.7, we obtain part 4.   ∎

For the reader who thinks the next proof is unnecessarily complicated, Exercise 6 presents most people's first attempt at the proof. It is intuitive but incorrect.

Theorem 5.2   (chain rule) Let I and J be intervals or rays in $\mathbb{R}$, let $f : I \to J$ and $g : J \to \mathbb{R}$, and let c be in I with f differentiable at c and g differentiable at $f(c)$. Then the composite function $g \circ f$ is differentiable at c and

$$(g \circ f)'(c) = g'(f(c))f'(c).$$

Proof   This proof is taken essentially from Apostol. Let $y_0 = f(c)$ and define $h : J \to \mathbb{R}$ by

$$h(y) = \begin{cases} \dfrac{g(y) - g(y_0)}{y - y_0} - g'(y_0) & \text{if } y \neq y_0 \\ 0 & \text{if } y = y_0. \end{cases} \tag{1}$$

By Definition 5.1, $\lim\limits_{y \to y_0} h(y) = 0 = h(y_0)$ and so h is continuous at y_0. Since f is continuous at c (Theorem 5.1) and h is continuous at $y_0 = f(c)$, $h \circ f$ is continuous at c. Thus,

$$\lim\limits_{x \to c} (h \circ f)(x) = (h \circ f)(c) = h(y_0) = 0.$$

Letting $y = f(x)$ in (1), we have

$$(h \circ f)(x) = \frac{g(f(x)) - g(y_0)}{f(x) - y_0} - g'(y_0) \text{ if } f(x) \neq y_0.$$

Therefore,

$$g(f(x)) - g(y_0) = [h(f(x)) + g'(y_0)][f(x) - y_0],$$

and this equation holds even if $f(x) = y_0$.

For x in I, $x \neq c$, we obtain

$$\frac{g(f(x)) - g(f(c))}{x - c} = [h(f(x)) + g'(y_0)]\left[\frac{f(x) - f(c)}{x - c}\right].$$

Thus,

$$\begin{aligned} (g \circ f)'(c) &= \lim\limits_{x \to c} \frac{g(f(x)) - g(f(c))}{x - c} \\ &= [\lim\limits_{x \to c} h(f(x)) + g'(y_0)] \cdot \lim\limits_{x \to c} \frac{f(x) - f(c)}{x - c} \\ &= [0 + g'(y_0)]f'(c) \\ &= g'(f(c))f'(c). \end{aligned}$$

$\blacksquare$

Example 5.2   Let

$$f(x) = \begin{cases} x^2 \sin \dfrac{1}{x} & \text{if } x \neq 0 \\ 0 & \text{if } x = 0. \end{cases}$$

By Proposition 5.1 and the chain rule, $f'(x) = 2x \sin(1/x) - \cos(1/x)$ for $x \neq 0$. $f'(0) = \lim\limits_{x \to 0} [f(x) - f(0)]/(x - 0) = \lim\limits_{x \to 0} x \sin(1/x) = 0$ (Example 4.14).

The graph of f follows.

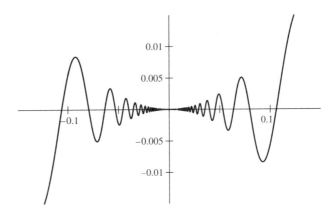

Thus, f is differentiable on $\mathbb{R}$. However, f' is not continuous at 0 since $\cos(1/x)$ in $f'(x)$ above does not have a limit as $x \to 0$.

Suppose that, in the definition of $f(x)$ in Example 5.2, we replace x^2 by x^n where n is in $\mathbb{N}$. In an analogous manner, the reader can show that f is differentiable everywhere and $f'(0) = 0$ if $n \geq 2$. For the case $n = 1$, see Exercise 5.

Remark   Given $f : I \to \mathbb{R}$, f' is a new function whose domain consists of those points in I where f is differentiable (points where f' is $\pm\infty$ are not allowed in order to keep f' real valued), and f' is also called the *first derivative* of f. If the domain of f' is an interval or ray in $\mathbb{R}$, then we can form the derivative of f', denoted by f'' and called the *second derivative* of f. In Example 5.2, $f'(0)$ exists in $\mathbb{R}$, but $f''(0)$ does not exist since f' is not continuous at 0 (Theorem 5.1).

In general, the *nth derivative* of f, denoted by $f^{(n)}$, is the first derivative of $f^{(n-1)}$, $n = 2, 3, 4, \ldots$. For $f^{(n)}(c)$ to exist in $\mathbb{R}$, $f^{(n-1)}$ must exist in an interval or ray containing c and $f^{(n-1)}$ must be differentiable at c.

Definition 5.2   Let $f : I \to \mathbb{R}$ with c in I. Then f has a *local maximum* (respectively, *local minimum*) at c if there is a neighborhood U of c such that $f(x) \leq f(c)$ [respectively, $f(x) \geq f(c)$] for all x in $U \cap I$.

Note that an absolute maximum or an absolute minimum, as defined in Chapter 4, is a local maximum or a local minimum. Recall that an interior point of I is a point in I that is not an endpoint of I.

Proposition 5.2   Suppose that f has a local maximum or a local minimum at an interior point c of I. If f is differentiable at c, then $f'(c) = 0$.

Proof   Let f have a local maximum at c. Since c is an interior point of I, by Definition 5.2, there is a $\delta > 0$ such that $f(x) \leq f(c)$ for all x such that

$c - \delta < x < c + \delta$. For $c < x < c + \delta$, $[f(x) - f(c)]/(x - c) \leq 0$. Since f is differentiable at c, by Definition 5.1 and Exercise 4 in Section 4.3, $f'(c) \leq 0$. For $c - \delta < x < c$, $[f(x) - f(c)]/(x - c) \geq 0$, and the same reasoning yields $f'(c) \geq 0$. Thus $f'(c) = 0$.

The proof for f having a local minimum at c is similar.   ∎

Although a derivative need not be continuous, the following theorem shows that a derivative satisfies the intermediate value property (Exercise 10 in Section 4.4). This will be useful in Section 6.5 in connection with the Fundamental Theorem of Calculus.

Theorem 5.3   If f is differentiable on I, then f' has the intermediate value property on I. That is, given a and b in I with $f'(a) \neq f'(b)$ and r between $f'(a)$ and $f'(b)$, there is a c between a and b such that $f'(c) = r$.

Proof   Assume that $a < b$ in I and $f'(a) < r < f'(b)$.

Define $F : [a, b] \to \mathbb{R}$ by $F(x) = f(x) - rx$. By Proposition 5.1, F is differentiable and hence continuous on $[a, b]$ (Theorem 5.1). Since $F'(x) = f'(x) - r$, it suffices to find a c in (a, b) such that $F'(c) = 0$.

By Theorem 4.2, F has an absolute maximum and an absolute minimum on $[a, b]$. If either of these occurs at an interior point c of $[a, b]$, then $F'(c) = 0$ by Proposition 5.2. So we must show that at least one of these occurs at an interior point.

If neither of these occurs at an interior point, then we have two cases.

Case 1 $F(a)$ is an absolute maximum and $F(b)$ is an absolute minimum.

Case 2 $F(a)$ is an absolute minimum and $F(b)$ is an absolute maximum.

In Case 1,

$$f'(b) - r = F'(b) = \lim_{\substack{x \to b \\ x < b}} \frac{F(x) - F(b)}{x - b} \leq 0,$$

and so $f'(b) \leq r$, which is a contradiction to $r < f'(b)$.

In Case 2,

$$f'(a) - r = F'(a) = \lim_{\substack{x \to a \\ x > a}} \frac{F(x) - F(a)}{x - a} \geq 0,$$

and so $f'(a) \geq r$, which is a contradiction to $f'(a) < r$. Thus, F has an absolute maximum or an absolute minimum at an interior point of $[a, b]$.

A similar proof holds for $f'(a) > r > f'(b)$.   ∎

Corollary 5.1   If f is differentiable on I, then all discontinuities of f' are of the second kind.

Proof   Referring to Section 4.6, if f' has a discontinuity of the first kind at a point c in I, then f' has a jump discontinuity at c. Thus f' cannot satisfy the intermediate value property in an interval around c.   ∎

The reader should note that in Example 5.2 the discontinuity of f' at 0 is of the second kind.

We end this section with a proposition concerning sequences and derivatives. We will use this proposition in Chapter 8 to obtain a surprising result—namely, an example of a continuous function on $\mathbb{R}$ that is nowhere differentiable. Intuitively, this is a continuous function with a "sawtooth" at every point.

Proposition 5.3 Let $f : I \to \mathbb{R}$ be differentiable at the interior point c of I. Let $(a_n)_{n\in\mathbb{N}}$ and $(b_n)_{n\in\mathbb{N}}$ be two sequences in I both converging to c with $a_n < c < b_n$ for all n in $\mathbb{N}$. Then

$$f'(c) = \lim_{n\to\infty} \frac{f(b_n) - f(a_n)}{b_n - a_n}.$$

Proof This proof is taken essentially from Rudin. For each n, let $\lambda_n = (b_n - c)/(b_n - a_n)$. Then, for each n, $0 < \lambda_n < 1$ and

$$\frac{f(b_n) - f(a_n)}{b_n - a_n} - f'(c) = \tag{2}$$

$$\lambda_n \left[\frac{f(b_n) - f(c)}{b_n - c} - f'(c) \right] + \tag{3}$$

$$(1 - \lambda_n) \left[\frac{f(a_n) - f(c)}{a_n - c} - f'(c) \right].$$

By Definition 5.1, the two expressions in brackets have limit 0 as $n \to \infty$. Since $(\lambda_n)_{n\in\mathbb{N}}$ and $(1 - \lambda_n)_{n\in\mathbb{N}}$ are bounded sequences, it follows that (3), and hence (2), approaches 0 as $n \to \infty$. ∎

═ Exercises ═

1. Let $f(x) = \sqrt{x}$ for $x \geq 0$. Find $f'(x)$ for $x \geq 0$ and note that the tangent line to f at 0 is vertical.

2. Complete the proof of Proposition 5.1.

3. Show that a polynomial is differentiable on $\mathbb{R}$, and that a rational function (the ratio of two polynomials) is differentiable wherever the denominator is not zero.

4. Let f and g be differentiable functions on I and suppose that $f(x)g(x) = 1$ for all x in I. Show that

$$\frac{f'(x)}{f(x)} + \frac{g'(x)}{g(x)} = 0.$$

5. Let

$$f(x) = \begin{cases} x \sin \dfrac{1}{x} & \text{if } x \neq 0 \\ 0 & \text{if } x = 0. \end{cases}$$

Show that f is continuous on $\mathbb{R}$, differentiable at $x \neq 0$, but not differentiable at $x = 0$.

6. Given the notation and hypothesis of Theorem 5.2, what is wrong with the following attempt to prove it?

$$\lim_{x \to c} \frac{g(f(x)) - g(f(c))}{x - c} = \lim_{x \to c} \left[\frac{g(f(x)) - g(f(c))}{f(x) - f(c)} \cdot \frac{f(x) - f(c)}{x - c} \right]$$

$$= g'(f(c)) \cdot f'(c).$$

7. Complete the proof of Proposition 5.2 when f has a local minimum at c.

8. Prove Rolle's Theorem: If f is continuous on $[a, b]$, f is differentiable on (a, b), and $f(a) = f(b) = 0$, then there is a c in (a, b) such that $f'(c) = 0$. [*Hint:* If f is nonzero, use Proposition 5.2.]

9. Let f be differentiable on I. Show that

(a) if f is monotone increasing on I, then $f'(x) \geq 0$ for all x in I;

(b) if f is monotone decreasing on I, then $f'(x) \leq 0$ for all x in I.

10. Suppose that f is differentiable on I and f' is monotone on I. Show that f' is continuous on I. [*Hint:* See Corollary 5.1.]

11. Let $h > 0$. Show that there does not exist a differentiable function on $[0, \infty)$ with $f'(0) = 0$ and $f'(x) \geq h$ for all $x > 0$.

12. Use mathematical induction to derive Leibnitz's formula for the nth derivative of a product of two differentiable functions f and g:

$$(fg)^{(n)}(x) = \sum_{k=0}^{n} \binom{n}{k} f^{(k)}(x) g^{(n-k)}(x),$$

where $f^{(0)} = f$ and

$$\binom{n}{k} = \frac{n!}{k!(n-k)!}.$$

5.2 Mean Value Theorems

The Mean Value Theorem, Theorem 5.5 below, is one of the most useful and important results in analysis as the corollaries, examples, and exercises will indicate. For example, Corollary 5.4 gives us an easy proof that $\sin x$ and $\cos x$ are uniformly continuous on $\mathbb{R}$. This theorem follows easily from the more general result, Theorem 5.4, which will also be used in the proof of L'Hôpital's rule in Section 5.4.

Theorem 5.4   (Generalized Mean Value Theorem) Let f and g be continuous functions on $[a, b]$ that are differentiable on (a, b). Then there is a c in (a, b) such that

$$[f(b) - f(a)]g'(c) = [g(b) - g(a)]f'(c).$$

Proof   Define $F : [a, b] \to \mathbb{R}$ by

$$F(x) = [f(b) - f(a)]g(x) - [g(b) - g(a)]f(x).$$

Then F is continuous on $[a, b]$ and differentiable on (a, b), and

$$F(a) = f(b)g(a) - g(b)f(a) = F(b).$$

We need to show that there is a c in (a, b) such that $F'(c) = 0$.

If F is a constant function, any c in (a, b) will work, so we can assume that F is not a constant function. By Theorem 4.2, F has an absolute maximum and an absolute minimum on $[a, b]$. Since F is nonconstant and $F(a) = F(b)$, at least one of these occurs at an interior point c of (a, b). By Proposition 5.2, $F'(c) = 0$. ■

Theorem 5.5 (Mean Value Theorem) If f is continuous on $[a, b]$ and differentiable on (a, b), then there is a c in (a, b) such that

$$f(b) - f(a) = f'(c)(b - a).$$

Proof   Let $g(x) = x$ in Theorem 5.4. ■

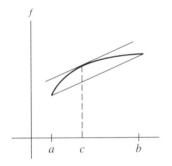

Figure 5.1

Geometrically (Figure 5.1), the Mean Value Theorem states that there is a c in (a, b) such that the slope of the tangent line to f at c is the same as the slope of the secant line connecting $(a, f(a))$ and $(b, f(b))$.

The Generalized Mean Value Theorem has the same geometric interpretation when we consider the curve as given parametrically by $x = g(t)$, $y = f(t)$ with t in $[a, b]$.

Corollary 5.2 Let f satisfy the hypothesis of Theorem 5.5.

1. If $f'(x) = 0$ for all x in (a, b), then f is a constant function.
2. If $f'(x) \geq 0$ for all x in (a, b), then f is monotone increasing.
3. If $f'(x) > 0$ for all x in (a, b), then f is strictly increasing.
4. If $f'(x) \leq 0$ for all x in (a, b), then f is monotone decreasing.
5. If $f'(x) < 0$ for all x in (a, b), then f is strictly decreasing.

All results for f are on $[a, b]$.

Proof   For any x_1 and x_2 in $[a, b]$ with $x_1 < x_2$, there exists, by Theorem 5.5, a c in (x_1, x_2) such that $f(x_2) - f(x_1) = f'(c)(x_2 - x_1)$. All parts 1 through 5 now follow. For example, if $f'(x) = 0$ for all x in (a, b), then $f(x_2) - f(x_1) = 0$ or $f(x_2) = f(x_1)$; if $f'(x) \geq 0$ for all x in (a, b), then $f(x_2) - f(x_1) \geq 0$ or $f(x_2) \geq f(x_1)$. ■

Example 5.3   Recall Exercise 9 in Section 4.4, and let $f(x) = x - \sin x$ for x in $[0, \pi/2]$. Since $f'(x) = 1 - \cos x > 0$ for x in $(0, \pi/2)$, f is strictly increasing on $[0, \pi/2]$ by part 3 of Corollary 5.2. Since $f(0) = 0$, $x > \sin x$ for x in $(0, \pi/2)$. Thus, $x > \sin x$ for all $x > 0$ since $\sin x \leq 1 < \pi/2$.

The following result will be used in Chapter 6.

Corollary 5.3   Suppose that f and g both satisfy the hypothesis of Theorem 5.5. If $f' = g'$ on (a, b), then $f - g$ is a constant function on $[a, b]$.

Proof   Let $h = f - g$. Since $h' = f' - g' = 0$ on (a, b), h is a constant function on $[a, b]$ by part 1 of Corollary 5.2. ■

Corollary 5.4   If f has a bounded derivative on I (an interval or ray in $\mathbb{R}$), then f is uniformly continuous on I.

Proof   Let $M > 0$ with $|f'(x)| \leq M$ for all x in I. Let $\varepsilon > 0$ and let $\delta = \varepsilon/M$. If x_1 and x_2 are in I with $0 < |x_1 - x_2| < \delta$, then, by Theorem 5.5, there is a c between x_1 and x_2 such that

$$|f(x_1) - f(x_2)| = |f'(c)| \, |x_1 - x_2|$$

$$\leq M \, |x_1 - x_2|$$

$$< M\delta = \varepsilon. \qquad \blacksquare$$

Example 5.4   From Corollary 5.4 it follows that $\sin x$ and $\cos x$ are uniformly continuous on $\mathbb{R}$.

Example 5.5   The Mean Value Theorem can be used to establish various inequalities. For instance, let $f(x) = e^x$ and let $a < b$. Then

$$f(b) - f(a) = f'(c)(b - a)$$

for some c in (a, b), and so

$$e^b - e^a = e^c(b - a),$$

where $e^a < e^c < e^b$. Thus,

$$e^a(b - a) < e^b - e^a < e^b(b - a).$$

Exercises

1. Let $f : [0, 2] \to \mathbb{R}$ be continuous on $[0, 2]$ and differentiable on $(0, 2)$, with $f(0) = 0$ and $f(1) = f(2) = 1$.

   (a) Show that there is a c_1 in $(0, 1)$ with $f'(c_1) = 1$.
   (b) Show that there is a c_2 in $(1, 2)$ with $f'(c_2) = 0$.
   (c) Show that there is a c_3 in $(0, 2)$ with $f'(c_3) = \frac{1}{3}$.

2. Let $f(x) = a_n x^n + a_{n-1} x^{n-1} + \cdots + a_1 x + a_0$, where the a_i's are in $\mathbb{R}$ and

   $$\frac{a_n}{n+1} + \frac{a_{n-1}}{n} + \cdots + \frac{a_1}{2} + a_0 = 0.$$

   Show that f has at least one root in $(0, 1)$.

3. Suppose that f is differentiable on $\mathbb{R}$ and that f has n distinct real roots. Show that f' has at least $n - 1$ distinct real roots. Show by example that f' can have more real roots than f.

4. The purpose of this exercise is to present another proof of the Mean Value Theorem. Let f satisfy the hypothesis of Theorem 5.5 and refer to Figure 5.1. Define $\varphi : [a, b] \to \mathbb{R}$ by $(\pm)\varphi(x) = $ the vertical distance between the secant line and the curve f at x. Show that φ satisfies Rolle's Theorem (Exercise 8 in Section 5.1).

5. Let $f : (0, 1] \to \mathbb{R}$ be differentiable on $(0, 1]$, with $|f'(x)| \leq 1$ for all x in $(0, 1]$. For each n in $\mathbb{N}$, let $a_n = f(1/n)$. Show that $(a_n)_{n \in \mathbb{N}}$ converges.

6. Let f be continuous on $[0, \infty)$ and differentiable on $(0, \infty)$, and let $f(0) = 0$. If f' is monotone increasing on $(0, \infty)$, show that $g(x) = f(x)/x$ is monotone increasing on $(0, \infty)$.

7. Let f be continuous on $[a, b]$ and differentiable on (a, b). If $\lim_{x \to a} f'(x)$ exists in $\mathbb{R}$, show that f is differentiable at a and $f'(a) = \lim_{x \to a} f'(x)$. A similar result holds for b.

8. In reference to Corollary 5.4, give an example of a uniformly continuous function on $[0, 1]$ that is differentiable on $(0, 1]$ but whose derivative is not bounded there.

9. Recall that a fixed point of a function f is a point c such that $f(c) = c$.
   (a) Show that if f is differentiable on $\mathbb{R}$ and $\left| f'(x) \right| < 1$ for all x in $\mathbb{R}$, then f has at most one fixed point.
   (b) Let $f(x) = x + (1 + e^x)^{-1}$. Show that f satisfies the hypothesis of part (a) but that f has no fixed point.

10. Show that $\sin x > x$ if $x < 0$.

11. Show that $(x - 1)/x < \ln x < x - 1$ for $x > 1$ and hence $\ln(1 + x) < x$ for $x > 0$.

12. For $0 < x < y$, show that $1 - (x/y) < \ln y - \ln x < (y/x) - 1$.

13. For $0 < x < \pi/2$, show that $x + \cos x < \pi/2$ and $x + \cot x > \pi/2$. (Thus, as $x \to \pi/2$ from the left, $\cos x$ is never large enough for $x + \cos x$ to be greater than $\pi/2$ and $\cot x$ is never small enough for $x + \cot x$ to be less than $\pi/2$.)

5.3 Taylor's Theorem

Given a differentiable function f on $[a, b]$ and $x \neq x_0$ in $[a, b]$, the Mean Value Theorem implies that there is a c between x and x_0 such that
$$f(x) = f(x_0) + f'(c)(x - x_0).$$
The tangent line to f at x_0 is given by the first degree polynomial
$$P(x) = f(x_0) + f'(x_0)(x - x_0).$$
Thus, $P(x)$ is an approximation to $f(x)$ with an error depending on how $f'(x_0)$ and $f'(c)$ differ.

Taylor's Theorem, under the appropriate hypotheses, tells us that f can be approximated by an nth degree polynomial and gives us an expression for measuring the error of the approximation.

Theorem 5.6 (Taylor's Theorem) Suppose that $f : [a, b] \to \mathbb{R}$, n is a positive integer, $f^{(n)}$ is continuous on $[a, b]$, and $f^{(n)}$ is differentiable on (a, b). For $x \neq x_0$ in $[a, b]$, there is a c between x and x_0 such that
$$f(x) = f(x_0) + f'(x_0)(x - x_0) + \frac{f''(x_0)}{2!}(x - x_0)^2 + \frac{f'''(x_0)}{3!}(x - x_0)^3 + \cdots +$$
$$\frac{f^{(n)}(x_0)}{n!}(x - x_0)^n + \frac{f^{(n+1)}(c)}{(n + 1)!}(x - x_0)^{n+1}.$$

Proof   Let $P_n(t) = \sum_{k=0}^{n} [f^{(k)}(x_0)/k!](t - x_0)^k$, where $f^{(0)} = f$ and t is in $[a, b]$ (x and x_0 are fixed and t is the variable). We want to show that

$$f(x) = P_n(x) + \frac{f^{(n+1)}(c)}{(n + 1)!}(x - x_0)^{n+1}$$

for some c between x and x_0. Letting

$$M = \frac{f(x) - P_n(x)}{(x - x_0)^{n+1}},$$

we want to show that $f^{(n+1)}(c) = M(n + 1)!$ for some c between x and x_0.

Define $g : [a, b] \to \mathbb{R}$ by

$$g(t) = f(t) - P_n(t) - M(t - x_0)^{n+1}.$$

For t in (a, b), $g^{(n+1)}(t) = f^{(n+1)}(t) - 0 - M(n + 1)!$, and so we want to show that $g^{(n+1)}(c) = 0$ for some c between x and x_0. Since $f^{(k)}(x_0) = P_n^{(k)}(x_0)$ for $k = 0, 1, 2, \ldots, n$,

$$g(x_0) = g'(x_0) = \cdots = g^{(n)}(x_0) = 0.$$

Also, $g(x) = 0$ by our choice of M. By the Mean Value Theorem (or by Rolle's Theorem, Exercise 8 in Section 5.1) applied to g on the closed interval with endpoints x and x_0, there is a c_1 between x and x_0 such that $g'(c_1) = 0$. Applying the Mean Value Theorem to g' on the closed interval with endpoints c_1 and x_0, there is a c_2 between c_1 and x_0 such that $g''(c_2) = 0$. Continuing, we get that there is a c between c_n and x_0, and hence between x and x_0, with $g^{(n+1)}(c) = 0$. ■

The conclusion of Taylor's Theorem is often written as

$$f(x) = P_n(x) + R_n(x),$$

where

$$P_n(x) = \sum_{k=0}^{n} \frac{f^{(k)}(x_0)}{k!}(x - x_0)^k$$

is the nth Taylor polynomial of f at x_0 and

$$R_n(x) = \frac{f^{(n+1)}(c)}{(n + 1)!}(x - x_0)^{n+1}$$

is the remainder term. To estimate the error in using $P_n(x)$ to approximate $f(x)$, we need to find a bound on $R_n(x)$ and hence a bound on $f^{(n+1)}(c)$.

Example 5.6   Let $f(x) = e^x$ and $x_0 = 0$. Then

$$P_n(x) = 1 + x + \frac{x^2}{2!} + \cdots + \frac{x^n}{n!},$$

which the reader may know is a partial sum of the Taylor series for f centered at 0. Also,

$$R_n(x) = \frac{e^c x^{n+1}}{(n + 1)!}$$

for some c between 0 and x.

If $|x| \leq 1$, then

$$|R_n(x)| \leq \frac{e}{(n+1)!} < \frac{3}{(n+1)!}.$$

Thus, for $|x| \leq 1$, to estimate e^x by $P_n(x)$ to within 10^{-6}, we need to choose n such that $(n+1)! > 3 \times 10^6$, and hence $n = 9$ will do.

Proposition 5.4 The number e is irrational.

Proof As in Example 5.6, by Taylor's Theorem, we have

$$e^x = 1 + x + \frac{x^2}{2!} + \cdots + \frac{x^n}{n!} + \frac{e^c x^{n+1}}{(n+1)!},$$

where n is in $\mathbb{N}$ and c is between 0 and x. Letting $x = 1$,

$$e = 2 + \frac{1}{2!} + \cdots + \frac{1}{n!} + \frac{e^c}{(n+1)!}, \tag{4}$$

with $0 < c < 1$.

Assume that e is rational and write $e = p/q$, where p and q are in $\mathbb{N}$ with p/q in lowest terms. Let n be in $\mathbb{N}$ with $n > \max\{q, e^c\}$. Multiplying (4) by $n!$, we have

$$n!\frac{p}{q} = \left[2 + \frac{1}{2!} + \cdots + \frac{1}{n!} \right] n! + \frac{e^c}{n+1}.$$

Since $n > q$, $n!(p/q)$ is a positive integer, as is $[2 + (1/2!) + \cdots + (1/n!)]n!$. Thus, $e^c/(n+1)$ is an integer. Since $n > e^c$, $0 < e^c/(n+1) < 1$. So we have produced a positive integer strictly between 0 and 1, which is a contradiction. Therefore, e is irrational. ∎

Proving that π is irrational is much harder than proving that e is irrational. The interested reader should see Simmons.

≡ Exercises ≡

In Exercises 1 through 3, let $x_0 = 0$ and calculate $P_7(x)$ and $R_7(x)$.

1. $f(x) = \sin x$, x in $\mathbb{R}$.
2. $f(x) = \cos x$, x in $\mathbb{R}$.
3. $f(x) = \ln(1 + x)$, $x \geq 0$.
4. In Exercises 1, 2, and 3, for $|x| \leq 1$, calculate a value of n such that $P_n(x)$ approximates $f(x)$ to within 10^{-6}.
5. Let $(a_n)_{n \in \mathbb{N}}$ be a sequence of positive real numbers such that $L = \lim_{n \to \infty} (a_{n+1}/a_n)$ exists in $\mathbb{R}$. If $L < 1$, show that $a_n \to 0$. [*Hint:* Let $L < r < 1$ and note that eventually $a_{n+1}/a_n < r$].
6. (a) Use Exercise 5 to show that for all x in $\mathbb{R}$, $R_n(x) \to 0$ for e^x in Example 5.6 and the functions in Exercises 1 and 2.
   (b) For the function in Exercise 3, show that $R_n(x) \to 0$ for $x \leq 1$.

7. If $x > 0$, show that
$$1 + x + \frac{x^2}{2} < e^x < 1 + x + \frac{x^2}{2}e^x.$$

8. The purpose of this exercise is to give a different proof of Taylor's Theorem, assuming a slightly stronger hypothesis, which produces a different form of the remainder. Assume that $f^{(n+1)}$ is continuous on $[a, b]$ and recall from calculus the integration by parts formula:
$$\int u \, dv = uv - \int v \, du.$$
For $x \neq x_0$ in $[a, b]$,
$$f(x) - f(x_0) = f(t) \mid_{x_0}^{x}$$
$$= \int_{x_0}^{x} f'(t) \, dt$$
$$= f'(t)(t - x) \mid_{x_0}^{x} - \int_{x_0}^{x} f''(t)(t - x) \, dt,$$
where in the integration by parts we used
$$u = f'(t), \quad du = f''(t) \, dt$$
$$dv = dt, \quad v = t - x.$$
[*Note:* We usually set $v = t$, but our choice here is also valid.] Hence,
$$f(x) = f(x_0) + f'(x_0)(x - x_0) - \int_{x_0}^{x} f''(t)(t - x) \, dt.$$
Continue to integrate by parts [$u = f''(t), dv = (t - x)dt$] to obtain
$$f(x) = f(x_0) + f'(x_0)(x - x_0) + \frac{f''(x_0)}{2!}(x - x_0)^2 + \cdots +$$
$$\frac{f^{(n)}(x_0)}{n!}(x - x_0)^n + (-1)^n \int_{x_0}^{x} \frac{f^{(n+1)}(t)(t - x)^n}{n!} \, dt.$$

5.4 L'Hôpital's Rule

The reader is probably aware that L'Hôpital's rule is used frequently in the evaluation of limits. It applies to indeterminate forms of the type $\frac{0}{0}$ or ∞/∞. Other indeterminate forms such as 0^0, 1^∞, and $\infty - \infty$ must first be reduced to the type $\frac{0}{0}$ or ∞/∞ in order to use L'Hôpital's rule. This is usually accomplished by taking exponentials or logarithms, or by performing algebraic manipulations; we illustrate this in Example 5.8. In Example 5.7 we show that L'Hôpital's rule cannot always be applied.

Theorem 5.7 (L'Hôpital's rule) Let f and g be differentiable on (a, b) where $-\infty \leq a < b \leq \infty$ and let $g'(x) \neq 0$ for all x in (a, b). Let L be in $\mathbb{R} \cup \{\pm\infty\}$ and suppose that
$$\frac{f'(x)}{g'(x)} \to L \text{ as } x \to a. \tag{5}$$

If
$$f(x) \to 0 \text{ and } g(x) \to 0 \text{ as } x \to a \tag{6}$$
or if
$$g(x) \to \infty \text{ as } x \to a, \tag{7}$$
then
$$\frac{f(x)}{g(x)} \to L \text{ as } x \to a. \tag{8}$$
A similar statement is true if $g(x) \to -\infty$ in (7) or $x \to b$.

Proof   This proof is taken essentially from Rudin.

We first let $-\infty \le L < \infty$. Let q and r be in $\mathbb{R}$, with $L < r < q$. By (5), there is a c in (a, b) such that $a < x < c$ implies
$$\frac{f'(x)}{g'(x)} < r.$$
For $x \ne y$ in (a, c), by Theorem 5.4, there is a t between x and y such that
$$\frac{f(y) - f(x)}{g(y) - g(x)} = \frac{f'(t)}{g'(t)} < r. \tag{9}$$
[Note that $g(y) \ne g(x)$, because otherwise the Mean Value Theorem would imply that $g'(u) = 0$ for some u in (a, b).]

Suppose that (6) holds. Letting $x \to a$ in (9), we have
$$\frac{f(y)}{g(y)} \le r < q \text{ for all } a < y < c.$$

Next, suppose that (7) holds. Keeping y fixed in (9), (7) implies that there is a c_1 in (a, c) such that $g(x) > g(y)$ and $g(x) > 0$ for $a < x < c_1$. Multiplying (9) by $[g(x) - g(y)]/g(x)$, we obtain
$$\frac{f(x)}{g(x)} < r - r\frac{g(y)}{g(x)} + \frac{f(y)}{g(x)} \text{ for } a < x < c_1. \tag{10}$$
Letting $x \to a$ in (10) and using (7), there is a c_2 in (a, c_1) such that
$$\frac{f(x)}{g(x)} \le r < q \text{ for } a < x < c_2.$$

If $-\infty < L \le \infty$, letting p be in $\mathbb{R}$ with $p < L$, we can show in the same manner as above that there is a c_3 in (a, b) such that
$$p < \frac{f(x)}{g(x)} \text{ for } a < x < c_3.$$
Hence, (8) holds.   ∎

The proof above shows that the right-hand limit of $f(x)/g(x)$ as x approaches a is L under the appropriate hypotheses. Combining this with the corresponding statement for the left-hand limit as x approaches b, we have that L'Hôpital's rule also applies to the limit as x approaches a real number.

Example 5.7   This example points out a nuance of L'Hôpital's rule:
$$\lim_{x \to \infty} \frac{4x + \sin x}{2x - \cos\sqrt{x}} = \lim_{x \to \infty} \frac{4 + \cos x}{2 + (\sin\sqrt{x})/2\sqrt{x}} \qquad \text{(by L'Hôpital).}$$

The first limit is of the form ∞/∞. The second limit does not exist, because the denominator approaches 2 and $\lim\limits_{x\to\infty} \cos x$ does not exist. Therefore, (5) in Theorem 5.7 does not hold, and thus L'Hôpital's rule does not apply.

We evaluate this limit as follows:

$$\lim_{x\to\infty} \frac{4x + \sin x}{2x - \cos \sqrt{x}} = \lim_{x\to\infty} \frac{4 + (\sin x)/x}{2 - (\cos \sqrt{x})/x}$$

$$= \frac{4 + 0}{2 - 0} = 2.$$

Notation   We will denote the left-hand (respectively, right-hand) limit of f as x approaches a by $\lim\limits_{x\to a^-} f(x)$ (respectively, $\lim\limits_{x\to a^+} f(x)$).

Example 5.8   We calculate $\lim\limits_{x\to 0^+} x^{\sin x}$. Note that this is a 0^0 indeterminate form and that L'Hôpital's rule applies to $\frac{0}{0}$ or ∞/∞ indeterminate forms. Since

$$x^{\sin x} = e^{\ln(x^{\sin x})} = e^{\sin x \ln x}$$

and e^x is continuous,

$$\lim_{x\to 0^+} x^{\sin x} = e^{\lim\limits_{x\to 0^+} \sin x \ln x}.$$

Also,

$$\lim_{\substack{x\to 0^+ \\ (0\cdot-\infty \text{ form})}} \sin x \ln x = \lim_{x\to 0^+} \frac{\ln x}{\csc x} \qquad \left(\tfrac{-\infty}{\infty} \text{form}\right)$$

$$= \lim_{x\to 0^+} \frac{1/x}{-\csc x \cot x} \qquad \text{(by L'Hôpital)}$$

$$= -\lim_{x\to 0^+} \frac{\sin x \tan x}{x} \qquad \left(\tfrac{0}{0} \text{ form}\right)$$

$$= -\lim_{x\to 0^+} \frac{\sin x \sec^2 x + \cos x \tan x}{1} \qquad \text{(by L'Hôpital)}$$

$$= 0.$$

Therefore,

$$\lim_{x\to 0^+} x^{\sin x} = e^0 = 1.$$

≡ Exercises ≡

Evaluate the following limits.

1. $\displaystyle\lim_{x\to 0^+} \cot x / \ln x$

2. $\displaystyle\lim_{x\to 0^+} x^2 \ln x$

3. $\displaystyle\lim_{x\to 0^+} x^x$

4. $\displaystyle\lim_{x\to 0^+} (\cos \sqrt{x})^{1/x}$

5. $\displaystyle\lim_{x\to 0} x^2 / (1 - \cos x)$

6. $\displaystyle\lim_{x\to\infty} e^x / \pi^x$

7. $\displaystyle\lim_{x\to\pi/2^-} (\sec x - \tan x)$

8. $\displaystyle\lim_{x\to\infty} [1 + (3/x)]^x$

6 Riemann Integration

In Section 6.1 we define the Riemann integral of a bounded function on a finite interval $[a, b]$ in terms of upper and lower sums. In the important Theorem 6.2 we obtain a necessary and sufficient condition for such a function to be Riemann integrable. In Section 6.2 we realize the Riemann integral as the limit of Riemann sums. Thus, Sections 6.1 and 6.2 establish the notation for the rest of the chapter and give us two ways to view a Riemann integral.

In Section 6.3 we derive the usual properties of the Riemann integral, such as linearity and monotonicity; and we obtain Theorems 6.5 and 6.6, which will help us show that various functions are integrable. In Section 6.4 we establish that continuous functions and monotone functions on $[a, b]$ are integrable, and we derive two Mean Value Theorems.

In Section 6.5 we obtain the Fundamental Theorem of Calculus, which shows the intimate connection between differentiation and integration. Thus, differentiation and integration are, in some sense, inverse operations. In Section 6.6 we evaluate improper integrals and establish comparison tests for deciding when an improper integral converges or diverges.

6.1 Existence of the Riemann Integral

Definition 6.1 Let $[a, b]$ be a closed bounded interval in $\mathbb{R}$ with $a < b$. A *partition* P of $[a, b]$ is a finite set of points $\{x_0, x_1, x_2, \ldots, x_n\}$ where n is in $\mathbb{N}$ and $a = x_0 < x_1 < x_2 < \cdots < x_{n-1} < x_n = b$. Let $\Delta x_i = x_i - x_{i-1}$ for $i = 1, 2, \ldots, n$ and let $\mathcal{P}$ or $\mathcal{P}[a, b]$ denote the set of all partitions of $[a, b]$.

Definition 6.2 Let f be a bounded function from $[a, b]$ into $\mathbb{R}$ (see Figure 6.1). For $P = \{x_i\}_{i=0}^{n}$ in $\mathcal{P}$, set

$$M_i = \sup\{f(x) : x_{i-1} \leq x \leq x_i\}$$

and

$$m_i = \inf\{f(x) : x_{i-1} \leq x \leq x_i\}.$$

(Note that each M_i and m_i are real numbers since f is bounded.) Set

$$U(P, f) = \sum_{i=1}^{n} M_i \Delta x_i$$

(the *upper sum* of f with respect to P) and

$$L(P, f) = \sum_{i=1}^{n} m_i \Delta x_i$$

(the *lower sum* of f with respect to P). Set

$$\overline{\int_a^b} f = \inf\{U(P, f) : P \in \mathcal{P}\}$$

(the *upper Riemann integral* of f over $[a, b]$) and

$$\underline{\int_a^b} f = \sup\{L(P, f) : P \in \mathcal{P}\}$$

(the *lower Riemann integral* of f over $[a, b]$).

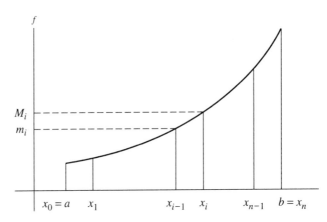

Figure 6.1

Remark The lower and upper Riemann integrals of f over $[a, b]$ always exist in $\mathbb{R}$. Since f is bounded, there exist real numbers m and M such that $m \le f(x) \le M$ for all x in $[a, b]$. For any P in $\mathcal{P}$, since $\sum_{i=1}^{n} \Delta x_i = (x_1 - x_0) + (x_2 - x_1) + \cdots + (x_{n-1} - x_{n-2}) + (x_n - x_{n-1}) = x_n - x_0 = b - a$, $m(b - a) \le L(P, f) \le U(P, f) \le M(b - a)$. Therefore, $\{L(P, f) : P \in \mathcal{P}\}$ and $\{U(P, f) : P \in \mathcal{P}\}$ are bounded sets of real numbers, and hence $\underline{\int_a^b} f$ and $\overline{\int_a^b} f$ are real numbers.

Our first goal is to establish that $\underline{\int_a^b} f \le \overline{\int_a^b} f$. Until we get to Section 6.6 our functions will be bounded.

Definition 6.3 For P and P^* in $\mathcal{P}$, P^* is a *refinement* of P if $P \subset P^*$.

Note that $P \cup P^*$ is a common refinement of both P and P^*. As will be seen below, this is an important observation.

Proposition 6.1   If P^* is a refinement of P, then

$$L(P, f) \le L(P^*, f) \text{ and } U(P^*, f) \le U(P, f).$$

Proof   We prove the first inequality, leaving the second one for the reader. Let $P = \{x_i\}_{i=0}^n$ be in $\mathcal{P}$ and suppose first that P^* contains just one more point than P. So $P^* = P \cup \{z\}$, where $x_{k-1} < z < x_k$ for some k in $\{1, 2, \ldots, n\}$. Then

$$L(P^*, f) = \sum_{i \ne k} m_i \Delta x_i + m'(z - x_{k-1}) + m''(x_k - z),$$

where

$$m' = \inf\{f(x) : x_{k-1} \le x \le z\}$$

and

$$m'' = \inf\{f(x) : z \le x \le x_k\}.$$

Since $m_k = \inf\{f(x) : x_{k-1} \le x \le x_k\}$, $m_k \le m'$ and $m_k \le m''$. Thus,

$$
\begin{aligned}
L(P^*, f) - L(P, f) &= m'(z - x_{k-1}) + m''(x_k - z) - m_k(x_k - x_{k-1}) \\
&= (m' - m_k)(z - x_{k-1}) + (m'' - m_k)(x_k - z) \ge 0.
\end{aligned}
$$

If P^* contains j more points than P, we repeat the reasoning above j times to arrive at $L(P, f) \le L(P^*, f)$.   ∎

Proposition 6.2   If P_1 and P_2 are in $\mathcal{P}$, then

$$L(P_1, f) \le U(P_2, f).$$

Proof   Let $P^* = P_1 \cup P_2$. Then P^* is a common refinement of P_1 and P_2. By Proposition 6.1,

$$L(P_1, f) \le L(P^*, f) \le U(P^*, f) \le U(P_2, f).$$ ∎

Theorem 6.1 $\underline{\int_a^b} f \le \overline{\int_a^b} f$.

Proof   Fix P_1 in $\mathcal{P}$. By Proposition 6.2, $U(P_1, f)$ is an upper bound of $\{L(P, f) : P \in \mathcal{P}\}$. Therefore, $\underline{\int_a^b} f = \sup\{L(P, f) : P \in \mathcal{P}\} \le U(P_1, f)$. Because P_1 is arbitrary, $\underline{\int_a^b} f$ is a lower bound of $\{U(P, f) : P \in \mathcal{P}\}$ and so

$$\underline{\int_a^b} f \le \inf\{U(P, f) : P \in \mathcal{P}\} = \overline{\int_a^b} f.$$ ∎

Definition 6.4 A bounded function f is *Riemann integrable* on $[a, b]$ if $\underline{\int_a^b} f = \overline{\int_a^b} f$, and this common value is the *Riemann integral* of f over $[a, b]$.

Notation and Terminology Let $\mathcal{R}$ or $\mathcal{R}[a, b]$ denote the set of Riemann integrable functions on $[a, b]$. For f in $\mathcal{R}$, we denote the Riemann integral of f over $[a, b]$ by $\int_a^b f$ or $\int_a^b f(x)dx$. Also, f is called the *integrand* and $[a, b]$ is called the *interval of integration*.

Example 6.1 Let
$$f(x) = \begin{cases} 1 & \text{if } x \in \mathbb{Q} \\ 0 & \text{if } x \notin \mathbb{Q}. \end{cases}$$
Then f is not Riemann integrable on any interval $[a, b]$ with $a < b$. To see this, let $P = \{x_i\}_{i=0}^n$ be in $\mathcal{P}$. Since $\mathbb{Q}$ and $\mathbb{R} \setminus \mathbb{Q}$ are dense in $\mathbb{R}$, $M_i = 1$ and $m_i = 0$ for all $i = 1, 2, \ldots, n$. Hence, $U(P, f) = \sum_{i=1}^n \Delta x_i = b - a$ and $L(P, f) = 0$. Since this holds for all P in $\mathcal{P}$, $\overline{\int_a^b} f = b - a$ and $\underline{\int_a^b} f = 0$. Therefore, f is not in $\mathcal{R}[a, b]$.

We next establish a necessary and sufficient condition for Riemann integrability. We then use this important theorem to show that certain discontinuous functions are Riemann integrable.

Theorem 6.2 A bounded function f is in $\mathcal{R}[a, b]$ if and only if for every $\varepsilon > 0$ there is a partition P of $[a, b]$ such that $U(P, f) - L(P, f) < \varepsilon$.

Proof   First suppose that f is in $\mathcal{R}[a, b]$ and let $\varepsilon > 0$. Since $\int_a^b f = \sup\{L(P, f) : P \in \mathcal{P}\}$, there is a P_1 in $\mathcal{P}$ such that $\int_a^b f - \varepsilon/2 < L(P_1, f)$. Since $\int_a^b f = \inf\{U(P, f) : P \in \mathcal{P}\}$, there is a P_2 in $\mathcal{P}$ such that $\int_a^b f + \varepsilon/2 > U(P_2, f)$. Let $P = P_1 \cup P_2$. By Proposition 6.1,
$$\int_a^b f - \frac{\varepsilon}{2} < L(P_1, f) \leq L(P, f) \leq U(P, f)$$
$$\leq U(P_2, f) < \int_a^b f + \frac{\varepsilon}{2},$$
and so $U(P, f) - L(P, f) < \varepsilon$.

Conversely, assume the statement after the if and only if. To show that f is in $\mathcal{R}[a, b]$, we need to show $\underline{\int_a^b} f = \overline{\int_a^b} f$. Let $\varepsilon > 0$. By hypothesis, there is a P in $\mathcal{P}$ with $U(P, f) - L(P, f) < \varepsilon$. Since $L(P, f) \leq \underline{\int_a^b} f \leq \overline{\int_a^b} f \leq U(P, f)$, $0 \leq \overline{\int_a^b} f - \underline{\int_a^b} f < \varepsilon$. Since ε is arbitrary, $\overline{\int_a^b} f - \underline{\int_a^b} f = 0$.   ∎

Example 6.2 Let
$$f(x) = \begin{cases} 1 & \text{if } 0 < x \leq 1 \\ 0 & \text{if } x = 0. \end{cases}$$

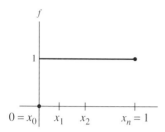

Figure 6.2

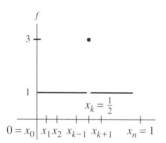

Figure 6.3

We use Theorem 6.2 to show that f is in $\mathcal{R}[0, 1]$. Let $P = \{x_i\}_{i=0}^n$ be any partition of $[0, 1]$ (Figure 6.2). Then $U(P, f) = \sum\limits_{i=1}^n 1\Delta x_i = 1 - 0 = 1$ and $L(P, f) = 0\Delta x_1 + \sum\limits_{i=2}^n 1\Delta x_i = 1 - x_1$. Hence, $U(P, f) - L(P, f) = x_1$.

Let $\varepsilon > 0$. Choose $P = \{x_i\}_{i=0}^n$ in $\mathcal{P}[0, 1]$ with $x_1 < \varepsilon$. Then $U(P, f) - L(P, f) < \varepsilon$. Thus, f is in $\mathcal{R}[0, 1]$ and $\int_a^b f = \inf\{U(P, f) : P \in \mathcal{P}\} = 1$.

Definition 6.5   For $P = \{x_i\}_{i=0}^n$ in $\mathcal{P}[a, b]$, the *mesh* (or *norm*) of P, denoted by $\|P\|$, is given by

$$\|P\| = \max\{\Delta x_i : i = 1, 2, \ldots, n\}.$$

In Example 6.2 we could have chosen P in $\mathcal{P}[0, 1]$ with $\|P\| < \varepsilon$ to get that $U(P, f) - L(P, f) = x_1 \leq \|P\| < \varepsilon$. Also note that if P^* is a refinement of P, then $\|P^*\| \leq \|P\|$.

Example 6.3   Let

$$f(x) = \begin{cases} 1 & \text{if } x \in [0, 1] \setminus \left\{\dfrac{1}{2}\right\} \\[2ex] 3 & \text{if } x = \dfrac{1}{2}. \end{cases}$$

Again, we use Theorem 6.2 to show that f is in $\mathcal{R}[0, 1]$. Let $P = \{x_i\}_{i=0}^n$ be any partition of $[0, 1]$ (Figure 6.3). By passing to a refinement, we may assume that $\frac{1}{2}$ is in P and that $x_k = \frac{1}{2}$ for some k in $\{2, \ldots, n-2\}$. (By adding points to P, we do not increase the mesh of P.) Then

$$L(P, f) = \sum_{i=1}^n 1\Delta x_i = 1 - 0 = 1$$

and

$$U(P, f) = \sum_{i=1}^{k-1} 1\Delta x_i + 3(x_k - x_{k-1}) + 3(x_{k+1} - x_k) + \sum_{i=k+2}^n 1\Delta x_i$$
$$= (x_{k-1} - 0) - 3x_{k-1} + 3x_{k+1} + (1 - x_{k+1})$$
$$= 1 + 2(x_{k+1} - x_{k-1}).$$

So,

$$U(P, f) - L(P, f) = 2(x_{k+1} - x_{k-1})$$
$$= 2[(x_{k+1} - x_k) + (x_k - x_{k-1})]$$
$$\leq 2[\|P\| + \|P\|] = 4\|P\|.$$

Given $\varepsilon > 0$, choose P in $\mathcal{P}[0, 1]$ with the properties given above and with $\|P\| < \varepsilon/4$. Then $U(P, f) - L(P, f) < \varepsilon$. Thus, f is in $\mathcal{R}[0, 1]$ and $\int_a^b f = \sup\{L(P, f) : P \in \mathcal{P}\} = 1$.

≡ Exercises ≡

Use Theorem 6.2 in Exercises 1 through 8.

1. Let $f(x) = c$ on $[a, b]$. Show that f is in $\mathcal{R}[a, b]$ and $\int_a^b f = c(b - a)$.

2. Let

$$f(x) = \begin{cases} 1 & \text{if} \quad 0 \leq x < 1 \\ 2 & \text{if} \quad x = 1. \end{cases}$$

Show that f is in $\mathcal{R}[0, 1]$ and find $\int_0^1 f$.

3. Let

$$f(x) = \begin{cases} 1 & \text{if} \quad 0 < x < 1 \\ 0 & \text{if} \quad x = 0 \text{ or } 1. \end{cases}$$

Show that f is in $\mathcal{R}[0, 1]$ and find $\int_0^1 f$.

4. Let

$$f(x) = \begin{cases} 1 & \text{if} \quad 0 \leq x < 1 \\ 3 & \text{if} \quad x = 1 \\ -1 & \text{if} \quad 1 < x \leq 2. \end{cases}$$

Show that f is in $\mathcal{R}[0, 2]$ and find $\int_0^2 f$.

5. Let

$$f(x) = \begin{cases} 0 & \text{if} \quad x = 1, \dfrac{1}{2}, \dfrac{1}{3}, \dfrac{1}{4}, \dots \\ 1 & \text{otherwise.} \end{cases}$$

Show that f is in $\mathcal{R}[0, 1]$ and find $\int_0^1 f$.

6. Let

$$f(x) = \begin{cases} 1 & \text{if} \quad x \in \mathbb{Q} \\ -1 & \text{if} \quad x \in \mathbb{R} \setminus \mathbb{Q}. \end{cases}$$

Show that f is not in $\mathcal{R}[a, b]$ for any $a < b$.

7. Let

$$f(x) = \begin{cases} x & \text{if} \quad x \in \mathbb{Q} \\ 0 & \text{if} \quad x \in \mathbb{R} \setminus \mathbb{Q}. \end{cases}$$

Show that f is not in $\mathcal{R}[0, 1]$.

8. Define $f : [0, 1] \to \mathbb{R}$ by

$$f(x) = \begin{cases} 0 & \text{if} \quad x \text{ is irrational} \\ \dfrac{1}{n} & \text{if} \quad x \text{ is rational and } x = \dfrac{m}{n} \\ & \quad \text{in lowest terms with } m \text{ and } n \text{ in } \mathbb{N} \\ 1 & \text{if} \quad x = 0. \end{cases}$$

From Example 4.8 and Exercise 8 in Section 4.2, f is continuous at every irrational number and discontinuous at every rational number. Show that f is in $\mathcal{R}[0, 1]$ and find $\int_0^1 f$. [*Hint:* Since each lower sum is 0, given $\varepsilon > 0$, we want a P such that $U(P, f) < \varepsilon$. In $[0, 1]$ there are only a finite number (say r) of points m/n with $1/n > \varepsilon/2$. Choose P with $\|P\| < \varepsilon/4r$.]

9. By considering upper and lower sums for $\int_1^n \ln x \, dx$, show that

$$\lim_{n \to \infty} \frac{n!}{n^n} = 0 \quad \text{and} \quad \lim_{n \to \infty} \frac{(n!)^{1/n}}{n} = \frac{1}{e}.$$

[*Hint:* Let $P = \{1, 2, 3, \ldots, n\}$.]

10. Prove the second inequality in Proposition 6.1.

6.2 Riemann Sums

In this section we realize $\int_a^b f$ as the limit of Riemann sums.

> **Definition 6.6**   Let $P = \{x_i\}_{i=0}^n$ be in $\mathcal{P}[a, b]$ and let f be a bounded function from $[a, b]$ into $\mathbb{R}$. For each $i = 1, 2, \ldots, n$ choose points $t_1, t_2, \ldots, t_n$ with $x_{i-1} \leq t_i \leq x_i$. A *Riemann sum* for f with respect to P, denoted by $S(P, f)$, is given by
>
> $$S(P, f) = \sum_{i=1}^n f(t_i) \, \Delta x_i.$$

Remark   For each P there are infinitely many Riemann sums that can be formed by varying the choice of the t_i's. In this sense, the notation $S(P, f)$ is ambiguous. If we keep in mind that $S(P, f)$ stands for infinitely many sums, one for each choice of the t_i's, then no difficulty should arise.

Note that $L(P, f)$ and $U(P, f)$ may not be Riemann sums since m_i and M_i may not be in the range of f. Of course, if f is continuous on $[a, b]$, then $L(P, f)$ and $U(P, f)$ are Riemann sums (Theorem 4.2). Since $m_i \leq f(t_i) \leq M_i$ for all i, we always have

$$L(P, f) \leq S(P, f) \leq U(P, f)$$

for all partitions P and all Riemann sums.

> **Definition 6.7**   Let I be in $\mathbb{R}$. The *limit of the Riemann sums* for f with respect to P as the mesh of P tends to 0, denoted by $\lim_{\|P\| \to 0} S(P, f)$, is I if for every $\varepsilon > 0$ there is a $\delta > 0$ such that if $P = \{x_i\}_{i=0}^n$ is in $\mathcal{P}$ with $\|P\| < \delta$, then $|S(P, f) - I| < \varepsilon$ for all possible choices of the t_i's in $[x_{i-1}, x_i]$.

Theorem 6.3   A bounded function f is in $\mathcal{R}[a, b]$ if and only if $\lim\limits_{\|P\|\to 0} S(P, f)$ exists in $\mathbb{R}$, and then $\int_a^b f = \lim\limits_{\|P\|\to 0} S(P, f)$.

Proof   The ideas of this proof are taken from Fridy. Suppose that f is in $\mathcal{R}[a, b]$ and let $\varepsilon > 0$. Since f is bounded, there is a $B > 0$ with $|f(x)| \le B$ for all x in $[a, b]$.

Since $\int_a^b f = \overline{\int_a^b} f$, there is a $P' = \{x_i\}_{i=0}^n$ in $\mathcal{P}$ with $U(P', f) < \int_a^b f + \varepsilon/2$. Let $\delta_1 = \varepsilon/(4nB)$ and let P be in $\mathcal{P}$ with $\|P\| < \delta_1$.

Consider $P^* = P \cup P'$. We first show that $U(P, f) - U(P^*, f) < \varepsilon/2$. In $U(P^*, f)$ we have two types of terms (see Figure 6.4):

$$\sum M_i \Delta z_i \text{ where } [z_{i-1}, z_i] \text{ contains no } x_j\text{'s and} \tag{1}$$

$$\sum M_i \Delta z_i \text{ where } z_{i-1} \text{ or } z_i, \text{ or both, are } x_j\text{'s.} \tag{2}$$

Since the type (1) terms are also terms of $U(P, f)$, $U(P, f) - U(P^*, f)$ consists only of type (2) terms. Since the number of type (2) terms is at most $2n$ and since each type (2) term is less than or equal to

$$B \, \|P^*\| \le B \, \|P\| < B\delta_1,$$

$$U(P, f) - U(P^*, f) < 2n(B\delta_1) = 2nB\varepsilon/(4nB) = \varepsilon/2.$$

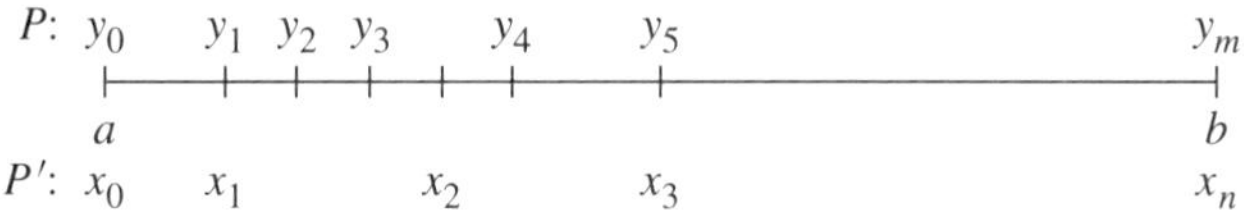

Figure 6.4

Using Proposition 6.1 and our choice of P',

$$U(P, f) < U(P^*, f) + \frac{\varepsilon}{2} \le U(P', f) + \frac{\varepsilon}{2} < \int_a^b f + \varepsilon$$

for all P in $\mathcal{P}$ with $\|P\| < \delta_1$.

Similarly, using the fact that $\int_a^b f = \underline{\int_a^b} f$, there is a $\delta_2 > 0$ such that $L(P, f) > \int_a^b f - \varepsilon$ for all P in $\mathcal{P}$ with $\|P\| < \delta_2$. Let $\delta = \min\{\delta_1, \delta_2\}$ and let P be in $\mathcal{P}$ with $\|P\| < \delta$. Then for any Riemann sum,

$$\int_a^b f - \varepsilon < L(P, f) \le S(P, f) \le U(P, f) < \int_a^b f + \varepsilon$$

and so $\left| S(P, f) - \int_a^b f \right| < \varepsilon$. Therefore, $\lim\limits_{\|P\|\to 0} S(P, f) = \int_a^b f$.

Now suppose that $\lim\limits_{\|P\|\to 0} S(P, f) = I$, where I is in $\mathbb{R}$. Let $\varepsilon > 0$. To show that f is in $R[a, b]$, by Theorem 6.2, we need to find a P in $\mathcal{P}$ with $U(P, f) - L(P, f) < \varepsilon$. By Definition 6.7 there is a $\delta > 0$ such that if P is in $\mathcal{P}$ with $\|P\| < \delta$, then $|S(P, f) - I| < \varepsilon/4$ for all possible choices of the t_i's.

Fix one $P = \{x_i\}_{i=0}^n$ in $\mathcal{P}$ with $\|P\| < \delta$. For each $i = 1, 2, \ldots, n$, since $M_i = \sup\{f(x) : x \in [x_{i-1}, x_i]\}$, there is a t_i in $[x_{i-1}, x_i]$ with

$$f(t_i) > M_i - \frac{\varepsilon}{4(b - a)}.$$

Then

$$U(P, f) = \sum_{i=1}^{n} M_i \Delta x_i$$

$$< \sum_{i=1}^{n} \left[f(t_i) + \frac{\varepsilon}{4(b-a)} \right] \Delta x_i$$

$$= \sum_{i=1}^{n} f(t_i) \Delta x_i + \frac{\varepsilon}{4(b-a)} \sum_{i=1}^{n} \Delta x_i$$

$$< I + \frac{\varepsilon}{4} + \frac{\varepsilon}{4}$$

$$= I + \frac{\varepsilon}{2}$$

since $\sum_{i=1}^{n} f(t_i) \Delta x_i$ is one particular $S(P, f)$.

Similarly, by approximating m_i for $i = 1, 2, \ldots, n$, $L(P, f) > I - \varepsilon/2$.

Therefore, $U(P, f) - L(P, f) < \varepsilon$, and so f is in $\mathcal{R}[a, b]$. Since $\int_a^b f$ and each $S(P, f)$ are in the closed interval $[L(P, f), U(P, f)]$, $|S(P, f) - \int_a^b f| < \varepsilon$ for all P with $\|P\| < \delta$, and so $I = \int_a^b f$.   ∎

Example 6.4   Let f be in $\mathcal{R}[0, 1]$ and for n in $\mathbb{N}$, let $P_n = \{i/n\}_{i=0}^{n}$ be a partition of $[0, 1]$. Then $(1/n) \sum_{i=1}^{n} f(i/n)$ is a Riemann sum for f. By Theorem 6.3 and Exercise 1,

$$\int_0^1 f = \lim_{n \to \infty} \frac{1}{n} \sum_{i=1}^{n} f\left(\frac{i}{n}\right).$$

In particular,

$$\lim_{n \to \infty} \frac{1}{n} \left[\left(\frac{1}{n}\right)^2 + \left(\frac{2}{n}\right)^2 + \cdots + \left(\frac{n}{n}\right)^2 \right] = \int_0^1 x^2 dx = \frac{1}{3}. \qquad \text{(by calculus)}$$

This can be written more provocatively as $\lim_{n \to \infty} 1/n^3 (1^2 + 2^2 + \cdots + n^2) = 1/3$. We can also do this algebraically as follows:

$$\frac{1}{n^3} \left(1^2 + 2^2 + \cdots + n^2\right) = \frac{1}{n^3} \frac{n(n+1)(2n+1)}{6} \qquad \text{(by Example 1.13)}$$

$$= \frac{1}{6} \left(\frac{n+1}{n}\right) \left(\frac{2n+1}{n}\right)$$

$$\xrightarrow[n]{} \frac{1}{6}(1)(2)$$

$$= \frac{1}{3}.$$

As the example above shows, sometimes a limit can be evaluated by realizing the limit as a Riemann sum for a particular function. We end this section with a definition that extends our interval of integration.

Definition 6.8 For f in $R[a, b]$, we set $\int_b^a f = -\int_a^b f$; and for any function f defined at a, we set $\int_a^a f = 0$.

═══ Exercises ═══

1. Suppose that $\lim\limits_{\|P\|\to 0} S(P, f) = I$, where I is in $\mathbb{R}$. Show that if $(P_n)_{n\in\mathbb{N}}$ is a sequence of partitions with $\|P_n\| \underset{n}{\to} 0$, then $\lim\limits_{n\to\infty} S(P_n, f) = I$.

   In Exercises 2–5, express the limit as a Riemann integral. (That the integrand is Riemann integrable on the appropriate interval will be shown in Section 6.4.)

2. $\lim\limits_{n\to\infty} (1/n) \sum\limits_{i=1}^{n} (i/n)$. Also calculate this limit algebraically.

3. $\lim\limits_{n\to\infty} \sum\limits_{i=1}^{n} 1/(i - 2n)$

4. $\lim\limits_{n\to\infty} \sum\limits_{i=0}^{n-1} 1/(n + i)$

5. $\lim\limits_{n\to\infty} (1/n) \sum\limits_{i=1}^{n} \sin(i\pi/2n)$

6. In the proof of Theorem 6.3, supply the details to show that there is a $\delta_2 > 0$ such that $L(P, f) > \int_a^b f - \varepsilon$ for all P with $\|P\| < \delta_2$.

6.3 Properties of the Riemann Integral

To show that a function is Riemann integrable, we use either Theorem 6.2 or Theorem 6.3, whichever is easier.

Proposition 6.3 Let f and g be in $\mathcal{R}[a, b]$ and let c be in $\mathbb{R}$. Then $f \pm g$ and cf are in $\mathcal{R}[a, b]$, and

$$\int_a^b (f \pm g) = \int_a^b f \pm \int_a^b g$$

and

$$\int_a^b cf = c \int_a^b f.$$

Proof We use Theorem 6.3. Let $\varepsilon > 0$. By Definition 6.7 there exist $\delta_1 > 0$ and $\delta_2 > 0$ such that for P in $\mathcal{P}[a, b]$, if $\|P\| < \delta_1$, then $\left|S(P, f) - \int_a^b f\right| < \varepsilon/2$ and if $\|P\| < \delta_2$, then $\left|S(P, g) - \int_a^b g\right| < \varepsilon/2$. Let $\delta = \min\{\delta_1, \delta_2\}$ and let $P = \{x_i\}_{i=0}^n$ be in $\mathcal{P}[a, b]$ with $\|P\| < \delta$. Since

$$S(P, f + g) = \sum_{i=1}^n [f(t_i) + g(t_i)] \Delta x_i$$

$$= \sum_{i=1}^n f(t_i) \Delta x_i + \sum_{i=1}^n g(t_i) \Delta x_i$$

$$= S(P, f) + S(P, g)$$

for all choices of the t_i's in $[x_{i-1}, x_i]$,

$$\left| S(P, f + g) - \left(\int_a^b f + \int_a^b g \right) \right|$$

$$= \left| S(P, f) + S(P, g) - \left(\int_a^b f + \int_a^b g \right) \right|$$

$$\leq \left| S(P, f) - \int_a^b f \right| + \left| S(P, g) - \int_a^b g \right|$$

$$< \frac{\varepsilon}{2} + \frac{\varepsilon}{2} = \varepsilon.$$

Therefore, $f + g$ is in $\mathcal{R}[a, b]$ and $\int_a^b (f + g) = \lim_{\|P\| \to 0} S(P, f + g) = \int_a^b f + \int_a^b g$.

If $c = 0$, then every Riemann sum for cf is zero and the result for cf follows. Assume that $c \neq 0$ and let $\varepsilon > 0$. By Definition 6.7 there exists a $\delta > 0$ such that if P is in $\mathcal{P}[a, b]$ with $\|P\| < \delta$, then $\left| S(P, f) - \int_a^b f \right| < \varepsilon/|c|$. Then for P in $\mathcal{P}[a, b]$ with $\|P\| < \delta$,

$$\left| S(P, cf) - c \int_a^b f \right| = \left| cS(P, f) - c \int_a^b f \right|$$

$$= |c| \left| S(P, f) - \int_a^b f \right|$$

$$< |c| \frac{\varepsilon}{|c|} = \varepsilon.$$

Thus, cf is in $\mathcal{R}[a, b]$ and $\int_a^b cf = c \int_a^b f$.

The result for $f - g$ follows by writing $f - g = f + cg$ with $c = -1$. ■

The next proposition tells us that the Riemann integral is monotone.

Proposition 6.4   Let f and g be in $\mathcal{R}[a, b]$, with $f(x) \leq g(x)$ for all x in $[a, b]$. Then $\int_a^b f \leq \int_a^b g$.

Proof   First suppose that h is in $\mathcal{R}[a, b]$, with $h(x) \geq 0$ for all x in $[a, b]$. Since $L(P, h) \geq 0$ for every P in $\mathcal{P}[a, b]$, $\int_a^b h = \sup\{L(P, h) : P \in \mathcal{P}[a, b]\} \geq 0$.

Since $(g - f)(x) \geq 0$ for all x in $[a, b]$, by Proposition 6.3,

$$0 \leq \int_a^b (g - f) = \int_a^b g - \int_a^b f$$

and so $\int_a^b f \leq \int_a^b g$. ■

As the reader is probably aware, it is often convenient to split the interval of integration. We accomplish this in the next proposition.

Proposition 6.5 Let f be in $\mathcal{R}[a, b]$ and let $a < c < b$. Then f is in $\mathcal{R}[a, c]$, f is in $\mathcal{R}[c, b]$, and $\int_a^b f = \int_a^c f + \int_c^b f$.

Proof We use Theorem 6.2. Let $\varepsilon > 0$, and let P be in $\mathcal{P}[a, b]$ with $U(P, f) - L(P, f) < \varepsilon$. By passing to a refinement and using Proposition 6.1, we can assume that c is in P. Then $P_1 = P \cap [a, c]$ is in $\mathcal{P}[a, c]$ and $P_2 = P \cap [c, b]$ is in $\mathcal{P}[c, b]$. Since $U(P, f) = U(P_1, f) + U(P_2, f)$ and $L(P, f) = L(P_1, f) + L(P_2, f)$, $U(P_1, f) - L(P_1, f) < \varepsilon$ and $U(P_2, f) - L(P_2, f) < \varepsilon$. By Theorem 6.2, f is in $\mathcal{R}[a, c]$ and f is in $\mathcal{R}[c, b]$.

Since $U(P_1, f) + U(P_2, f) - [L(P_1, f) + L(P_2, f)] < \varepsilon$,

$$\int_a^b f \leq U(P, f)$$
$$= U(P_1, f) + U(P_2, f)$$
$$< L(P_1, f) + L(P_2, f) + \varepsilon$$
$$\leq \int_a^c f + \int_c^b f + \varepsilon.$$

Since $\varepsilon > 0$ is arbitrary, $\int_a^b f \leq \int_a^c f + \int_c^b f$.

To obtain $\int_a^b f \geq \int_a^c f + \int_c^b f$, note that

$$\int_a^b f \geq L(P, f)$$
$$= L(P_1, f) + L(P_2, f)$$
$$> U(P_1, f) + U(P_2, f) - \varepsilon$$
$$\geq \int_a^c f + \int_c^b f - \varepsilon. \qquad \blacksquare$$

Remark Suppose that $a < b < c$ and f is in $\mathcal{R}[a, c]$. By Proposition 6.5, $\int_a^c f = \int_a^b f + \int_b^c f$ or, equivalently, $\int_a^b f = \int_a^c f - \int_b^c f$. By Definition 6.8, $\int_c^b f = -\int_b^c f$, and so $\int_a^b f = \int_a^c f + \int_c^b f$.

Example 6.5 The purpose of this example is to show that the composition of two Riemann integrable functions need not be Riemann integrable. Let f be the function defined in Exercise 8 in Section 6.1, and let g be the function in Example 6.2. Then both f and g are in $\mathcal{R}[0, 1]$. However, $g \circ f : [0, 1] \to [0, 1]$ is given by

$$(g \circ f)(x) = \begin{cases} 0 & \text{if} \quad x \text{ is irrational} \\ 1 & \text{if} \quad x \text{ is rational} \end{cases}$$

and, by Example 6.1, $g \circ f$ is not in $\mathcal{R}[0, 1]$.

Although the following result takes some work to establish, it is well worth the effort as Corollary 6.1 and Exercise 1 indicate.

Theorem 6.4 Let f be in $\mathcal{R}[a, b]$, with the range of f contained in the closed interval $[m, M]$. Let $\phi : [m, M] \to \mathbb{R}$ be continuous on $[m, M]$. Then $\phi \circ f$ is in $\mathcal{R}[a, b]$.

Proof   We use Theorem 6.2. Let $\varepsilon > 0$. By Proposition 4.8, there is a $K > 0$ with $|\phi(y)| \leq K$ for all y in $[m, M]$. Let $\varepsilon' = \varepsilon/(b-a+2K)$. By Theorem 4.4, ϕ is uniformly continuous on $[m, M]$; so there exists a $\delta > 0$ such that $\delta < \varepsilon'$, and if y_1 and y_2 are in $[m, M]$ with $|y_1 - y_2| < \delta$, then $|\phi(y_1) - \phi(y_2)| < \varepsilon'$.

Since f is in $\mathcal{R}[a, b]$, by Theorem 6.2, there is a partition $P = \{x_i\}_{i=0}^n$ of $[a, b]$ with

$$U(P, f) - L(P, f) < \delta^2. \tag{3}$$

Let M_i and m_i have the same meaning as in Definition 6.2 for f, and let M_i^* and m_i^* be the analogous numbers for $\phi \circ f$. Let

$$A = \{i \in \{1, 2, \ldots, n\} : M_i - m_i < \delta\} \text{ and}$$
$$B = \{i \in \{1, 2, \ldots, n\} : M_i - m_i \geq \delta\}.$$

For i in A, our choice of δ implies that $M_i^* - m_i^* \leq \varepsilon'$, and hence

$$\sum_{i \in A} \left(M_i^* - m_i^*\right) \Delta x_i \leq \varepsilon' \sum_{i \in A} \Delta x_i \leq \varepsilon'(b - a).$$

By (3),

$$\delta \sum_{i \in B} \Delta x_i \leq \sum_{i \in B} (M_i - m_i)\Delta x_i < \delta^2.$$

Hence, $\sum_{i \in B} \Delta x_i < \delta$. For i in B, $M_i^* - m_i^* \leq 2K$, and so

$$\sum_{i \in B} \left(M_i^* - m_i^*\right) \Delta x_i \leq 2K \sum_{i \in B} \Delta x_i < 2K\delta < 2K\varepsilon'.$$

Therefore,

$$U(P, \phi \circ f) - L(P, \phi \circ f) = \sum_{i=1}^n (M_i^* - m_i^*)\Delta x_i$$
$$= \sum_{i \in A}(M_i^* - m_i^*)\Delta x_i + \sum_{i \in B}(M_i^* - m_i^*)\Delta x_i$$
$$< \varepsilon'(b - a) + 2K\varepsilon'$$
$$= \varepsilon'(b - a + 2K)$$
$$= \varepsilon.$$

By Theorem 6.2, $\phi \circ f$ is in $\mathcal{R}[a, b]$.   ■

Corollary 6.1   If f and g are in $\mathcal{R}[a, b]$, then

1. fg is in $\mathcal{R}[a, b]$;

2. $|f|$ is in $\mathcal{R}[a, b]$ and $\left|\int_a^b f\right| \leq \int_a^b |f|$;

3. $\max\{f, g\}$ and $\min\{f, g\}$ are in $\mathcal{R}[a, b]$.

Proof

1. Letting $\phi(y) = y^2$ in Theorem 6.4, we have that h^2 is in $\mathcal{R}[a, b]$ whenever h is in $\mathcal{R}[a, b]$. Using Proposition 6.3, since

$$fg = \frac{1}{4}\left[(f + g)^2 - (f - g)^2\right],$$

fg is in $\mathcal{R}[a, b]$.

2. Letting $\phi(y) = |y|$, Theorem 6.4 shows that $|f|$ is in $\mathcal{R}[a, b]$. Since $-|f(x)| \leq f(x) \leq |f(x)|$ for all x in $[a, b]$, by Proposition 6.4,

$$-\int_a^b |f| \leq \int_a^b f \leq \int_a^b |f|,$$

and so $\left| \int_a^b f \right| \leq \int_a^b |f|$.

3. Using Proposition 6.3 and Part 2, this follows by expressing

$$\max(f, g) = \frac{1}{2}[|f - g| + f + g]$$

and

$$\min(f, g) = \frac{1}{2}[-|f - g| + f + g].$$ ∎

Exercise 7 asks for an example showing that $|f|$ Riemann integrable does not imply that f is Riemann integrable.

We end this section with two theorems that the reader should apply to Examples 6.2 and 6.3 and Exercises 2, 3, and 4 in Section 6.1. These two theorems will help us enlarge our collection of Riemann integrable functions.

Theorem 6.5   Suppose that f is in $\mathcal{R}[a, b]$ and that $g = f$ except at a finite number of points in $[a, b]$. Then g is in $\mathcal{R}[a, b]$ and $\int_a^b g = \int_a^b f$.

Proof   First assume that $g = f$ on $[a, b]$ except at exactly one point z in $[a, b]$. Let $P = \{x_i\}_{i=0}^n$ be in $\mathcal{P}[a, b]$ and let $\{t_i\}_{i=1}^n$ be any collection of points with $x_{i-1} \leq t_i \leq x_i$ for each $i = 1, 2, \ldots, n$. Since $S(P, f) = \sum_{i=1}^n f(t_i)\Delta x_i$ and $S(P, g) = \sum_{i=1}^n g(t_i)\Delta x_i$, the only way these Riemann sums can differ is if some $t_k = z$. Since z could be a partition point, say $z = x_k$, the worst that could happen would be for both t_k and t_{k+1} to equal z. Then

$$|S(P, g) - S(P, f)| \leq |g(z) - f(z)| \Delta x_k + |g(z) - f(z)| \Delta x_{k+1}$$
$$\leq |g(z) - f(z)| 2 \|P\|.$$

Let $\varepsilon > 0$. Since f is in $\mathcal{R}[a, b]$, by Theorem 6.3, choose $0 < \delta < \varepsilon/(4|g(z) - f(z)|)$ such that if P is in $\mathcal{P}[a, b]$ with $\|P\| < \delta$, then $|S(P, f) - \int_a^b f| < \varepsilon/2$. Then $\|P\| < \delta$ implies that

$$\left| S(P, g) - \int_a^b f \right| \leq |S(P, g) - S(P, f)| + \left| S(P, f) - \int_a^b f \right|$$

$$\leq |g(z) - f(z)| 2 \|P\| + \left| S(P, f) - \int_a^b f \right|$$

$$< |g(z) - f(z)| 2\delta + \frac{\varepsilon}{2}$$

$$< \frac{\varepsilon}{2} + \frac{\varepsilon}{2}$$

$$= \varepsilon.$$

By Theorem 6.3, g is in $\mathcal{R}[a, b]$ and $\int_a^b g = \lim_{\|P\| \to 0} S(P, g) = \int_a^b f$.

Next, if g differs from f on $[a, b]$ at m points, we repeat the procedure above m times to show that g is in $\mathcal{R}[a, b]$ and $\int_a^b g = \int_a^b f$.   ∎

Exercise 2 states that if f is in $\mathcal{R}[a, b]$, then f is in $\mathcal{R}[c, d]$ for all closed subintervals $[c, d]$ of $[a, b]$. The following theorem reverses this situation.

Theorem 6.6   If f is bounded on $[a, b]$ and f is Riemann integrable on every closed subinterval of (a, b), then f is in $\mathcal{R}[a, b]$.

Proof   We use Theorem 6.2. Let $\varepsilon > 0$ and choose $B > 0$ such that $|f(x)| \leq B$ for all x in $[a, b]$. Let $a < c < d < b$. Since f is in $\mathcal{R}[c, d]$, there is a P in $\mathcal{P}[c, d]$ with $U(P, f) - L(P, f) < \varepsilon/3$. Let $P^* = \{a\} \cup P \cup \{b\}$. Then P^* is in $\mathcal{P}[a, b]$, and

$$U(P^*, f) - L(P^*, f) = \left[\sup_{x \in [a,c]} f(x) - \inf_{x \in [a,c]} f(x) \right] (c - a) +$$

$$[U(P, f) - L(P, f)] +$$

$$\left[\sup_{x \in [d,b]} f(x) - \inf_{x \in [d,b]} f(x) \right] (b - d) <$$

$$2B(c - a) + \frac{\varepsilon}{3} + 2B(b - d).$$

If we choose c and d with $a < c < d < b$, $c - a < \varepsilon/6B$, and $b - d < \varepsilon/6B$, then $U(P^*, f) - L(P^*, f) < \varepsilon$ and so f is in $\mathcal{R}[a, b]$.   ∎

Example 6.6   Recall Exercise 5 in Section 6.1. Let $f : [0, 1] \to [0, 1]$ be given by

$$f(x) = \begin{cases} 0 & \text{if } x = 1, \dfrac{1}{2}, \dfrac{1}{3}, \ldots \\ 1 & \text{otherwise.} \end{cases}$$

If $[c, d]$ is any closed subinterval of $(0, 1)$, then on $[c, d]$, f differs from the constant function 1 at only a finite number of points. By Theorem 6.5 and Exercise 1 in Section 6.1, f is in $\mathcal{R}[c, d]$. So, by Theorem 6.6, f is in $\mathcal{R}[0, 1]$. By Exercise 8,

$$\int_0^1 f = \lim_{c \to 0^+} \int_c^1 f = \lim_{c \to 0^+} (1 - c) = 1.$$

≡ Exercises ≡

1. Let f be in $\mathcal{R}[a, b]$.
   (a) If $f \geq 0$ on $[a, b]$, show that $\sqrt{f}$ is in $\mathcal{R}[a, b]$.
   (b) For n in $\mathbb{N}$, show that f^n is in $\mathcal{R}[a, b]$, where $f^n(x) = (f(x))^n$.
   (c) If there is a $\delta > 0$ such that $f(x) \geq \delta$ for all x in $[a, b]$, then $1/f$ is in $\mathcal{R}[a, b]$. A similar result holds if $f(x) \leq -\delta$ for all x in $[a, b]$.

2. Show that if f is in $\mathcal{R}[a, b]$, then f is in $\mathcal{R}[c, d]$ for all closed subintervals $[c, d]$ of $[a, b]$.

3. Let $a < c < b$. If f is in $\mathcal{R}[a, c]$ and $\mathcal{R}[c, b]$, show that f is in $\mathcal{R}[a, b]$.

4. If $f_1, f_2, \ldots, f_m$ are in $\mathcal{R}[a, b]$, show that $f_1 + f_2 + \cdots + f_m$ is in $\mathcal{R}[a, b]$ and $\int_a^b (f_1 + f_2 + \cdots + f_m) = \sum_{i=1}^m \int_a^b f_i$.

5. Let $P = \{x_i\}_{i=0}^n$ be in $\mathcal{P}[a, b]$. Let $f : [a, b] \to \mathbb{R}$ be such that f is constant on each open subinterval (x_{i-1}, x_i)—say, $f(x) = y_i$ for $x_{i-1} < x < x_i$ for $i = 1, 2, \ldots, n$. Such a function f is a *step function*. Show that f is in $\mathcal{R}[a, b]$ and $\int_a^b f = \sum_{i=1}^n y_i \Delta x_i$.

6. Let f be in $\mathcal{R}[a, b]$ with $m \le f(x) \le M$ for all x in $[a, b]$. Show that there is a λ in $[m, M]$ with $\int_a^b f = \lambda(b - a)$.

7. Give an example in which $|f|$ is in $\mathcal{R}[a, b]$ but f is not.

8. If f is in $\mathcal{R}[a, b]$, show that $\int_a^b f = \lim_{c \to a^+} \int_c^b f$.

9. Let f be in $\mathcal{R}[-a, a]$, where $a > 0$. Show that
   (a) if f is an even function $[f(-x) = f(x)$ for all $x]$, then $\int_{-a}^a f = 2 \int_0^a f$;
   (b) if f is an odd function $[f(-x) = -f(x)$ for all $x]$, then $\int_{-a}^a f = 0$.

10. For f and g in $\mathcal{R}[a, b]$, show that

$$\left| \int_a^b fg \right| \le \left[\left(\int_a^b f^2 \right) \left(\int_a^b g^2 \right) \right]^{1/2}.$$

This is the Cauchy-Bunyakovsky-Schwarz inequality for integrals. [*Hint:* For any real number x, expand $\int_a^b (xf + g)^2$ into a quadratic in x and use the discriminant from the quadratic formula.]

11. Assuming that $\sqrt{x}$ and $\sqrt{\cos x}$ are in $\mathcal{R}[0, \pi/2]$, show that

$$\int_0^{\pi/2} \sqrt{x \cos x}\, dx \le \frac{\pi}{2\sqrt{2}}.$$

12. For f and g in $\mathcal{R}[a, b]$, show that

$$\left[\int_a^b (f + g)^2 \right]^{1/2} \le \left(\int_a^b f^2 \right)^{1/2} + \left(\int_a^b g^2 \right)^{1/2}.$$

This is Minkowski's inequality. [*Hint:* Expand the integral on the left and use Exercise 10.]

13. Assuming that $\sqrt{x}$ and $\sqrt{\sin x}$ are in $\mathcal{R}[0, \pi]$, show that

$$\int_0^\pi \left(\sqrt{\sin x} + \sqrt{x} \right)^2 dx \le \left(\sqrt{2} + \frac{\pi}{\sqrt{2}} \right)^2.$$

6.4 Families of Riemann Integrable Functions

In this section we show that continuous functions and monotone functions are Riemann integrable on $[a, b]$. We also derive two Mean Value Theorems.

Continuous Functions

Theorem 6.7   If f is continuous on $[a, b]$, then f is in $\mathcal{R}[a, b]$.

Proof   We use Theorem 6.2. Let $\varepsilon > 0$. Since f is uniformly continuous on $[a, b]$ (Theorem 4.4), there is a $\delta > 0$ such that if x' and x'' are in $[a, b]$ with $|x' - x''| < \delta$, then $|f(x') - f(x'')| < \varepsilon/(b - a)$.

Let $P = \{x_i\}_{i=0}^n$ be in $\mathcal{P}[a, b]$ with $\|P\| < \delta$. For each $i = 1, 2, \ldots, n$, by Theorem 4.2, there are points μ_i' and μ_i'' in $[x_{i-1}, x_i]$ such that $f(\mu_i') = M_i$ and $f(\mu_i'') = m_i$ (where M_i and m_i are as in Definition 6.2). Since $|\mu_i' - \mu_i''| \leq \|P\| < \delta$, $M_i - m_i = |f(\mu_i') - f(\mu_i'')| < \varepsilon/(b - a)$, and so

$$U(P, f) - L(P, f) = \sum_{i=1}^n (M_i - m_i)\, \Delta x_i$$

$$< \frac{\varepsilon}{b - a} \sum_{i=1}^n \Delta x_i$$

$$= \varepsilon.$$

By Theorem 6.2, f is in $\mathcal{R}[a, b]$.   ∎

Example 6.7   Let

$$f(x) = \begin{cases} \dfrac{1 - \cos x}{x} & \text{if } 0 < x \leq 1 \\ 13 & \text{if } x = 0. \end{cases}$$

By L'Hôpital, $\lim\limits_{x \to 0} (1 - \cos x)/x = 0$. Therefore, f agrees with a continuous function on $[0, 1]$ except at one point. Since this continuous function is in $\mathcal{R}[0, 1]$, f is in $\mathcal{R}[0, 1]$ by Theorem 6.5.

Example 6.8   Let

$$f(x) = \begin{cases} \sin \dfrac{1}{x} & \text{if } 0 < x \leq 1 \\ 0 & \text{if } x = 0. \end{cases}$$

Since $\lim\limits_{x \to 0} \sin(1/x)$ does not exist, unlike Example 6.7, f cannot be extended continuously to 0. However, since f is continuous on every closed subinterval of $(0, 1)$ and f is bounded on $[0, 1]$, f is in $\mathcal{R}[0, 1]$ by Theorem 6.6.

Theorem 6.8   (First Mean Value Theorem) If f is continuous on $[a, b]$, then there is a c in $[a, b]$ such that $\int_a^b f = f(c)(b - a)$.

Proof   By Theorem 4.2, $m = \inf\{f(x) : x \in [a, b]\}$ and $M = \sup\{f(x) : x \in [a, b]\}$ are in $f([a, b])$. Since $m \leq f(x) \leq M$ on $[a, b]$, by Proposition 6.4,

$$m(b - a) = \int_a^b m \leq \int_a^b f \leq \int_a^b M = M(b - a).$$

Therefore, the number $[1/(b - a)] \int_a^b f$ is in $[m, M]$. By the Intermediate Value Theorem (Theorem 4.3), there is a c in $[a, b]$ such that $f(c) = [1/(b - a)] \int_a^b f$.   ∎

Remark   The reader should compare Theorem 6.8 with Exercise 6 in Section 6.3. For f continuous and nonnegative on $[a, b]$, the geometric interpretation of Theorem 6.8 is that the area under f from a to b lying above the horizontal axis is the same as the area of the rectangle over $[a, b]$ of height $f(c)$. That the number c guaranteed by Theorem 6.8 may not be unique can be seen by considering $f(x) = x^2$ on $[-1, 0.5]$.

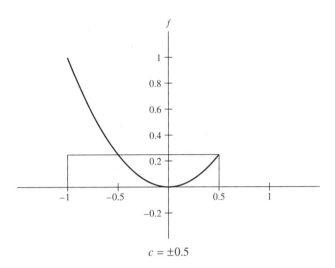

$$c = \pm 0.5$$

Corollary 6.2   If f is continuous and nonnegative on $[a, b]$, with $\int_a^b f = 0$, then $f(x) = 0$ for all x in $[a, b]$.

Proof   Suppose that $f(x_0) > 0$ for some x_0 in $[a, b]$. By Exercise 4 in Section 4.1, there exist $u < v$ with x_0 in $[u, v] \subset [a, b]$ and $f(x) > 0$ for all x in $[u, v]$. By Theorem 6.8, there is a c in $[u, v]$ such that $\int_u^v f = f(c)(v - u) > 0$. Since $f \geq 0$ on $[a, b]$, by Proposition 6.5,

$$\int_a^b f = \int_a^u f + \int_u^v f + \int_v^b f \geq \int_u^v f > 0,$$

which is a contradiction. Therefore, f is identically 0 on $[a, b]$.   ∎

Remark   Note that for f continuous and nonnegative on $[a, b]$, the contrapositive of Corollary 6.2 states that if f is positive anywhere on $[a, b]$, then $\int_a^b f > 0$. If

$$f(x) = \begin{cases} 1 & \text{if } x = 1, \dfrac{1}{2}, \dfrac{1}{3}, \ldots \\ 0 & \text{otherwise,} \end{cases}$$

then $\int_0^1 f = 0$, which implies that continuity is necessary in Corollary 6.2.

Monotone Functions

Theorem 6.9   If f is monotone on $[a, b]$, then f is in $\mathcal{R}[a, b]$.

Proof   First let f be monotone increasing on $[a, b]$. If $f(a) = f(b)$, then f is a constant function and hence in $\mathcal{R}[a, b]$; so we assume that $f(a) < f(b)$. If $P = \{x_i\}_{i=0}^n$ is any partition of $[a, b]$, then $M_i = f(x_i)$ and $m_i = f(x_{i-1})$

for each $i = 1, 2, \ldots, n$. Therefore,

$$U(P, f) - L(P, f) = \sum_{i=1}^{n} (M_i - m_i)\Delta x_i$$

$$= \sum_{i=1}^{n} [f(x_i) - f(x_{i-1})]\Delta x_i$$

$$\leq \|P\| \sum_{i=1}^{n} [f(x_i) - f(x_{i-1})]$$

$$= \|P\| [f(x_n) - f(x_0)]$$

$$= \|P\| [f(b) - f(a)].$$

Given $\varepsilon > 0$, choose a P in $\mathcal{P}[a, b]$ with $\|P\| < \varepsilon/[f(b) - f(a)]$. Then $U(P, f) - L(P, f) < \varepsilon$, and so f is in $\mathcal{R}[a, b]$ by Theorem 6.2.

If f is monotone decreasing on $[a, b]$, then $-f$ is monotone increasing on $[a, b]$ (Exercise 5 in Section 4.6), and so $f = -(-f)$ is in $\mathcal{R}[a, b]$ by Proposition 6.3. ∎

Example 6.9   Let f be bounded on $[a, b]$. Define g and h from $[a, b]$ into $\mathbb{R}$ by

$$g(x) = \sup\{f(y) : a \leq y \leq x\}$$

and

$$h(x) = \inf\{f(y) : a \leq y \leq x\}.$$

Then g is monotone increasing and h is monotone decreasing on $[a, b]$, and so both g and h are in $\mathcal{R}[a, b]$.

Theorem 6.10   (Second Mean Value Theorem) If f is monotone on $[a, b]$, then there is a c in $[a, b]$ such that

$$\int_a^b f = f(a)(c - a) + f(b)(b - c).$$

Proof   Assume that f is monotone increasing on $[a, b]$. Since $f(a) \leq f(x) \leq f(b)$ for all x in $[a, b]$, Proposition 6.4 implies that $f(a)(b - a) \leq \int_a^b f \leq f(b)(b - a)$.

Define $g : [a, b] \to \mathbb{R}$ by $g(x) = f(a)(x - a) + f(b)(b - x)$. Then g is continuous on $[a, b]$ and $g(a) = f(b)(b - a)$ and $g(b) = f(a)(b - a)$. By the Intermediate Value Theorem (Theorem 4.3), there is a c in $[a, b]$ with $g(c) = \int_a^b f$. Hence,

$$\int_a^b f = f(a)(c - a) + f(b)(b - c).$$

If f is monotone decreasing on $[a, b]$, apply the equation above to $-f$, which is monotone increasing, and then multiply by -1. ∎

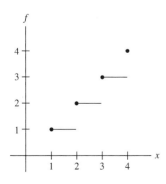

Figure 6.5

Example 6.10   To interpret Theorem 6.10 geometrically, let $f(x) = [x]$ denote the largest integer less than or equal to x (Figure 6.5). Then $\int_1^4 f = 6$.

By Theorem 6.10, there is a c in $[1, 4]$ with $6 = 1(c - 1) + 4(4 - c)$, and so $c = 3$. Thus, the area under $[x]$ from 1 to 4 is split into two rectangles, one from 1 to 3 with height $1 = f(1)$ and the other from 3 to 4 with height $4 = f(4)$.

Exercises

1. Show that
$$f(x) = \begin{cases} \dfrac{\sin x}{x} & \text{if} \quad x \neq 0 \\ 0 & \text{if} \quad x = 0 \end{cases}$$
is in $\mathcal{R}[0, 1]$.

2. For n in $\mathbb{N}$, show that
$$f(x) = \begin{cases} x^n \sin \dfrac{1}{x} & \text{if} \quad x \neq 0 \\ 0 & \text{if} \quad x = 0 \end{cases}$$
is in $\mathcal{R}[0, 1]$.

3. Show that
$$f(x) = \begin{cases} \cos \dfrac{1}{x} & \text{if} \quad x \neq 0 \\ 13 & \text{if} \quad x = 0 \end{cases}$$
is in $\mathcal{R}[0, 1]$.

4. Show that a bounded function with a finite number of discontinuities is in $\mathcal{R}[a, b]$.

5. Let m_k and b_k be in $\mathbb{R}$ for each $k = 1, 2, \ldots, n$, where n is in $\mathbb{N}$. Show that
$$f(x) = \begin{cases} m_k x + b_k & \text{if} \quad x \in [k - 1, k) \\ 47 & \text{if} \quad x = n \end{cases}$$
is in $\mathcal{R}[0, n]$.

6. Let f be continuous on $[a, b]$, with $f(x) \leq 0$ for all x in $[a, b]$. If f is not identically 0, show that $\int_a^b f < 0$.

7. Let f and g be continuous on $[a, b]$, with $\int_a^b f = \int_a^b g$. Show that $f(c) = g(c)$ for some c in $[a, b]$.

8. Let f be continuous and nonnegative on $[a, b]$. Show that there is a c in $[a, b]$ with
$$f(c) = \left(\frac{1}{b - a} \int_a^b f^2 \right)^{1/2}.$$

9. Let f be continuous on $[a, b]$, and let g be in $\mathcal{R}[a, b]$ with g nonnegative. Show that there is a c in $[a, b]$ such that $\int_a^b fg = f(c) \int_a^b g$.

10. Find the value of c guaranteed by Theorem 6.10 for

$$f(x) = \begin{cases} \dfrac{|x|}{x} & \text{if } x \neq 0 \\ 0 & \text{if } x = 0 \end{cases}$$

on $[-a, a]$ where $a > 0$.

11. Let f be continuous and nonnegative on $[0, 1]$ with $M = \sup\{f(x) : x \in [0, 1]\}$. Show that

$$M = \lim_{n \to \infty} \left(\int_0^1 f^n \right)^{1/n}.$$

[*Hint:* First show that $\left(\int_0^1 f^n \right)^{1/n} \leq M$ for each n. For $\varepsilon > 0$, there is an interval $[c, d] \subset [0, 1]$ with $f(x) > M - \varepsilon/2$ on $[c, d]$. Conclude that, for large n, $\left(\int_0^1 f^n \right)^{1/n} > M - \varepsilon.$]

6.5 Fundamental Theorem of Calculus

Theorems 6.11 and 6.12 below are sometimes both called Fundamental Theorems of Calculus. When the integrand is continuous, some books call Theorem 6.11 a corollary of Theorem 6.12. These are very important theorems relating integration and differentiation.

> **Definition 6.9**   Let f and φ be two functions defined on any interval I. If $\varphi'(x) = f(x)$ for all x in I, then φ is an *antiderivative* (or *primitive*) of f on I.

By Corollary 5.3, any two antiderivatives of f differ by a constant on I.

Theorem 6.11   (Fundamental Theorem of Calculus) If f is in $\mathcal{R}[a, b]$ and φ is an antiderivative of f on $[a, b]$, then

$$\int_a^b f = \varphi(b) - \varphi(a).$$

Proof   Let $P = \{x_i\}_{i=0}^n$ be a partition of $[a, b]$. By the Mean Value Theorem (Theorem 5.5), for each $i = 1, 2, \ldots, n$, choose t_i in (x_{i-1}, x_i) with

$$\varphi(x_i) - \varphi(x_{i-1}) = \varphi'(t_i)(x_i - x_{i-1}) = f(t_i)\Delta x_i.$$

Then

$$\varphi(b) - \varphi(a) = \sum_{i=1}^n [\varphi(x_i) - \varphi(x_{i-1})] = \sum_{i=1}^n f(t_i)\Delta x_i.$$

Since f is in $\mathcal{R}[a, b]$, Theorem 6.3 implies that the last sum tends to $\int_a^b f$ as

$\|P\| \to 0$, and so

$$\varphi(b) - \varphi(a) = \int_a^b f. \qquad \blacksquare$$

Sometimes $\varphi(b) - \varphi(a)$ is denoted by $\varphi(x)|_a^b$.

Example 6.11   Let

$$f(x) = \begin{cases} -1 & \text{if} \quad -1 \le x \le 0 \\ 1 & \text{if} \quad 0 < x \le 1. \end{cases}$$

Then f is in $\mathcal{R}[-1, 1]$ since f is monotone. However, f has no antiderivative on $[-1, 1]$ since f does not satisfy the intermediate value property there (Theorem 5.3).

The following theorem shows that a continuous function on $[a, b]$ always has an antiderivative.

Theorem 6.12   Let f be in $\mathcal{R}[a, b]$ and define $F : [a, b] \to \mathbb{R}$ by $F(x) = \int_a^x f$. Then F is continuous on $[a, b]$. Moreover, if f is continuous at a point c in $[a, b]$, then F is differentiable at c and $F'(c) = f(c)$.

Proof   Let $\varepsilon > 0$ and let $M > 0$ such that $|f(t)| \le M$ for all t in $[a, b]$. By Proposition 6.5 and part 2 of Corollary 6.1, for $x < y$ in $[a, b]$,

$$|F(y) - F(x)| = \left| \int_a^y f - \int_a^x f \right| = \left| \int_x^y f \right|$$

$$\le \int_x^y |f|$$

$$\le M(y - x).$$

Hence, $|F(y) - F(x)| < \varepsilon$ if $|y - x| < \varepsilon/M$, and so F is continuous (in fact, uniformly continuous) on $[a, b]$.

Now suppose that f is continuous at c in $[a, b]$. Let $\varepsilon > 0$ and choose $\delta > 0$ such that if t is in $[a, b]$ with $|t - c| < \delta$, then $|f(t) - f(c)| < \varepsilon$. For x in $[a, b]$ with $0 < |x - c| < \delta$,

$$\left| \frac{F(x) - F(c)}{x - c} - f(c) \right| = \left| \frac{1}{x - c} \int_c^x f(t)\, dt - f(c) \right|$$

$$= \left| \frac{1}{x - c} \int_c^x [f(t) - f(c)]\, dt \right|$$

$$\le \frac{1}{|x - c|} \left| \int_c^x |f(t) - f(c)|\, dt \right|$$

$$< \frac{1}{|x - c|} \varepsilon \, |x - c|$$

$$= \varepsilon.$$

Hence,

$$F'(c) = \lim_{x \to c} \frac{F(x) - F(c)}{x - c} = f(c).$$ ■

Remark   If f is continuous on $[a, b]$, then Theorem 6.12 can be used to prove Theorem 6.11 as follows. Let φ be any antiderivative of f on $[a, b]$ and let $F(x) = \int_a^x f$ for x in $[a, b]$. By Theorem 6.12, F is an antiderivative of f on $[a, b]$ and so $\varphi(x) - F(x) = C$ on $[a, b]$, where C is a constant. Then $C = \varphi(a) - F(a) = \varphi(a)$ and so $\varphi(b) - F(b) = \varphi(a)$. Therefore $\varphi(b) - \varphi(a) = F(b) = \int_a^b f$.

Example 6.12   Let f be the function in Example 6.11. For $-1 \le x \le 0$, $F(x) = \int_{-1}^x (-1)dt = -(x + 1)$ and, for $0 \le x \le 1$, $F(x) = \int_{-1}^0 (-1)dt + \int_0^x 1dt = x - 1$. Thus,

$$F(x) = \begin{cases} -(x + 1) & \text{if} & -1 \le x \le 0 \\ x - 1 & \text{if} & 0 \le x \le 1. \end{cases}$$

Draw F and note that F is continuous on $[-1, 1]$, F is differentiable on $[-1, 1]$ except at 0, and $F'(x) = f(x)$ for $x \ne 0$.

The results in the remainder of this section are probably familiar to the reader from calculus.

Proposition 6.6   Let f be continuous on $[a, b]$, let I be an interval or ray in $\mathbb{R}$, and let v be a differentiable function from I into $[a, b]$. If $G(x) = \int_a^{v(x)} f$ for x in I, then $G'(x) = f(v(x))v'(x)$ for x in I.

Proof   Let $F(x) = \int_a^x f$ on $[a, b]$. Then $G = F \circ v$ on I and, by the chain rule (Theorem 5.2),

$$G'(x) = F'(v(x))v'(x) = f(v(x))v'(x)$$

for all x in I.   ■

Variations on Proposition 6.6 will be considered in Exercise 4.

Proposition 6.7   (integration by parts) Let f and g be differentiable functions on $[a, b]$, with f' and g' both Riemann integrable on $[a, b]$. Then

$$\int_a^b fg' = f(b)g(b) - f(a)g(a) - \int_a^b f'g.$$

Proof   Since differentiability implies continuity, f and g are continuous on $[a, b]$. By part 1 of Corollary 6.1, fg' and $f'g$ are in $\mathcal{R}[a, b]$. Using the product rule for derivatives (part 2 of Proposition 5.1), we have

$$\int_a^b fg' + \int_a^b f'g = \int_a^b (fg)' = f(b)g(b) - f(a)g(a)$$

by Theorem 6.11 since fg is an antiderivative of $(fg)'$ on $[a, b]$.   ■

We end this section with a proposition that shows how to substitute or change variables in a Riemann integral. Recall that the continuous image of a closed interval is again a closed interval by Corollary 4.3.

Proposition 6.8 (change of variable) Let g be a differentiable function on $[a, b]$ such that g' is in $\mathcal{R}[a, b]$. If f is continuous on the range of g, then

$$\int_a^b f(g(t))g'(t)\,dt = \int_{g(a)}^{g(b)} f(x)\,dx.$$

[Formally, this is obtained by letting $x = g(t)$, $dx = g'(t)dt$.]

Proof   First note that g is continuous on $[a, b]$ by Theorem 5.1, and so $(f \circ g)g'$ is in $\mathcal{R}[a, b]$ by part 1 of Corollary 6.1.

Define F on the range of g by $F(x) = \int_{g(a)}^x f$. By Exercise 4(a), $F'(x) = f(x)$ on the range of g. By the chain rule (Theorem 5.2),

$$(F \circ g)'(t) = F'(g(t))g'(t) = f(g(t))g'(t)$$

for all t in $[a, b]$. Hence, by Theorem 6.11,

$$\int_a^b f(g(t))g'(t)dt = (F \circ g)(b) - (F \circ g)(a)$$

$$= F(g(b)) - 0$$

$$= \int_{g(a)}^{g(b)} f. \qquad \blacksquare$$

≡ Exercises ≡

1. Let

$$f(x) = \begin{cases} 0 & \text{if} \quad 0 \le x \le 1 \\ 1 & \text{if} \quad 1 < x \le 2. \end{cases}$$

Show that f is in $\mathcal{R}[0, 2]$, f has no antiderivative on $[0, 2]$, and find $F(x) = \int_0^x f$ for x in $[0, 2]$.

2. Let f be the function in Exercise 5 in Section 6.1. Find $F(x) = \int_0^x f$ for x in $[0, 1]$, and note that F is differentiable even though f is not continuous on $[0, 1]$. [*Hint:* First let $h = 1 - f$ and find $\int_0^x h$.]

3. The purpose of this exercise is to show that a function that has an antiderivative is not necessarily Riemann integrable. Let

$$f(x) = \begin{cases} x^2 \sin \dfrac{1}{x^2} & \text{if} \quad 0 < x \le 1 \\ 0 & \text{if} \quad x = 0. \end{cases}$$

Find f'. Then f' has an antiderivative but f' is not in $\mathcal{R}[0, 1]$ since f' is unbounded there.

4. Let f be continuous on $[a, b]$.

   (a) For c in $[a, b]$, show that $\frac{d}{dx} \int_c^x f = f(x)$ on $[a, b]$.

   (b) Show that $\frac{d}{dx} \int_x^b f = -f(x)$ on $[a, b]$.

   (c) Show that $\frac{d}{dx} \int_{u(x)}^{v(x)} f = f(v(x))v'(x) - f(u(x))u'(x)$ for all x in I, where u and v are differentiable functions on an interval or ray I with values in $[a, b]$.

5. Find $\frac{d}{dx} \int_{x^2}^{x^3} \cos(t^2)\, dt$.

6. Let f be continuous on $[a, b]$ such that $\int_a^x f = \int_x^b f$ for all x in $[a, b]$. Show that $f(x) = 0$ on $[a, b]$.

7. If f is continuous and strictly positive on $[a, b]$, show that $F(x) = \int_a^x f$ is strictly increasing on $[a, b]$.

8. Let f be continuous and monotone on $[a, b]$ and let g be Riemann integrable and nonnegative on $[a, b]$. Show that there is a c in $[a, b]$ such that

$$\int_a^b fg = f(a) \int_a^c g + f(b) \int_c^b g.$$

   [*Hint:* First consider f to be monotone increasing. By Exercise 9 in Section 6.4, there is a u in $[a, b]$ such that $\int_a^b fg = f(u) \int_a^b g$. Let $G(x) = [f(b) - f(a)] \int_x^b g$ and use the monotonicity of f.]

9. Let f be continuous on $[0, \infty)$, with $f(x) > 0$ for all $x > 0$. If $f^2(x) = 2 \int_0^x f$ for all $x \geq 0$, show that $f(x) = x$ for all $x \geq 0$.

10. Let f' be continuous on $[0, \infty)$, with $f(0) = 1$. If $\int_0^x f' = \int_0^x f$ for all $x > 0$, show that $f(x) = e^x$ for all $x \geq 0$.

6.6 Improper Integrals

Previously, all integrals involved a bounded integrand on a finite interval $[a, b]$. These are often referred to as *proper* Riemann integrals. *Improper Riemann integrals of the first kind* occur when a or b is $\pm\infty$, and *improper Riemann integrals of the second kind* occur when the integrand is unbounded on $[a, b]$. Of course, various combinations of these integrals can occur. The general technique is to split an improper integral into a combination of improper integrals (and sometimes proper integrals), each of which is improper only at one place ($+\infty$, $-\infty$, or a point where the integrand is unbounded). Then each of the latter improper integrals is written as a limit of proper Riemann integrals. This process allows us to use our already developed theory for proper Riemann integrals and limits to calculate improper Riemann integrals. We point out that the density functions of continuous random variables in probability and statistics often lead to improper Riemann integrals of the first kind.

As examples, $\int_1^\infty (1/x)\, dx$ is improper only at ∞; $\int_0^1 \ln x\, dx$ is improper only at 0; $\int_0^\infty (1/\sqrt{x})\, dx$ is improper at both 0 and ∞; but $\int_0^1 [(\sin x)/x]\, dx$ is a proper integral. Letting $f(x) = (\sin x)/x$ for $x \neq 0$ and letting $f(0)$ equal any real number, f is bounded on $[0, 1]$ since $\lim_{x \to 0+} [(\sin x)/x] = 1$. Also, f is in $\mathcal{R}[0, 1]$ by Theorem 6.5 or Theorem 6.6.

In this section, a, b, and t denote real numbers while f and g denote functions.

Improper Integrals of the First Kind

Definition 6.10 Let f be in $\mathcal{R}[a, t]$ for all $t \geq a$. *The improper Riemann integral of f on $[a, \infty)$, denoted by $\int_a^\infty f$ or $\int_a^\infty f(x)dx$, is* $\lim_{t \to \infty} \int_a^t f(x)dx$. *Similarly, if f is in $\mathcal{R}[t, b]$ for all $t \leq b$, the improper Riemann integral of f on $(-\infty, b]$, denoted by $\int_{-\infty}^b f$ or $\int_{-\infty}^b f(x)dx$,* is $\lim_{t \to -\infty} \int_t^b f(x)dx$.

The improper integral is *convergent* (or *converges*) if the corresponding limit is a real number; otherwise it is *divergent* (or *diverges*).

Example 6.13 $\int_{-\infty}^0 e^x = \lim_{t \to -\infty} \int_t^0 e^x = \lim_{t \to -\infty} e^x \big|_t^0 = \lim_{t \to -\infty} (1 - e^t) = 1 - 0 = 1$.

Remark Some people, including the authors, simplify (or abuse) the notation in working with improper integrals by not using the limit. For example,

$$\int_{-\infty}^0 e^x = e^x \big|_{-\infty}^0 = e^0 - e^{-\infty} = 1 - 0 = 1.$$

If one keeps in mind that this integral is improper at $-\infty$ and that $e^{-\infty}$ is actually a limit, then no problem should arise.

Example 6.14 Let $a > 0$. Then $\int_a^\infty x^{-1}\, dx = \lim_{t \to \infty} \ln x \big|_a^t = \lim_{t \to \infty} (\ln t - \ln a) = \infty$. For $p \neq 1$,

$$\int_a^\infty x^{-p}\, dx = \lim_{t \to \infty} \frac{x^{-p+1}}{-p+1} \bigg|_a^t$$

$$= \lim_{t \to \infty} \frac{t^{1-p} - a^{1-p}}{1 - p} = \begin{cases} \infty & \text{if } p < 1 \\ \dfrac{a^{1-p}}{p - 1} & \text{if } p > 1. \end{cases}$$

Hence, for $a > 0$, $\int_a^\infty x^{-p}$ converges if and only if $p > 1$.

Although we were able to calculate the values of the improper integrals in the examples above, many times this is not feasible. However, in these instances we may be able to determine whether the improper integral converges or diverges. We next develop tests for convergence of improper integrals of the form $\int_a^\infty f$. Similar tests can be developed for improper integrals of the form $\int_{-\infty}^b f$, but such tests are unnecessary since $\int_{-\infty}^b f(x)dx = \int_{-b}^\infty f(-y)dy$ by the change of variable $y = -x$.

Theorem 6.13 (Comparison test) Let f be in $\mathcal{R}[a, t]$ for all $t \geq a$. If $0 \leq f(x) \leq g(x)$ for all $x \geq a$ and if $\int_a^\infty g$ converges, then $\int_a^\infty f$ converges.

Proof   Let $F(t) = \int_a^t f$ for $t \geq a$. Since f is nonnegative, F is monotone increasing on $[a, \infty)$. By Proposition 6.4, for each $t \geq a$, $F(t) \leq \int_a^t g \leq \int_a^\infty g$, which is a real number since $\int_a^\infty g$ converges. Therefore,

$$\int_a^\infty f = \lim_{t \to \infty} F(t) = \sup\{F(t) : t \geq a\}$$

is a real number by the Completeness Axiom for $\mathbb{R}$.   ■

Example 6.15 $\int_1^\infty \sin^2(1/x)\,dx$ converges since $\sin^2(1/x) \leq 1/x^2$ for $x > 0$ and $\int_1^\infty x^{-2}\,dx$ converges by Example 6.14. $\int_2^\infty (1/\ln x)\,dx$ diverges since $1/\ln x > 1/x$ and $\int_2^\infty (1/x)\,dx$ diverges by Example 6.14 (we are using the contrapositive of Theorem 6.13).

Theorem 6.14 (Limit Comparison test) Let f be nonnegative and let g be positive on $[a, \infty)$, with both f and g in $\mathcal{R}[a, t]$ for all $t \geq a$. Suppose that

$$\lim_{x \to \infty} \frac{f(x)}{g(x)} = c.$$

1. For $0 < c < \infty$, $\int_a^\infty f$ and $\int_a^\infty g$ either both converge or both diverge.

2. For $c = 0$, if $\int_a^\infty g$ converges, then $\int_a^\infty f$ converges.

3. For $c = \infty$, if $\int_a^\infty g$ diverges, then $\int_a^\infty f$ diverges.

Proof   If $0 < c < \infty$, then there is an $\alpha > 0$ such that $c/2 < f(x)/g(x) < \frac{3}{2}c$ for all $x > \alpha$. Thus, $x > \alpha$ implies $(c/2)g(x) < f(x) < (3c/2)g(x)$. Applying Theorem 6.13 twice to this last inequality proves part 1.

For part 2, note that there is an $M > a$ such that $f(x)/g(x) < 1$ or, equivalently, $f(x) < g(x)$ for all $x \geq M$. Since $\int_a^\infty f = \int_a^M f + \int_M^\infty f$ and $\int_M^\infty f$ converges by Theorem 6.13, $\int_a^\infty f$ converges.

For part 3, note that for large x, $f(x)/g(x) > 1$ or $f(x) > g(x)$. Divergence of $\int_a^\infty f$ now follows from the divergence of $\int_a^\infty g$ and Theorem 6.13.   ■

Example 6.16 $\int_1^\infty \sin(1/x)\,dx$ diverges since

$$\lim_{x \to \infty} \frac{\sin(1/x)}{1/x} = \lim_{y \to 0^+} \frac{\sin y}{y} = 1$$

and $\int_1^\infty (1/x)$ diverges. For $p > 1$, $\int_2^\infty [1/(x^p \ln x)]\,dx$ converges since

$$\lim_{x \to \infty} \frac{1/(x^p \ln x)}{x^{-p}} = \lim_{x \to \infty} \frac{1}{\ln x} = 0$$

and $\int_2^\infty x^{-p}\,dx$ converges.

Definition 6.11   An improper integral of f is *absolutely convergent* (or *converges absolutely*) if the improper integral of $|f|$ converges. If an improper integral is convergent but not absolutely convergent, then the improper integral is *conditionally convergent*.

Note that Theorems 6.13 and 6.14 are actually tests of absolute convergence. As one would expect, absolute convergence implies convergence as the following theorem shows for integrals of the form $\int_a^\infty f$.

Theorem 6.15 Let f be in $\mathcal{R}[a, t]$ for all $t \geq a$. If $\int_a^\infty |f|$ converges, then $\int_a^\infty f$ converges.

Proof   For $x \geq a$, $-|f(x)| \leq f(x) \leq |f(x)|$ and so $0 \leq f(x) + |f(x)| \leq 2|f(x)|$. By Theorem 6.13, $\int_a^\infty (f + |f|)$ converges. For $t > a$, by Proposition 6.3, $\int_a^t f = \int_a^t [(f + |f|) - |f|] = \int_a^t (f + |f|) - \int_a^t |f|$. As $t \to \infty$, both limits on the right exist in $\mathbb{R}$, and so $\int_a^\infty f$ converges. ∎

The next example shows that the converse of Theorem 6.15 is false.

Example 6.17   We show that $\int_1^\infty x^{-p} \sin x \, dx$ is absolutely convergent if $p > 1$ and conditionally convergent if $0 < p \leq 1$. In particular, $\int_1^\infty [(\sin x)/x] \, dx$ converges.

Since $|x^{-p} \sin x| \leq x^{-p}$ for $x \geq 1$ and $\int_1^\infty x^{-p} dx$ converges for $p > 1$ (Example 6.14), $\int_1^\infty x^{-p} \sin x \, dx$ converges absolutely for $p > 1$.

Let $p > 0$ and let $t > 1$. Integration by parts yields $\int_1^t x^{-p} \sin x \, dx = -x^{-p} \cos x \big|_1^t - p \int_1^t x^{-(p+1)} \cos x \, dx$. Since $\lim_{t \to \infty} [(\cos t)/t^p] = 0$, $\int_1^\infty x^{-p} \sin x \, dx$ will converge if $\int_1^\infty x^{-(p+1)} \cos x \, dx$ converges. This last integral is actually absolutely convergent since $|x^{-(p+1)} \cos x| \leq x^{-(p+1)}$ and $p + 1 > 1$.

Now let $0 < p \leq 1$. We have left to show that $\int_1^\infty x^{-p} |\sin x| \, dx$ does not converge. For n in $\mathbb{N}$, $n > 1$,

$$\int_\pi^{n\pi} x^{-p} |\sin x| \, dx = \sum_{k=2}^n \int_{(k-1)\pi}^{k\pi} x^{-p} |\sin x| \, dx$$

$$\geq \sum_{k=2}^n \frac{1}{k\pi} \int_{(k-1)\pi}^{k\pi} |\sin x| \, dx$$

$$= \frac{2}{\pi} \sum_{k=2}^n \frac{1}{k}.$$

The inequality on the second line follows from the fact that $x^{-p} \geq (1/k\pi)^p \geq 1/k\pi$. The third line follows since $\int_{(k-1)\pi}^{k\pi} |\sin x| \, dx = 2$. From calculus, the harmonic series $\sum_{k=1}^\infty 1/k$ diverges and so $\lim_{n \to \infty} \int_\pi^{n\pi} x^{-p} |\sin x| \, dx = \infty$. Therefore, $\int_1^\infty x^{-p} |\sin x| \, dx$ diverges.

Definition 6.12   For f defined on $\mathbb{R}$ and c in $\mathbb{R}$, $\int_{-\infty}^\infty f$ converges and

$$\int_{-\infty}^\infty f = \int_{-\infty}^c f + \int_c^\infty f \qquad (4)$$

provided both integrals on the right of (4) converge. If either of the integrals on the right of (4) diverges, then $\int_{-\infty}^\infty f$ diverges.

Remark   From Proposition 6.5 it follows that the choice of c is immaterial. The *Cauchy principal value* of $\int_{-\infty}^{\infty} f$ is $\lim\limits_{t \to \infty} \int_{-t}^{t} f$. By writing the integrals on the right of (4) in terms of limits, it should be clear that if $\int_{-\infty}^{\infty} f$ converges, then its value is the same as the Cauchy principal value. However, $\int_{-\infty}^{\infty} f$ may diverge but its Cauchy principal value may exist. For example, the Cauchy principal value of $\int_{-\infty}^{\infty} x \, dx$ is 0 but $\int_{-\infty}^{\infty} x \, dx$ diverges.

Improper Integrals of the Second Kind

Definition 6.13   Let f be in $\mathcal{R}[a, t]$ for all t in $[a, b)$ with f unbounded at b. The *improper Riemann integral of f on $[a, b]$* is $\lim\limits_{t \to b^-} \int_a^t f$. Similarly, if f is in $\mathcal{R}[t, b]$ for all t in $(a, b]$ with f unbounded at a, the *improper Riemann integral of f on $[a, b]$* is $\lim\limits_{t \to a^+} \int_t^b f$.

The improper integral is *convergent* (or *converges*) if the corresponding limit is a real number; otherwise it is *divergent* (or *diverges*).

Notation   Although some authors use the notation $\int_{a+}^{b} f$ to denote that this integral is improper at a, we will simply use $\int_a^b f$. For f in $\mathcal{R}[a, b]$, this is justified by Exercise 8 in Section 6.3. It is incumbent on the reader to decide first whether $\int_a^b f$ is proper or improper.

Example 6.18   By the symmetry (draw both graphs on the same axes) between e^x and $\ln x$, it follows from Example 6.13 that $\int_0^1 \ln x = -1$. Doing this directly involves integration by parts and L'Hôpital's rule to evaluate the limit.

Example 6.19   Let $b > 0$. Then $\int_0^b x^{-1} dx = \lim\limits_{t \to 0^+} \ln x \big|_t^b = \lim\limits_{t \to 0^+} (\ln b - \ln t) = \infty$. For $p \neq 1$,

$$\int_0^b x^{-p} dx = \lim_{t \to 0^+} \frac{b^{1-p} - t^{1-p}}{1 - p} = \begin{cases} \dfrac{b^{1-p}}{1 - p} & \text{if} \quad p < 1 \\ \infty & \text{if} \quad p > 1. \end{cases}$$

Hence, for $b > 0$, $\int_0^b x^{-p}$ converges if and only if $p < 1$.

Remark   In Example 6.19 changing variable $y = 1/x$ gives that $\int_0^b x^{-p} \, dx = \int_{1/b}^{\infty} y^{-(2-p)} \, dy$ which, by Example 6.14, converges if and only if $2 - p > 1$ or, equivalently, $p < 1$.

Similarly, an improper integral of the second kind can be changed into an improper integral of the first kind by the appropriate substitution. Thus, one would expect theorems analogous to Theorems 6.13, 6.14, and 6.15 to hold for improper integrals of the second kind. Since their proofs are basically the same, we simply state these theorems for integrals that are improper at b.

Comparison test   If f is in $\mathcal{R}[a, t]$ for all t in $[a, b)$, with $0 \leq f(x) \leq g(x)$ for all x in $[a, b)$, then the convergence of $\int_a^b g$ implies the convergence of $\int_a^b f$.

Limit Comparison test   If f is nonnegative on $[a, b)$, g is positive on $[a, b)$, with both f and g in $\mathcal{R}[a, t]$ for all t in $[a, b)$, and $\lim\limits_{x \to b^-} f(x)/g(x) = c$, then

1. for $0 < c < \infty$, $\int_a^b f$ and $\int_a^b g$ either both converge or both diverge;

2. for $c = 0$, $\int_a^b g$ converges implies that $\int_a^b f$ converges;

3. for $c = \infty$, if $\int_a^b g$ diverges, then $\int_a^b f$ diverges.

Absolute convergence implies convergence   If f is in $\mathcal{R}[a, t]$ for all t in $[a, b)$, then $\int_a^b |f|$ converges implies that $\int_a^b f$ converges.

Example 6.20 $\int_0^1 e^{-x} x^{p-1}\, dx$ is proper if $p \geq 1$. For $0 < p < 1$, $\int_0^1 e^{-x} x^{p-1}\, dx$ converges since $e^{-x} x^{p-1} \leq x^{p-1}$, and $\int_0^1 x^{p-1} dx$ converges if $1 - p < 1$ or, equivalently, $p > 0$.

Example 6.21   Consider $\int_0^1 x^{-p} \sin x\, dx$. This is a proper integral if $p \leq 1$ for $\lim\limits_{x \to 0^+} (\sin x)/x^p$ is 1 if $p = 1$ and is 0 if $0 < p < 1$ (use L'Hôpital's rule). If $p > 1$, $\lim\limits_{x \to 0^+} (x^{-p} \sin x)/x^{-p+1} = 1$ and since $\int_0^1 x^{-p+1} dx$ converges if and only if $p - 1 < 1$, $\int_0^1 x^{-p} \sin x\, dx$ converges if and only if $p < 2$.

Definition 6.14   If f is unbounded at an interior point c of (a, b), then $\int_a^b f$ *converges* and

$$\int_a^b f = \int_a^c f + \int_c^b f \tag{5}$$

provided both integrals on the right of (5) converge. Otherwise, $\int_a^b f$ *diverges*.

The expression in (5) is valid even when f is not unbounded at c. For example,

$$\int_0^\infty \frac{\sin x}{x} dx = \int_0^1 \frac{\sin x}{x} dx + \int_1^\infty \frac{\sin x}{x} dx$$

converges because the first integral on the right side is proper, and $\int_1^\infty [(\sin x)/x]\, dx$ converges by Example 6.17.

≡ Exercises ≡

1. Show that $\int_1^\infty [1/(x\sqrt{x+1})]\,dx$ converges but $\int_0^\infty [1/(x\sqrt{x+1})]\,dx$ diverges.

2. Show that $\int_1^\infty \sin(x^2)\,dx$ converges. [*Hint:* Let $y = x^2$.]

3. Show that $\int_0^\infty e^{-x}\sin(1/x)\,dx$ converges absolutely.

4. Show that $\int_0^1 [(1/x)\sin(1/x)]\,dx$ is conditionally convergent.

5. (a) Show that $\int_0^1 |\ln x|\,dx$ converges.

   (b) Show that $\int_0^1 [x\ln x/(1+x^2)]\,dx$ converges.

6. Show that $\int_0^1 x^{-p}\cos x\,dx$ converges if and only if $p < 1$. (This integral is proper if $p \le 0$.)

7. Show that $\int_1^\infty x^{-p}\cos x\,dx$ converges for $p > 0$ and converges absolutely if $p > 1$.

8. Show that $\int_2^\infty x^{-1}(\ln x)^{-p}\,dx$ converges if and only if $p > 1$.

9. Show that $\int_1^\infty e^{-x}x^p\,dx$ converges for every p in $\mathbb{R}$. [*Hint:* Use the Limit Comparison test with x^{-2}.]

10. For $p > 0$, the *Gamma function* is defined by

$$\Gamma(p) = \int_0^\infty e^{-x}x^{p-1}\,dx.$$

   (a) Show that $\Gamma(p)$ is a real number for every $p > 0$. [*Hint:* Combine Example 6.20 with Exercise 9.]

   (b) Use integration by parts to show that $\Gamma(p) = (p-1)\Gamma(p-1)$ for $p > 1$.

   (c) Calculate $\Gamma(1)$ and then use part (b) to show that $\Gamma(n) = (n-1)!$ for n in $\mathbb{N}$.

11. Show that $\int_{-\infty}^\infty [1/(x^2+1)]\,dx = \pi$.

12. For this exercise, assume that $\int_0^\infty [(\sin x)/x]\,dx = \pi/2$. We have not built up the machinery to make this calculation. The interested reader can find this machinery in Apostol, page 444, and most of this exercise on page 456.

   (a) Use $\sin 2x = 2\sin x\cos x$ to obtain $\int_0^\infty [(\sin x\cos x)/x]\,dx = \pi/4$.

   (b) Integrate the integral in part (a) by parts to obtain $\int_0^\infty [(\sin^2 x)/x^2]\,dx = \pi/2$.

   (c) Use $\sin^2 x + \cos^2 x = 1$ with part (b) to obtain $\int_0^\infty [(\sin^4 x)/x^2]\,dx = \pi/4$.

   (d) Use $\sin^2 x = (1 - \cos 2x)/2$ along with part (b) to obtain $\int_0^\infty [(1 - \cos x)/x^2]\,dx = \pi/2$.

   (e) Use two integrations by parts ($u = \sin^4 x$ in the first one) with part (c) to obtain $\int_0^\infty [(\sin^4 x)/x^4]\,dx = \pi/3$.

Infinite Series

7

The reader may wonder why infinite series were not covered earlier in this text, especially since we touched on geometric series in Chapters 2 and 3, mentioned other series in Examples 3.16 and 4.26, and alluded to the harmonic series in two of the exercises in Chapter 3. Our purpose was twofold. We first wanted to show that infinite series can be useful. Second, by this point in the text, the reader should have sufficient mathematical maturity to appreciate this chapter without finding it overwhelming.

The first one and one-half sections of this chapter contain material usually seen in calculus, and the reader may be surprised by how straightforward it now seems. Even though most of the remaining material may be new, the reader should find it interesting and enjoyable.

7.1 Convergence and Divergence

Definition 7.1 Let $(a_n)_{n \in \mathbb{N}}$ be a sequence in $\mathbb{R}$. The symbol $\sum_{n=1}^{\infty} a_n$ or, equivalently, the symbol $\sum_{k=1}^{\infty} a_k$, represents an *infinite series* or simply a *series*. For each n in $\mathbb{N}$, a_n is the *nth term* of the series $\sum_{k=1}^{\infty} a_k$; and letting

$$s_n = a_1 + a_2 + \cdots + a_n = \sum_{k=1}^{n} a_k,$$

the number s_n is the *nth partial sum* and $(s_n)_{n \in \mathbb{N}}$ is the *sequence of partial sums* of the series $\sum_{k=1}^{\infty} a_k$.

Comment The reader may notice that Definition 7.1 does not say what an infinite series is; it says that a certain symbol represents an infinite series. A more precise definition can be given as follows: if $(a_n)_{n \in \mathbb{N}}$ is a sequence in $\mathbb{R}$ and $s_n = \sum_{k=1}^{n} a_k$ for each n in $\mathbb{N}$, then an *infinite series* is the pair $\langle (a_n)_{n \in \mathbb{N}}, (s_n)_{n \in \mathbb{N}} \rangle$. The authors feel that this is one of those rare times in mathematics where it is better to be slightly imprecise but totally understandable.

Notation It is often convenient to start a series at 0 or at some other nonnegative integer. If p is a nonnegative integer, the symbol $\sum_{n=p}^{\infty} b_n$ means $\sum_{n=1}^{\infty} a_n$ where $a_n = b_{p+n-1}$. When no confusion can arise, we will write Σb_n for $\sum_{n=p}^{\infty} b_n$.

Definition 7.2 Given the series $\sum\limits_{n=1}^{\infty} a_n$ with partial sums $(s_n)_{n \in \mathbb{N}}$, the series $\sum\limits_{n=1}^{\infty} a_n$ *converges* (or is *convergent*) if the sequence $(s_n)_{n \in \mathbb{N}}$ converges, and the series $\sum\limits_{n=1}^{\infty} a_n$ *diverges* (or is *divergent*) if the sequence $(s_n)_{n \in \mathbb{N}}$ diverges. If $(s_n)_{n \in \mathbb{N}}$ converges to s, then s is the *sum* of the series $\sum\limits_{n=1}^{\infty} a_n$, and we write $\sum\limits_{n=1}^{\infty} a_n = s$.

Remark For convergent series the same symbol, $\sum\limits_{n=1}^{\infty} a_n$, is used to denote both the series and its sum. Also note that, as in Chapter 3, "converges" means converges to a real number. When $\lim\limits_{n \to \infty} s_n = \infty$ (or $-\infty$), we will also write $\sum\limits_{n=1}^{\infty} a_n = \infty$ (or $-\infty$).

Example 7.1 $\sum\limits_{n=1}^{\infty} 1 = \infty$ since $s_n = n \to \infty$, and $\sum\limits_{n=1}^{\infty} (-1)^{n+1} = 1 - 1 + 1 - 1 + \cdots$ diverges since $s_{2n} = 0$ and $s_{2n-1} = 1$ for each n in $\mathbb{N}$.

Example 7.2 (a telescoping series) $\sum\limits_{n=1}^{\infty} 1/[n(n+1)] = 1$ for

$$s_n = \sum_{k=1}^{n} \frac{1}{k(k+1)}$$

$$= \sum_{k=1}^{n} \left(\frac{1}{k} - \frac{1}{k+1} \right)$$

$$= \left(1 - \frac{1}{2} \right) + \left(\frac{1}{2} - \frac{1}{3} \right) + \left(\frac{1}{3} - \frac{1}{4} \right) + \cdots + \left(\frac{1}{n} - \frac{1}{n+1} \right)$$

$$= 1 - \frac{1}{n+1} \to 1.$$

Proposition 7.1 If $\sum\limits_{n=1}^{\infty} a_n$ converges, then $\lim\limits_{n \to \infty} a_n = 0$. Thus, if $\lim\limits_{n \to \infty} a_n \neq 0$, then $\sum\limits_{n=1}^{\infty} a_n$ diverges.

Proof Let $s_n = \sum\limits_{k=1}^{n} a_k$ for each n in $\mathbb{N}$ and let $\sum\limits_{n=1}^{\infty} a_n = A$. Since $a_n = s_n - s_{n-1}$ for $n \geq 2$, $\lim\limits_{n \to \infty} a_n = \lim\limits_{n \to \infty} s_n - \lim\limits_{n \to \infty} s_{n-1} = A - A = 0$. ∎

Note that Proposition 7.1 easily implies that the two series in Example 7.1 diverge.

Example 7.3 (geometric series) $\sum\limits_{n=1}^{\infty} x^{n-1} = 1 + x + x^2 + x^3 + \cdots$ converges if and only if $|x| < 1$ and then $\sum\limits_{n=1}^{\infty} x^{n-1} = 1/(1-x)$. If $|x| \geq 1$, then $\lim\limits_{n \to \infty} x^{n-1} \neq 0$ and so $\sum\limits_{n=1}^{\infty} x^{n-1}$ diverges. For $|x| < 1$, $s_n = \sum\limits_{k=1}^{n} x^{k-1} = 1 + x + x^2 + \cdots + x^{n-1} = (1 - x^n)/(1-x) \xrightarrow[n]{} 1/(1-x)$.

Remark   Note that $\sum_{n=1}^{\infty} x^{n-1} = \sum_{n=0}^{\infty} x^n$, which can be obtained by the substitution $m = n - 1$ in $\sum_{n=1}^{\infty} x^{n-1}$. Also in this context, when $x = 0$, we interpret 0^0 as 1.

Theorem 7.1   (Cauchy condition) The series $\sum_{n=1}^{\infty} a_n$ converges if and only if for every $\varepsilon > 0$ there exists an n_0 in $\mathbb{N}$ such that if $n > m \geq n_0$, then $\left| \sum_{k=m+1}^{n} a_k \right| < \varepsilon$.

Proof   The series $\sum_{n=1}^{\infty} a_n$ converges

if and only if the sequence $(s_n)_{n \in \mathbb{N}}$ of partial sums converges (Definition 7.2)

if and only if $(s_n)_{n \in \mathbb{N}}$ is Cauchy (Theorem 3.12)

if and only if for all $\varepsilon > 0$ there is an n_0 in $\mathbb{N}$ such that if $n \geq n_0$ and $m \geq n_0$, then $|s_n - s_m| < \varepsilon$ (Definition 3.11)

if and only if for all $\varepsilon > 0$ there is an n_0 in $\mathbb{N}$ such that if $n > m \geq n_0$, then $\left| \sum_{k=m+1}^{n} a_k \right| < \varepsilon$ (since $s_n - s_m = \sum_{k=m+1}^{n} a_k$). ∎

Example 7.4   (harmonic series) $\sum_{n=1}^{\infty} 1/n$ diverges. To see this, observe that for m in $\mathbb{N}$,

$$\sum_{k=2^m+1}^{2^m+2^m} \frac{1}{k} = \frac{1}{2^m + 1} + \frac{1}{2^m + 2} + \cdots + \frac{1}{2^m + 2^m}$$

$$\geq \frac{1}{2^m + 2^m} + \frac{1}{2^m + 2^m} + \cdots + \frac{1}{2^m + 2^m}$$

$$= \frac{2^m}{2^m + 2^m} = \frac{1}{2}.$$

Hence, the Cauchy condition cannot be satisfied when $\varepsilon < \frac{1}{2}$.

The two basic questions about an infinite series are whether or not it converges and, if so, to what (that is, what is its sum). By evaluating the limit of the partial sums, we are actually answering the latter question, which is the harder question. We now develop some elementary tests to answer the first question. This will increase the collection of series that we know to be convergent or divergent.

Theorem 7.2   Let $a_n \geq 0$ for each n in $\mathbb{N}$. Then $\sum_{n=1}^{\infty} a_n$ converges if and only if the sequence of partial sums is bounded above.

Proof   Since $a_n \geq 0$ for each n, the sequence $(s_n)_{n \in \mathbb{N}}$ of partial sums is monotone increasing. By Proposition 3.2 and Theorem 3.7, $(s_n)_{n \in \mathbb{N}}$ converges if and only if $(s_n)_{n \in \mathbb{N}}$ is bounded above. ∎

Remark   If $a_n \geq 0$ for each n in $\mathbb{N}$, Theorem 7.2 implies that $\sum\limits_{n=1}^{\infty} a_n$ diverges if and only if the limit of the partial sums is ∞. Combining this with Example 7.4, we have that $\lim\limits_{n\to\infty} [1 + \frac{1}{2} + \frac{1}{3} + \cdots + (1/n)] = \infty$, which is not obvious.

Theorem 7.3   (Integral test) Suppose that $f : [1, \infty) \to \mathbb{R}$ satisfies $f(x) \geq 0$ for all x in $[1, \infty)$ and that f is monotone decreasing on $[1, \infty)$. Let $a_n = f(n)$ for each n in $\mathbb{N}$. Then the series $\sum\limits_{n=1}^{\infty} a_n$ converges if and only if the improper integral $\int_1^{\infty} f$ converges.

Proof   Consider Figure 7.1 and note that f is integrable on $[1, n]$ for each n in $\mathbb{N}$ since f is monotone (Theorem 6.9).

For $n \geq 2$, $\sum\limits_{k=1}^{n-1} a_k$ is an upper sum and $\sum\limits_{k=2}^{n} a_k$ is a lower sum of f on $[1, n]$, and so

$$\sum_{k=1}^{n-1} a_k \geq \int_1^n f \geq \sum_{k=2}^n a_k. \tag{1}$$

If $\int_1^{\infty} f$ converges, then the second inequality in (1) shows that the partial sums of $\sum\limits_{n=2}^{\infty} a_n$ are bounded above by $\int_1^{\infty} f$. By Theorem 7.2, $\sum\limits_{n=2}^{\infty} a_n$ converges and so $\sum\limits_{n=1}^{\infty} a_n$ converges by Exercise 4.

If $\int_1^{\infty} f$ diverges, then the first inequality in (1) shows that the partial sums of $\sum\limits_{n=1}^{\infty} a_n$ are unbounded above, and so $\sum\limits_{n=1}^{\infty} a_n$ diverges by Theorem 7.2.   ∎

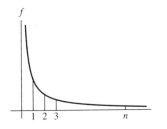

Figure 7.1

Example 7.5   (p-series) $\sum\limits_{n=1}^{\infty} 1/n^p$ converges if and only if $p > 1$. This follows from the Integral test and Example 6.14, where we showed that $\int_1^{\infty} 1/x^p$ converges if and only if $p > 1$.

The reader will find that the p-series is very useful in determining whether other series converge or diverge. This will be demonstrated in the remainder of this section.

Theorem 7.4   (Comparison test) Suppose that $0 \leq a_n \leq b_n$ for all n in $\mathbb{N}$. Then

1. if $\sum\limits_{n=1}^{\infty} b_n$ converges, $\sum\limits_{n=1}^{\infty} a_n$ converges;

2. if $\sum\limits_{n=1}^{\infty} a_n$ diverges, $\sum\limits_{n=1}^{\infty} b_n$ diverges.

Proof   Clearly, part 2 is the contrapositive of part 1. To show part 1, note that for each n in $\mathbb{N}$, $\sum\limits_{k=1}^{n} a_k \leq \sum\limits_{k=1}^{n} b_k$. If $\sum\limits_{n=1}^{\infty} b_n$ converges, then the partial sums of $\sum\limits_{n=1}^{\infty} a_n$ are bounded above by $\sum\limits_{n=1}^{\infty} b_n$; and so, by Theorem 7.2, $\sum\limits_{n=1}^{\infty} a_n$ converges.   ∎

Example 7.6 $\sum\limits_{n=2}^{\infty} 1/\ln n$ diverges since $1/n < 1/\ln n$ for $n \geq 2$ and $\sum\limits_{n=2}^{\infty} 1/n$ diverges. Also, $\sum\limits_{n=1}^{\infty} 1/(n^2+3)$ converges since $1/(n^2+3) \leq 1/n^2$ and $\sum\limits_{n=1}^{\infty} 1/n$ converges by Example 7.5.

Theorem 7.5 (Limit Comparison test) Let $a_n \geq 0$ and $b_n > 0$ for each n in $\mathbb{N}$ and suppose that $\lim\limits_{n\to\infty} a_n/b_n = L$.

1. If $0 < L < \infty$, then $\sum\limits_{n=1}^{\infty} a_n$ and $\sum\limits_{n=1}^{\infty} b_n$ either both converge or both diverge.

2. If $L = 0$ and $\sum\limits_{n=1}^{\infty} b_n$ converges, then $\sum\limits_{n=1}^{\infty} a_n$ converges.

3. If $L = \infty$ and $\sum\limits_{n=1}^{\infty} b_n$ diverges, then $\sum\limits_{n=1}^{\infty} a_n$ diverges.

Proof   The proof is very similar to the proof of Theorem 6.14 and is left as Exercise 10.   ∎

Example 7.7 $\sum\limits_{n=2}^{\infty} 1/\sqrt{n^3 - 2}$ converges by Theorem 7.5 because

$$\lim_{n\to\infty} \frac{1/\sqrt{n^3-2}}{1/n^{3/2}} = \lim_{n\to\infty} \sqrt{\frac{n^3}{n^3-2}}$$

$$= \lim_{n\to\infty} \sqrt{\frac{1}{1 - (2/n^3)}}$$

$$= 1,$$

and $\sum\limits_{n=2}^{\infty} 1/n^{3/2}$ converges by Example 7.5.

Note that to apply the Comparison test in this example, one would have to choose a convergent dominating series $\sum\limits_{n=2}^{\infty} b_n$. Since the choice of such a dominating series is not clear, the Limit Comparison test is easier to apply in this case. We leave it to the reader to decide which test is easier to apply to the series $\sum\limits_{n=1}^{\infty} (\ln n)/(2n^3 - 1)$, which (recalling that $\ln n < n$) converges by comparison with $\sum\limits_{n=1}^{\infty} 1/n^2$.

Exercise 4 implies that the deletion or insertion of a finite number of terms in a series does not alter the convergence or divergence of the series. Of course, if the series is convergent, deletion or insertion of terms may change the value of the sum. Thus, many of our theorems could be generalized by having the hypotheses hold for all n beyond some fixed positive integer. This, together with Example 7.4, also justifies why $\sum\limits_{n=2}^{\infty} 1/n$ diverges, as stated in Example 7.6.

One final comment. The reader has probably noticed a connection between infinite series and improper integrals. This analogy will be made precise in Chapter 10.

≡ Exercises ≡

1. Let $\sum_{n=1}^{\infty} a_n$ and $\sum_{n=1}^{\infty} b_n$ be convergent series, with $\sum_{n=1}^{\infty} a_n = A$ and $\sum_{n=1}^{\infty} b_n = B$, and let c be in $\mathbb{R}$. Show that $\sum_{n=1}^{\infty} (a_n + b_n)$ and $\sum_{n=1}^{\infty} c a_n$ both converge and that $\sum_{n=1}^{\infty} (a_n + b_n) = A + B$ and $\sum_{n=1}^{\infty} c a_n = cA$.

2. Show that $\sum_{n=1}^{\infty} (\sqrt{n+1} - \sqrt{n})$ diverges.

3. Find the sum of $\sum_{n=1}^{\infty} 1/(n^2 + 3n + 2)$.

4. Let $(a_n)_{n \in \mathbb{N}}$ and $(b_n)_{n \in \mathbb{N}}$ be two sequences such that $a_n = b_n$ for all $n \geq n_0$, where n_0 is in $\mathbb{N}$. Show that $\sum_{n=1}^{\infty} a_n$ and $\sum_{n=1}^{\infty} b_n$ either both converge or both diverge. Conclude that the insertion or deletion of a finite number of terms in a series does not affect the convergence or divergence of the series.

5. Find the sum of $\sum_{n=1}^{\infty} \left(\frac{1}{2}\right)^n$.

6. Show that $\sum_{n=2}^{\infty} 1/[n(\ln n)^p]$ converges if and only if $p > 1$.

7. Let $a_n \geq 0$ for all n in $\mathbb{N}$. Show that if $\sum_{n=1}^{\infty} a_n$ converges, then $\sum_{n=1}^{\infty} a_n^2$ converges.

8. Let $a_n \geq 0$ for all n in $\mathbb{N}$. Show that if $\sum_{n=1}^{\infty} a_n$ converges, then $\sum_{n=1}^{\infty} \sqrt{a_n}/n$ converges. [*Hint:* Expand $[\sqrt{a_n} - (1/n)]^2$.]

9. Use Theorem 7.5 to show that
   (a) $\sum_{n=1}^{\infty} (n-2)/(3n^2 - n + 6)$ diverges
   and
   (b) $\sum_{n=1}^{\infty} (\ln n)/(n^2 + 4)$ converges.

10. Prove Theorem 7.5.

7.2 Absolute and Conditional Convergence

In this section we first consider alternating series. We next distinguish between absolute and conditional convergence, and then we develop tests for each type of convergence.

Alternating Series

Definition 7.3   An *alternating series* is one whose successive terms are alternately positive and negative.

If $a_n > 0$ for each n in $\mathbb{N}$, then

$$\sum_{n=1}^{\infty}(-1)^{n+1}a_n = a_1 - a_2 + a_3 - a_4 + \cdots$$

and

$$\sum_{n=1}^{\infty}(-1)^{n}a_n = -a_1 + a_2 - a_3 + a_4 - \cdots$$

are alternating series. Since the second series is the constant -1 times the first series, in view of Exercise 1 in Section 7.1, we consider the first series.

Theorem 7.6   (Alternating Series test) If $(a_n)_{n\in\mathbb{N}}$ is a monotone decreasing sequence of positive real numbers with $\lim_{n\to\infty} a_n = 0$, then $\sum_{n=1}^{\infty}(-1)^{n+1}a_n$ converges. In this case, if A and $(s_n)_{n\in\mathbb{N}}$ are, respectively, the sum and partial sums of this series, then $|A - s_n| \leq a_{n+1}$ for each n in $\mathbb{N}$.

Proof   For n in $\mathbb{N}$,

$$s_{2n} = (a_1 - a_2) + (a_3 - a_4) + \cdots + (a_{2n-1} - a_{2n}) \geq 0$$

since $(a_n)_{n\in\mathbb{N}}$ is monotone decreasing (each term in parentheses is nonnegative). Since $s_{2n+2} - s_{2n} = a_{2n+1} - a_{2n+2} \geq 0$, the sequence $(s_{2n})_{n\in\mathbb{N}}$ is monotone increasing. Also, since $(a_n)_{n\in\mathbb{N}}$ is monotone decreasing,

$$s_{2n} = a_1 - (a_2 - a_3) - (a_4 - a_5) - \cdots - (a_{2n-2} - a_{2n-1}) - a_{2n} \leq a_1.$$

Therefore, $(s_{2n})_{n\in\mathbb{N}}$ is a bounded monotone increasing sequence and by Theorem 3.7 must converge to a real number, say A.

For n in $\mathbb{N}$,

$$\lim_{n\to\infty} s_{2n+1} = \lim_{n\to\infty} (s_{2n} + a_{2n+1})$$

$$= \lim_{n\to\infty} s_{2n} + \lim_{n\to\infty} a_{2n+1}$$

$$= A + 0$$

$$= A.$$

Hence, $s_n \to A$ and so $\sum_{n=1}^{\infty}(-1)^{n+1}a_n = A$ with $A \leq a_1$.

For n in $\mathbb{N}$, $A - s_n = (-1)^n(a_{n+1} - a_{n+2} + a_{n+3} - \cdots)$. Since $a_{n+1} - a_{n+2} + a_{n+3} - \cdots$ is an alternating series with the same properties as $\sum_{n=1}^{\infty}(-1)^{n+1}a_n$, by the first part of the proof, it converges and $0 \leq \sum_{k=1}^{\infty}(-1)^{k+1}a_{n+k} \leq a_{n+1}$. Thus, $|A - s_n| \leq a_{n+1}$. ∎

Example 7.8   The alternating harmonic series $\sum_{n=1}^{\infty}(-1)^{n+1}(1/n) = 1 - \frac{1}{2} + \frac{1}{3} - \frac{1}{4} + \cdots$ converges by the Alternating Series test.

Absolute Versus Conditional Convergence

Definition 7.4   The series $\sum\limits_{n=1}^{\infty} a_n$ *converges absolutely* (or is *absolutely convergent*) if $\sum\limits_{n=1}^{\infty} |a_n|$ converges. The series $\sum\limits_{n=1}^{\infty} a_n$ *converges conditionally* (or is *conditionally convergent*) if $\sum\limits_{n=1}^{\infty} a_n$ converges but not absolutely (that is, if $\sum\limits_{n=1}^{\infty} a_n$ converges and $\sum\limits_{n=1}^{\infty} |a_n|$ diverges).

Example 7.9   The alternating harmonic series $\sum\limits_{n=1}^{\infty} (-1)^{n+1}(1/n)$ is conditionally convergent by Examples 7.8 and 7.4.

Remark   For series of nonnegative terms, absolute convergence is the same as convergence. Also note that the Integral test and both comparison tests from Section 7.1 are tests of absolute convergence. In Sections 7.3 and 7.4, we will see that absolutely convergent series behave very much like finite sums, whereas conditionally convergent series are completely different.

Proposition 7.2   Absolute convergence implies convergence.

Proof   Let $\sum\limits_{n=1}^{\infty} a_n$ converge absolutely, and let $\varepsilon > 0$. By the Cauchy condition (Theorem 7.1), there is an n_0 in $\mathbb{N}$ such that $\sum\limits_{k=m+1}^{n} |a_k| < \varepsilon$ whenever $n > m \geq n_0$.

Then, for $n > m \geq n_0$, $\left| \sum\limits_{k=m+1}^{n} a_k \right| \leq \sum\limits_{k=m+1}^{n} |a_k| < \varepsilon$, and so $\sum\limits_{n=1}^{\infty} a_n$ converges by the Cauchy condition.   ∎

Tests for Absolute Convergence

We next develop three more tests for absolute convergence.

Theorem 7.7   (Ratio test) If $\sum\limits_{n=1}^{\infty} a_n$ is a series of nonzero real numbers with $L = \lim\limits_{n \to \infty} |a_{n+1}/a_n|$, then

1. $\sum\limits_{n=1}^{\infty} a_n$ converges absolutely if $L < 1$;

2. $\sum\limits_{n=1}^{\infty} a_n$ diverges if $L > 1$;

3. the test is inconclusive if $L = 1$.

Proof

1. First choose a real number r such that $L < r < 1$. Then there is an n_0 in $\mathbb{N}$ with $|a_{n+1}/a_n| \leq r$ for all $n \geq n_0$ or, equivalently, $|a_{n+1}| \leq r |a_n|$ for all $n \geq n_0$. Hence,

$$|a_{n_0+1}| \leq r |a_{n_0}|,$$
$$|a_{n_0+2}| \leq r |a_{n_0+1}| \leq r^2 |a_{n_0}|,$$
$$\vdots$$
$$|a_{n_0+k}| \leq r^k |a_{n_0}|.$$

Since $\sum\limits_{k=1}^{\infty} r^k$ is a convergent geometric series, $\sum\limits_{m=n_0+1}^{\infty} |a_m|$ converges by the Comparison test. By Exercise 4 in Section 7.1, $\sum\limits_{n=1}^{\infty} |a_n|$ converges.

2. Since $L > 1$, there is an n_0 in $\mathbb{N}$ such that $|a_{n+1}| > |a_n|$ for all $n \geq n_0$. Hence $\lim\limits_{n\to\infty} a_n \neq 0$, and Proposition 7.1 implies that $\sum\limits_{n=1}^{\infty} a_n$ diverges.

3. The divergent series $\sum\limits_{n=1}^{\infty} 1/n$ and the convergent series $\sum\limits_{n=1}^{\infty} 1/n^2$ both have $L = 1$. ∎

Example 7.10 $\sum\limits_{n=1}^{\infty} n!/n^n$ converges by the Ratio test. First note that

$$\frac{a_{n+1}}{a_n} = \frac{(n+1)!}{(n+1)^{n+1}} \cdot \frac{n^n}{n!} = \frac{n^n}{(n+1)^n}.$$

By L'Hôpital's rule or Example 3.6, $\lim\limits_{n\to\infty} [n/(n+1)]^n = 1/e < 1$ (this is a good exercise for the reader).

Theorem 7.8   (Root test) Given the series $\sum\limits_{n=1}^{\infty} a_n$, let $L = \lim\limits_{n\to\infty} \sqrt[n]{|a_n|}$. Then

1. $\sum\limits_{n=1}^{\infty} a_n$ converges absolutely if $L < 1$;

2. $\sum\limits_{n=1}^{\infty} a_n$ diverges if $L > 1$;

3. the test is inconclusive if $L = 1$.

Proof

1. First choose a number r such that $L < r < 1$. Then there is an n_0 in $\mathbb{N}$ with $\sqrt[n]{|a_n|} \leq r$ for all $n \geq n_0$ or, equivalently, $|a_n| \leq r^n$ for all $n \geq n_0$. Hence, $\sum\limits_{n=n_0}^{\infty} |a_n|$ converges by the Comparison test and so $\sum\limits_{n=1}^{\infty} |a_n|$ converges by Exercise 4 in Section 7.1.

2. Since $L > 1$, $|a_n| \geq 1$ for all large n and so $\lim\limits_{n\to\infty} a_n \neq 0$.

3. By an application of L'Hôpital's rule, $\sum\limits_{n=1}^{\infty} 1/n$ and $\sum\limits_{n=1}^{\infty} 1/n^2$ both have $L = 1$. ∎

Example 7.11   Our purpose here is to give an example in which the Ratio test fails but the Root test works, thus giving the correct impression that the Root test is more powerful than the Ratio test. (More precisely, if the Ratio test shows convergence, then so does the Root test; and if the Root test is inconclusive, then so is the Ratio test. See Rudin, p. 59.)

Consider $\sum\limits_{n=1}^{\infty} a_n = 2r + r^2 + 2r^3 + r^4 + 2r^5 + \cdots$, which clearly converges if $r = 0$. For $r \neq 0$,

$$\left| \frac{a_{n+1}}{a_n} \right| = \begin{cases} 2|r| & \text{if } n \text{ is even} \\ \dfrac{1}{2}|r| & \text{if } n \text{ is odd.} \end{cases}$$

Therefore, $\lim\limits_{n\to\infty} |a_{n+1}/a_n|$ does not exist for $r \neq 0$ and so the Ratio test fails.

However,

$$\sqrt[n]{|a_n|} = \begin{cases} |r^n|^{1/n} = |r| & \text{if } n \text{ is even} \\ |2r^n|^{1/n} = 2^{1/n}\,|r| & \text{if } n \text{ is odd.} \end{cases}$$

Since $\lim_{n\to\infty} 2^{1/n} = 1$, $\lim_{n\to\infty} \sqrt[n]{|a_n|} = |r|$. By the Root test, $\sum_{n=1}^{\infty} a_n$ converges absolutely if $|r| < 1$ and diverges if $|r| > 1$.

If $r = \pm 1$, then $\lim_{n\to\infty} a_n \neq 0$ and so this series diverges if $|r| \geq 1$.

Our next test is often used when the Ratio and Root tests fail.

Theorem 7.9   (Raabe's test) If $\sum_{n=1}^{\infty} a_n$ is a series of nonzero real numbers with $L = \lim_{n\to\infty} n(1 - |a_{n+1}/a_n|)$, then

1. $\sum_{n=1}^{\infty} a_n$ converges absolutely if $L > 1$;

2. $\sum_{n=1}^{\infty} a_n$ diverges or converges conditionally if $L < 1$;

3. the test is inconclusive if $L = 1$.

Proof

1. First choose a number r such that $1 < r < L$. Then there is an $n_0 \geq 2$ in $\mathbb{N}$ with $n(1 - |a_{n+1}/a_n|) \geq r$ for all $n \geq n_0$ or, equivalently, $|a_{n+1}|/|a_n| \leq 1 - (r/n)$ for all $n \geq n_0$. By manipulation it follows that

$$(n-1)\,|a_n| - n\,|a_{n+1}| \geq (r-1)\,|a_n| \qquad (2)$$

for all $n \geq n_0$. Let $m > n_0$ and rewrite (2) for $n = n_0, n_0 + 1, \ldots, m$:

$$\begin{aligned}
(n_0 - 1)\,\big|a_{n_0}\big| &- n_0\,\big|a_{n_0+1}\big| &\geq& \quad (r-1)\,\big|a_{n_0}\big| \\[4pt]
n_0\,\big|a_{n_0+1}\big| &- (n_0+1)\,\big|a_{n_0+2}\big| &\geq& \quad (r-1)\,\big|a_{n_0+1}\big| \\[4pt]
(n_0+1)\,\big|a_{n_0+2}\big| &- (n_0+2)\,\big|a_{n_0+3}\big| &\geq& \quad (r-1)\,\big|a_{n_0+2}\big| \\[2pt]
\vdots & \qquad \vdots & & \qquad \vdots \\[2pt]
(m-1)\,|a_m| &- m\,|a_{m+1}| &\geq& \quad (r-1)\,|a_m|.
\end{aligned}$$

Adding these inequalities, we get

$$(n_0 - 1)\,\big|a_{n_0}\big| - m\,\big|a_{m+1}\big| \geq (r-1)\big(\big|a_{n_0}\big| + \big|a_{n_0+1}\big| + \cdots + |a_m|\big).$$

Setting $s_n = \sum_{k=1}^{n} |a_k|$, we obtain

$$(n_0 - 1)\,\big|a_{n_0}\big| - m\,\big|a_{m+1}\big| \geq (r-1)[s_m - s_{n_0-1}].$$

Hence,

$$\begin{aligned}
(r-1)s_m &\leq (n_0 - 1)\,\big|a_{n_0}\big| + (r-1)s_{n_0-1} - m\,\big|a_{m+1}\big| \\[4pt]
&\leq (n_0 - 1)\,\big|a_{n_0}\big| + (r-1)s_{n_0-1}.
\end{aligned}$$

Therefore, $(s_m)_{m=n_0+1}^{\infty}$ is a bounded sequence and so $(s_n)_{n\in\mathbb{N}}$ is bounded. By Theorem 7.2, $\sum_{n=1}^{\infty} |a_n|$ converges.

2. Since $L < 1$, there is an $n_0 \geq 2$ in $\mathbb{N}$ such that $n(1 - |a_{n+1}/a_n|) \leq 1$ for all $n \geq n_0$ or, equivalently,

$$\frac{|a_{n+1}|}{|a_n|} \geq \frac{n-1}{n}$$

for all $n \geq n_0$. Hence, $n\,|a_{n+1}| \geq (n-1)\,|a_n|$ for all $n \geq n_0$, and so

$$n\,|a_{n+1}| \geq (n_0 - 1)\,|a_{n_0}|$$

for all $n \geq n_0$. Letting $c = (n_0 - 1)\,|a_{n_0}| > 0$, we have $|a_{n+1}| \geq c/n$ for all $n \geq n_0$. Since $\sum\limits_{n=n_0}^{\infty} c/n$ diverges, $\sum\limits_{n=n_0+1}^{\infty} |a_n|$ diverges by the Comparison test, and so $\sum\limits_{n=1}^{\infty} |a_n|$ diverges.

3. The divergent harmonic series and the conditionally convergent alternating harmonic series both have $L = 1$. If $a_2 = 2$ and

$$a_{n+1} = \frac{n \ln n}{(n+1)\ln n + 2} a_n$$

for $n \geq 2$, then $\sum\limits_{n=2}^{\infty} a_n$ has $L = 1$ and converges absolutely. The easiest way to show the absolute convergence is to use Kummer's test (see Fridy, p. 157), which we have not developed.   ■

Example 7.12   Consider

$$\sum_{n=1}^{\infty} \frac{2 \cdot 5 \cdot 8 \cdot \,\cdots\, \cdot (3n-1)}{9 \cdot 12 \cdot 15 \cdot \,\cdots\, \cdot (3n+6)} = \frac{2}{9} + \frac{2 \cdot 5}{9 \cdot 12} + \frac{2 \cdot 5 \cdot 8}{9 \cdot 12 \cdot 15} + \cdots.$$

By calculation, $a_{n+1}/a_n = (3n+2)/(3n+9) \to 1$, and so the Ratio test fails. (The Root test does not look good either.) Since

$$n\left(1 - \left|\frac{a_{n+1}}{a_n}\right|\right) = \frac{7n}{3n+9} \to \frac{7}{3} > 1,$$

this series converges by Raabe's test.

Tests for Conditional Convergence

Although there are many more tests for absolute convergence, we now turn our attention to tests for determining convergence when the series may not converge absolutely. So far we have only one of these, the Alternating Series test. We develop one such test and leave a second one as Exercise 7.

Lemma 7.1   (Abel's partial summation formula) Let $(a_n)_{n\in\mathbb{N}}$ and $(b_n)_{n\in\mathbb{N}}$ be two sequences in $\mathbb{R}$, and let $A_n = \sum\limits_{k=1}^{n} a_k$ for each n in $\mathbb{N}$. Then

$$\sum_{k=1}^{n} a_k b_k = A_n b_{n+1} - \sum_{k=1}^{n} A_k(b_{k+1} - b_k). \tag{3}$$

In particular, $\sum\limits_{k=1}^{\infty} a_k b_k$ converges if both the series $\sum\limits_{k=1}^{\infty} A_k(b_{k+1} - b_k)$ and the sequence $(A_n b_{n+1})_{n\in\mathbb{N}}$ converge.

Proof   Letting $A_0 = 0$, we have

$$\sum_{k=1}^{n} a_k b_k = \sum_{k=1}^{n} (A_k - A_{k-1}) b_k = \sum_{k=1}^{n} A_k b_k - \sum_{k=1}^{n} A_k b_{k+1} + A_n b_{n+1},$$

and so (3) follows.   ■

Theorem 7.10   (Dirichlet's test) Let $\sum_{n=1}^{\infty} a_n$ be a series with bounded partial sums, and let $(b_n)_{n\in\mathbb{N}}$ be a monotone decreasing sequence with $b_n \underset{n}{\to} 0$. Then $\sum_{n=1}^{\infty} a_n b_n$ converges.

Proof   Let $A_n = \sum_{k=1}^{n} a_k$, and let $M > 0$ with $|A_n| \le M$ for all n in $\mathbb{N}$. Since $b_n \underset{n}{\to} 0$, $\lim_{n\to\infty} A_n b_{n+1} = 0$. By Lemma 7.1, it suffices to show that $\sum_{k=1}^{\infty} A_k(b_{k+1} - b_k)$ converges. Since $(b_n)_{n\in\mathbb{N}}$ is monotone decreasing,

$$|A_k(b_{k+1} - b_k)| \le M(b_k - b_{k+1})$$

for each k in $\mathbb{N}$. Since

$$\sum_{k=1}^{n}(b_k - b_{k+1}) = (b_1 - b_2) + (b_2 - b_3) + \cdots + (b_n - b_{n+1})$$
$$= b_1 - b_{n+1} \underset{n}{\to} b_1,$$

$\sum_{k=1}^{\infty}(b_k - b_{k+1})$ converges. By the Comparison test, $\sum_{k=1}^{\infty} A_k(b_{k+1} - b_k)$ converges absolutely.   ∎

Note that the Alternating Series test follows directly from Dirichlet's test, because given the hypothesis of Theorem 7.6, $\sum_{n=1}^{\infty} (-1)^{n+1}$ has bounded partial sums and the a_n's (which are the b_n's in Theorem 7.10) decrease monotonically to 0.

Example 7.13   For $x \ne 0, \pm 2\pi, \pm 4\pi, \ldots$, the trigonometric identity

$$\sum_{k=1}^{n} \sin kx = \frac{\cos \frac{1}{2}x - \cos(n + \frac{1}{2})x}{2 \sin \frac{1}{2}x}$$

implies that $\left| \sum_{k=1}^{n} \sin kx \right| \le 1/|\sin \frac{1}{2}x|$. Of course, $\sin(2m\pi) = 0$ for any integer m. Thus, the partial sums of $\sum_{n=1}^{\infty} \sin nx$ are bounded for all x in $\mathbb{R}$. So if $(b_n)_{n\in\mathbb{N}}$ is a monotone decreasing sequence with limit 0, Dirichlet's test implies that $\sum_{n=1}^{\infty} b_n \sin nx$ converges for all x in $\mathbb{R}$. In particular, $\sum_{n=1}^{\infty} (\sin n)/n$ converges.

Exercises

1. Show that the following series converge absolutely.

   (a) $\sum_{n=1}^{\infty} r^n/n!$ where r is in $\mathbb{R}$ (b) $\sum_{n=1}^{\infty} (n + 2^n)/3^n$

2. Show that the following series converge.

   (a) $\sum_{n=2}^{\infty} 1/(\ln n)^n$ (b) $\sum_{n=2}^{\infty} (\sqrt[n]{n} - 1)^n$

3. Show that

$$\sum_{n=1}^{\infty}(-1)^{n+1}\cdot\frac{1\cdot3\cdot5\cdot\;\cdots\;\cdot(2n-1)}{2\cdot4\cdot6\cdot\;\cdots\;\cdot(2n)}=\frac{1}{2}-\frac{1\cdot3}{2\cdot4}+\frac{1\cdot3\cdot5}{2\cdot4\cdot6}-\cdots.$$

is conditionally convergent. [*Hint:* Use Raabe's test (the Ratio test fails) to show nonabsolute convergence; then use the Alternating Series test. To show $a_n \to 0$, show that $a_n^2 < 1/(2n+1)$ by induction.]

4. Suppose that $\sum_{n=1}^{\infty} a_n$ converges with each $a_n > 0$, and let $r_n = \sum_{k=n}^{\infty} a_k$ for each n in $\mathbb{N}$. Show that

(a) $(r_n)_{n\in\mathbb{N}}$ is strictly decreasing with limit 0;

(b) $\sum_{n=1}^{\infty} a_n r_n$ converges;

(c) $\sum_{n=1}^{\infty} a_n/r_n$ diverges. [*Hint:* Use the Cauchy condition.]

5. Suppose that $\sum_{n=1}^{\infty} a_n$ converges and x is in $[0, 1]$. Show that $\sum_{n=1}^{\infty} a_n x^n$ converges.

6. Use the trigonometric identity

$$\sum_{k=1}^{n}\cos kx=\frac{\sin(n+\tfrac{1}{2})x-\sin\tfrac{1}{2}x}{2\sin\tfrac{1}{2}x}$$

for $x \neq 2m\pi$ (where m is an integer) to show that $\sum_{n=1}^{\infty} b_n \cos nx$ converges for all $x \neq 2m\pi$ whenever $(b_n)_{n\in\mathbb{N}}$ is a monotone decreasing sequence with limit 0.

7. This is Abel's test. Show that $\sum_{n=1}^{\infty} a_n b_n$ converges if $\sum_{n=1}^{\infty} a_n$ converges and the sequence $(b_n)_{n\in\mathbb{N}}$ is monotone and bounded. [*Hint:* Use Dirichlet's test.]

8. Suppose that $\sum_{n=1}^{\infty} a_n$ converges with each $a_n > 0$, and let $s_n = \sum_{k=1}^{n} a_k$. Show that $\sum_{n=1}^{\infty} a_n s_n$ and $\sum_{n=1}^{\infty} a_n/s_n$ both converge.

9. Suppose that $\sum_{n=1}^{\infty} a_n$ diverges with each $a_n > 0$, and let $s_n = \sum_{k=1}^{n} a_k$.

(a) Show that $\sum_{n=1}^{\infty} a_n/s_n$ diverges. [*Hint:* Use the Cauchy condition.]

(b) Show that $\sum_{n=1}^{\infty} a_n/s_n^2$ converges. [*Hint:* $a_n/s_n^2 = (s_n - s_{n-1})/s_n^2 \leq 1/s_{n-1} - 1/s_n$ for $n \geq 2$ and $\sum_{n=2}^{\infty} (1/s_{n-1} - 1/s_n)$ converges.]

10. Consider

$$\sum_{n=1}^{\infty}\left[\frac{1\cdot3\cdot5\cdot\;\cdots\;\cdot(2n-1)}{2\cdot4\cdot6\cdot\;\cdots\;\cdot(2n)}\right]^{p}=\left(\frac{1}{2}\right)^{p}+\left(\frac{1\cdot3}{2\cdot4}\right)^{p}+\left(\frac{1\cdot3\cdot5}{2\cdot4\cdot6}\right)^{p}+\cdots,$$

where p is in $\mathbb{R}$.

(a) Use Raabe's test to show that this series converges if $p > 2$ and diverges if $p < 2$.

(b) When $p = 2$, Raabe's test is inconclusive. To show that this series diverges when $p = 2$, first do long division to obtain

$$\frac{a_{n+1}}{a_n} = 1 - \frac{4n + 3}{4n^2 + 8n + 4} = 1 - \frac{1}{n} + \frac{5n + 4}{n(4n^2 + 8n + 4)} > 1 - \frac{1}{n}.$$

It follows that $a_{n+1} > (1/n)a_2$ for $n \geq 2$.

7.3 Regrouping and Rearranging Series

Regrouping

The first question we consider is what happens to the convergence or divergence of a series if the terms of the series are regrouped by inserting parentheses. For example, what happens when we regroup the alternating harmonic series

$$1 - \frac{1}{2} + \frac{1}{3} - \frac{1}{4} + \cdots$$

as

$$\left(1 - \frac{1}{2}\right) + \left(\frac{1}{3} - \frac{1}{4}\right) + \left(\frac{1}{5} - \frac{1}{6}\right) + \cdots ?$$

The next definition makes the insertion of parentheses precise.

Definition 7.5   Let ψ be a function from $\mathbb{N}$ into $\mathbb{N}$ such that $\psi(m) < \psi(n)$ if $m < n$, and let $\sum_{n=1}^{\infty} a_n$ be a given series. Define a new series $\sum_{n=1}^{\infty} b_n$ as follows:

$$b_1 = a_1 + a_2 + \cdots + a_{\psi(1)},$$
$$b_2 = a_{\psi(1)+1} + a_{\psi(1)+2} + \cdots + a_{\psi(2)},$$
$$\vdots$$
$$b_{n+1} = a_{\psi(n)+1} + a_{\psi(n)+2} + \cdots + a_{\psi(n+1)}.$$
$$\vdots$$

Then $\sum_{n=1}^{\infty} b_n$ is a *regrouping* of $\sum_{n=1}^{\infty} a_n$ *formed by inserting parentheses*. If we start with the series $\sum_{n=1}^{\infty} b_n$ above, then the series $\sum_{n=1}^{\infty} a_n$ is a *regrouping* of $\sum_{n=1}^{\infty} b_n$ *formed by removing parentheses*.

In Definition 7.5, note that

$$\sum_{n=1}^{\infty} b_n = \left(a_1 + a_2 + \cdots + a_{\psi(1)}\right) + \left(a_{\psi(1)+1} + a_{\psi(1)+2} + \cdots + a_{\psi(2)}\right) + \cdots.$$

In the regrouped series above Definition 7.5, $\psi(n) = 2n$.

Theorem 7.11   If $\sum\limits_{n=1}^{\infty} a_n$ converges to A, then every regrouping of $\sum\limits_{n=1}^{\infty} a_n$ formed by inserting parentheses also converges to A.

Proof   Let $\sum\limits_{n=1}^{\infty} a_n$ and $\sum\limits_{n=1}^{\infty} b_n$ be related as in Definition 7.5, and let $s_n = \sum\limits_{k=1}^{n} a_k$ and $t_n = \sum\limits_{k=1}^{n} b_k$. Then $(t_n)_{n\in\mathbb{N}}$ is a subsequence of $(s_n)_{n\in\mathbb{N}}$; in fact, $t_n = s_{\psi(n)}$ for each n in $\mathbb{N}$. Since $s_n \to A$, $t_n \to A$ by Theorem 3.6.   ∎

Example 7.14   Our purpose here is to show that a regrouping of a divergent series may converge or diverge. For the divergent series $\sum\limits_{n=1}^{\infty} (-1)^{n+1} = 1 - 1 + 1 - 1 + 1 - 1 + \cdots$, the regrouping $(1 - 1) + (1 - 1) + (1 - 1) + \cdots = 0 + 0 + 0 + \cdots$ converges, whereas the regrouping $(1 - 1) + 1 - 1 + 1 - 1 + \cdots [\psi(n) = n + 1]$ diverges.

Remark   Going backward in Example 7.14 from the first regrouping to the original series shows that removing parentheses may destroy convergence. Also, a series is a regrouping of itself [let $\psi(n) = n$]; hence, the original series could have been used for the second regrouping in Example 7.14.

Rearrangements

Definition 7.6   Let φ be a one-to-one function from $\mathbb{N}$ onto $\mathbb{N}$ and let $\sum\limits_{n=1}^{\infty} a_n$ be a given series. Define a new series $\sum\limits_{n=1}^{\infty} b_n$ by $b_n = a_{\varphi(n)}$ for each n in $\mathbb{N}$. Then $\sum\limits_{n=1}^{\infty} b_n$ is a *rearrangement* of $\sum\limits_{n=1}^{\infty} a_n$.

Remark   First note that every term of $\sum\limits_{n=1}^{\infty} a_n$ appears once and only once in $\sum\limits_{n=1}^{\infty} b_n$, but the order in which the terms appear may be changed. Second, $\sum\limits_{n=1}^{\infty} a_n$ is also a rearrangement of $\sum\limits_{n=1}^{\infty} b_n$ since $a_n = b_{\varphi^{-1}(n)}$. Third, a series is a rearrangement of itself [let $\varphi(n) = n$].

The question we now consider is whether a rearrangement of a convergent series converges or diverges; and if it converges, whether or not it will have the same sum.

Example 7.15   Let A be the sum of the conditionally convergent alternating harmonic series

$$1 - \frac{1}{2} + \frac{1}{3} - \frac{1}{4} + \frac{1}{5} - \frac{1}{6} + \frac{1}{7} - \frac{1}{8} + \cdots.$$

Consider the rearrangement of this series formed by having one positive term followed by two negative terms:

$$1 - \frac{1}{2} - \frac{1}{4} + \frac{1}{3} - \frac{1}{6} - \frac{1}{8} + \frac{1}{5} - \frac{1}{10} - \frac{1}{12} + \cdots.$$

Let s_n be a partial sum of this rearrangement of the form

$$s_n = 1 - \frac{1}{2} - \frac{1}{4} + \frac{1}{3} - \frac{1}{6} - \frac{1}{8} + \cdots + \frac{1}{2n-1} - \frac{1}{4n-2} - \frac{1}{4n}.$$

Then

$$s_n = \left(1 - \frac{1}{2}\right) - \frac{1}{4} + \left(\frac{1}{3} - \frac{1}{6}\right) - \frac{1}{8} + \cdots + \left(\frac{1}{2n-1} - \frac{1}{4n-2}\right) - \frac{1}{4n}$$

$$= \frac{1}{2} - \frac{1}{4} + \frac{1}{6} - \frac{1}{8} + \cdots + \frac{1}{2(2n-1)} - \frac{1}{2(2n)}$$

$$= \frac{1}{2}\left(1 - \frac{1}{2} + \frac{1}{3} - \frac{1}{4} + \cdots + \frac{1}{2n-1} - \frac{1}{2n}\right).$$

As $n \to \infty$, $s_n \to (1/2)A$. If r_n and t_n are partial sums of the rearranged series whose last terms are of the form $1/(2n-1)$ and $-1/(4n-2)$, respectively, then $|s_n - r_n| = 1/(4n-2) + 1/4n \to 0$ and $|s_n - t_n| = 1/4n \to 0$. Therefore,

$$\lim_{n\to\infty} r_n = \lim_{n\to\infty} t_n = \lim_{n\to\infty} s_n = (1/2)A,$$

and so our rearranged series converges and has sum $(1/2)A$.

The next theorem, which we used in Example 4.26, is probably what one would expect. However, Theorem 7.13 and Example 7.16 are surprising.

Theorem 7.12   If $\sum_{n=1}^{\infty} a_n$ is absolutely convergent with sum A, then every rearrangement is absolutely convergent and has sum A.

Proof   Let $\sum_{n=1}^{\infty} b_n$ be a rearrangement of $\sum_{n=1}^{\infty} a_n$ as given in Definition 7.6. Let $\varepsilon > 0$. By the Cauchy condition, there is an n_0 in $\mathbb{N}$ such that $\sum_{k=m+1}^{n} |a_k| < \varepsilon/2$ whenever $n > m \geq n_0$. Since $\sum_{k=n_0+1}^{\infty} |a_k|$ converges and has all of its partial sums less than $\varepsilon/2$, $\sum_{k=n_0+1}^{\infty} |a_k| \leq \varepsilon/2$. Choose n_1 in $\mathbb{N}$ such that $\{1, 2, \ldots, n_0\} \subset \{\varphi(1), \varphi(2), \ldots, \varphi(n_1)\}$. Then $k > n_1$ implies $\varphi(k) > n_0$, and hence, for $n > m \geq n_1$,

$$\sum_{k=m+1}^{n} |b_k| = \sum_{k=m+1}^{n} |a_{\varphi(k)}| \leq \sum_{k=n_0+1}^{\infty} |a_k| \leq \frac{\varepsilon}{2} < \varepsilon.$$

By the Cauchy condition, $\sum_{n=1}^{\infty} b_n$ converges absolutely.

We now show that $\sum_{n=1}^{\infty} b_n = A$. Let $s_n = \sum_{k=1}^{n} a_k$ and $t_n = \sum_{k=1}^{n} b_k$. Since $s_n \to A$ and $(t_n)_{n\in\mathbb{N}}$ converges, it suffices to show that $\lim_{n\to\infty} (t_n - s_n) = 0$. Let $\varepsilon > 0$ and choose n_0 and n_1 as in the first part of the proof. For $n > n_1$,

$$|t_n - s_n| = |b_1 + \cdots + b_n - (a_1 + \cdots + a_n)|$$

$$= |a_{\varphi(1)} + \cdots + a_{\varphi(n)} - (a_1 + \cdots + a_n)|$$

$$\leq \sum_{k=n_0+1}^{\infty} |a_k| < \varepsilon,$$

and so $\lim\limits_{n \to \infty} (t_n - s_n) = 0$. (The first inequality follows since $a_1, \ldots, a_{n_0}$ cancel out in the subtraction.) ∎

To simplify the proof of the next theorem, we first consider the following lemma.

Lemma 7.2 Suppose that $\sum\limits_{n=1}^{\infty} a_n$ converges conditionally. Let $P_n = (|a_n| + a_n)/2$ and let $Q_n = (|a_n| - a_n)/2$ for each n in $\mathbb{N}$. Then $\sum\limits_{n=1}^{\infty} P_n$ and $\sum\limits_{n=1}^{\infty} Q_n$ both diverge.

Proof First note that $a_n = P_n - Q_n$, $|a_n| = P_n + Q_n$, $P_n \geq 0$, and $Q_n \geq 0$ for each n in $\mathbb{N}$. If both $\sum\limits_{n=1}^{\infty} P_n$ and $\sum\limits_{n=1}^{\infty} Q_n$ converge, then $\sum\limits_{n=1}^{\infty} P_n + \sum\limits_{n=1}^{\infty} Q_n = \sum\limits_{n=1}^{\infty} |a_n|$ would converge, contrary to the hypothesis. Since $\sum\limits_{k=1}^{n} a_k = \sum\limits_{k=1}^{n} P_k - \sum\limits_{k=1}^{n} Q_k$, if $\sum\limits_{n=1}^{\infty} P_n$ converges and $\sum\limits_{n=1}^{\infty} Q_n$ diverges (or vice versa), then $\sum\limits_{n=1}^{\infty} a_n$ would diverge, again contrary to the hypothesis. ∎

Theorem 7.13 Suppose that $\sum\limits_{n=1}^{\infty} a_n$ converges conditionally and L is in $\mathbb{R}$. Then there is a rearrangement of $\sum\limits_{n=1}^{\infty} a_n$ that has sum L. (See Exercise 7 for a similar result when $L = \pm\infty$.)

Proof Let $p_1, p_2, p_3, \ldots$ be the nonnegative terms of $\sum\limits_{n=1}^{\infty} a_n$ in the order in which they occur; and let $-q_1, -q_2, -q_3, \ldots$ denote the negative terms of $\sum\limits_{n=1}^{\infty} a_n$ in the order in which they occur. Since $\sum\limits_{n=1}^{\infty} p_n$ and $\sum\limits_{n=1}^{\infty} q_n$ differ from $\sum\limits_{n=1}^{\infty} P_n$ and $\sum\limits_{n=1}^{\infty} Q_n$ of Lemma 7.2 only by zero terms, both $\sum\limits_{n=1}^{\infty} p_n$ and $\sum\limits_{n=1}^{\infty} q_n$ diverge. Hence, the partial sums of these last two series both have limit ∞.

To construct the desired rearrangement, we do the following: we take just enough nonnegative terms until we exceed L, then just enough negative terms until we are less than L, then just enough nonnegative terms until we exceed L, and so on.

Let n_1 be the smallest positive integer such that

$$p_1 + \cdots + p_{n_1} > L.$$

Let m_1 be the smallest positive integer such that

$$p_1 + \cdots + p_{n_1} - q_1 - \cdots - q_{m_1} < L.$$

Next, let n_2 be the smallest positive integer such that

$$p_1 + \cdots + p_{n_1} - q_1 - \cdots - q_{m_1} + p_{n_1+1} + \cdots + p_{n_2} > L$$

and let m_2 be the smallest positive integer such that

$$p_1 + \cdots + p_{n_1} - q_1 - \cdots - q_{m_1} + p_{n_1+1} + \cdots + p_{n_2} -$$
$$q_{m_1+1} - \cdots - q_{m_2} < L.$$

These steps are possible since $\sum\limits_{n=1}^{\infty} p_n$ and $\sum\limits_{n=1}^{\infty} q_n$ are divergent series of nonnegative terms. Continuing this process (by induction), we obtain a rearrangement of $\sum\limits_{n=1}^{\infty} a_n$.

To see that this rearrangement converges to L, let s_n be a partial sum whose last term is p_{n_j} or $-q_{m_j}$. Then $|L - s_n| \leq \max\{p_{n_j}, q_{m_j}\}$ (by the way p_{n_j} and $-q_{m_j}$ were chosen). Since $\sum\limits_{n=1}^{\infty} a_n$ converges, both $p_{n_j} \to 0$ and $q_{m_j} \to 0$ as $j \to \infty$, and so $s_n \to L$. ∎

Example 7.16   The behavior of a rearrangement of a conditionally convergent series can be stranger than that indicated in Theorem 7.13. Let $\sum\limits_{n=1}^{\infty} a_n$ be a conditionally convergent series and let $-\infty \leq x < y \leq \infty$. Let $(x_n)_{n\in\mathbb{N}}$ and $(y_n)_{n\in\mathbb{N}}$ be two sequences in $\mathbb{R}$ with $x_n \to x$, $y_n \to y$, $x_n < y_n$, and $x_n < y_{n+1}$ for each n, and $y_1 > 0$.

In the notation of the proof of Theorem 7.13, choose just enough nonnegative terms so that $p_1 + \cdots + p_{n_1} > y_1$, followed by just enough negative terms so that $p_1 + \cdots + p_{n_1} - q_1 - \cdots - q_{m_1} < x_1$, followed by just enough nonnegative terms so that $p_1 + \cdots + p_{n_1} - q_1 - \cdots - q_{m_1} + p_{n_1+1} + \cdots + p_{n_2} > y_2$, followed by just enough negative terms so that $p_1 + \cdots + p_{n_1} - q_1 - \cdots - q_{m_1} + p_{n_1+1} + \cdots + p_{n_2} - q_{m_1+1} - \cdots - q_{m_2} < x_2$, and so on.

If r_n and t_n are partial sums of this rearrangement whose last terms are p_{n_j} and $-q_{m_j}$, respectively, then $|r_n - y_n| \leq p_{n_j}$ and $|t_n - x_n| \leq q_{m_j}$. Since $p_{n_j} \to 0$ and $q_{m_j} \to 0$ as $j \to \infty$, $r_n \to y$ and $t_n \to x$ as $n \to \infty$. Recall Definition 3.14 and let $(s_n)_{n\in\mathbb{N}}$ denote the sequence of partial sums of this rearrangement. Then $\liminf s_n = x$ and $\limsup s_n = y$ since no number smaller than x or larger than y can be a subsequential limit of $(s_n)_{n\in\mathbb{N}}$.

≡ Exercises ≡

1. Show that $\sum\limits_{n=1}^{\infty} 1/[2n(2n-1)]$ converges and has the same sum as the alternating harmonic series.

2. Suppose that $\sum\limits_{n=1}^{\infty} a_n = \infty$ (or $-\infty$) [that is, the limit of the partial sums of $\sum\limits_{n=1}^{\infty} a_n$ is ∞ (or $-\infty$)]. Let $\sum\limits_{n=1}^{\infty} b_n$ be a regrouping of $\sum\limits_{n=1}^{\infty} a_n$ formed by inserting parentheses. Show that $\sum\limits_{n=1}^{\infty} b_n = \infty$ (or $-\infty$).

3. (a) Does the convergence of $\sum\limits_{n=1}^{\infty} a_n$ imply the convergence of $\sum\limits_{n=1}^{\infty} (a_n + a_{n+1})$?

   (b) Does the convergence of $\sum\limits_{n=1}^{\infty} (a_n + a_{n+1})$ imply the convergence of $\sum\limits_{n=1}^{\infty} a_n$?

   (c) Does the convergence of $\sum\limits_{n=1}^{\infty} (|a_n| + |a_{n+1}|)$ imply the convergence of $\sum\limits_{n=1}^{\infty} |a_n|$?

4. If $\sum\limits_{n=1}^{\infty} a_n$ converges absolutely and $\sum\limits_{n=1}^{\infty} b_n$ is a rearrangement of $\sum\limits_{n=1}^{\infty} a_n$, show that $\sum\limits_{n=1}^{\infty} |b_n| = \sum\limits_{n=1}^{\infty} |a_n|$.

5. Can a conditionally convergent series be rearranged to form an absolutely convergent series?

6. In the notation of Lemma 7.2, show that if $\sum\limits_{n=1}^{\infty} a_n$ is absolutely convergent, then $\sum\limits_{n=1}^{\infty} P_n$ and $\sum\limits_{n=1}^{\infty} Q_n$ both converge and $\sum\limits_{n=1}^{\infty} a_n = \sum\limits_{n=1}^{\infty} P_n - \sum\limits_{n=1}^{\infty} Q_n$.

7. If $\sum\limits_{n=1}^{\infty} a_n$ is conditionally convergent, show that there is a rearrangement with sum ∞. (A similar result holds for $-\infty$.)

8. Give an example showing that a rearrangement of a divergent series may diverge or converge.

9. Let $\sum\limits_{n=1}^{\infty} a_n$ be a divergent series of nonnegative terms. Show that every rearrangement of $\sum\limits_{n=1}^{\infty} a_n$ diverges and has sum ∞.

7.4 Multiplication of Series

In this section we start our series at 0.

Definition 7.7 The *Cauchy product* of the two series $\sum\limits_{n=0}^{\infty} a_n$ and $\sum\limits_{n=0}^{\infty} b_n$ is the series $\sum\limits_{n=0}^{\infty} c_n$, where

$$c_n = \sum_{k=0}^{n} a_k b_{n-k} \qquad \text{for } n = 0, 1, 2, 3, \ldots.$$

If we think of the following series as "big" polynomials and multiply them as polynomials, we get

$$\left(\sum_{n=0}^{\infty} a_n x^n\right)\left(\sum_{n=0}^{\infty} b_n x^n\right)$$
$$= (a_0 + a_1 x + a_2 x^2 + \cdots) \cdot$$
$$(b_0 + b_1 x + b_2 x^2 + \cdots)$$
$$= a_0 b_0 + (a_0 b_1 + a_1 b_0)x +$$
$$(a_0 b_2 + a_1 b_1 + a_2 b_0)x^2 + \cdots$$
$$= c_0 + c_1 x + c_2 x^2 + \cdots.$$

Replacing x with 1 is the motivation for Definition 7.7.

The question we consider is whether the Cauchy product of two convergent series converges; and if it does, whether the sum of the Cauchy product is equal to the product of the sums of the two convergent series. That is, does

$$\sum_{n=0}^{\infty} c_n = \left(\sum_{n=0}^{\infty} a_n \right) \left(\sum_{n=0}^{\infty} b_n \right)?$$

Example 7.17   The series

$$\sum_{n=0}^{\infty} \frac{(-1)^n}{\sqrt{n+1}} = 1 - \frac{1}{\sqrt{2}} + \frac{1}{\sqrt{3}} - \frac{1}{\sqrt{4}} + \cdots$$

is conditionally convergent by Theorem 7.6 and Example 7.5. We show that the Cauchy product of this series with itself diverges. First note that

$$c_n = \sum_{k=0}^{n} a_k a_{n-k}$$

$$= \sum_{k=0}^{n} \frac{(-1)^k}{\sqrt{k+1}} \cdot \frac{(-1)^{n-k}}{\sqrt{n-k+1}}$$

$$= (-1)^n \sum_{k=0}^{n} \frac{1}{\sqrt{(n-k+1)(k+1)}}.$$

Since

$$(n - k + 1)(k + 1) = \left(\frac{n}{2} + 1 \right)^2 - \left(\frac{n}{2} - k \right)^2$$

$$\leq \left(\frac{n}{2} + 1 \right)^2$$

$$= \left(\frac{n+2}{2} \right)^2,$$

it follows that

$$|c_n| \geq \sum_{k=0}^{n} \frac{2}{n+2} = \frac{2(n+1)}{n+2} \to 2.$$

Hence, $\lim_{n \to \infty} c_n \neq 0$ and $\sum_{n=0}^{\infty} c_n$ diverges.

Exercise 1 provides us with an example of a convergent Cauchy product of two conditionally convergent series. The following is our main theorem concerning Cauchy products.

Theorem 7.14   (Mertens' Theorem) Suppose that $\sum_{n=0}^{\infty} a_n$ converges absolutely and has sum A, while $\sum_{n=0}^{\infty} b_n$ converges and has sum B. Then the Cauchy product of these two series converges and has sum AB. (That is, the Cauchy product of two convergent series converges, and to the appropriate value, if at least one of the series is absolutely convergent.)

Proof   This proof is taken from Rudin. For n in $\mathbb{N} \cup \{0\}$, let

$$A_n = \sum_{k=0}^{n} a_k, \quad B_n = \sum_{k=0}^{n} b_k, \quad C_n = \sum_{k=0}^{n} c_k, \quad \beta_n = B_n - B.$$

Then

$$C_n = a_0 b_0 + (a_0 b_1 + a_1 b_0) + \cdots + (a_0 b_n + a_1 b_{n-1} + \cdots + a_n b_0)$$
$$= a_0 B_n + a_1 B_{n-1} + \cdots + a_n B_0$$
$$= a_0 (B + \beta_n) + a_1 (B + \beta_{n-1}) + \cdots + a_n (B + \beta_0)$$
$$= A_n B + a_0 \beta_n + a_1 \beta_{n-1} + \cdots + a_n \beta_0.$$

Let $\gamma_n = a_0 \beta_n + a_1 \beta_{n-1} + \cdots + a_n \beta_0$. We want to show that $C_n \to AB$. Since $\sum_{n=0}^{\infty} a_n = A$, $A_n B \to AB$, and so it suffices to show that $\gamma_n \to 0$.

Let $\varepsilon > 0$. Since $\sum_{n=0}^{\infty} a_n$ is absolutely convergent, let $\alpha = \sum_{n=0}^{\infty} |a_n|$. Since $\sum_{n=0}^{\infty} b_n = B$, $\beta_n \to 0$, and so there is an n_0 in $\mathbb{N}$ such that $|\beta_n| < \varepsilon/(\alpha + 1)$ for all $n \geq n_0$. For $n \geq n_0$,

$$|\gamma_n| \leq |a_n \beta_0 + a_{n-1} \beta_1 + \cdots + a_{n-n_0+1} \beta_{n_0-1}| +$$
$$|a_{n-n_0} \beta_{n_0} + a_{n-n_0-1} \beta_{n_0+1} + \cdots + a_0 \beta_n|$$
$$\leq |a_n \beta_0 + a_{n-1} \beta_1 + \cdots + a_{n-n_0+1} \beta_{n_0-1}| + \frac{\varepsilon}{\alpha + 1} \cdot \alpha$$
$$\leq |a_n \beta_0 + a_{n-1} \beta_1 + \cdots + a_{n-n_0+1} \beta_{n_0-1}| + \varepsilon.$$

Letting $n \to \infty$ and keeping n_0 fixed, since $a_k \to 0$ as $k \to \infty$, we obtain that $\limsup |\gamma_n| \leq \varepsilon$. Since ε is arbitrary, $\lim_{n \to \infty} \gamma_n = 0$.   ∎

In Chapter 9 we will prove one further result about Cauchy products— namely, if $\sum_{n=0}^{\infty} a_n$, $\sum_{n=0}^{\infty} b_n$, and their Cauchy product $\sum_{n=0}^{\infty} c_n$ converge to A, B, and C, respectively, then $C = AB$.

Example 7.18   Consider

$$\sum_{n=0}^{\infty} \frac{n+1}{2^n} = 1 + 1 + \frac{3}{2^2} + \frac{4}{2^3} + \frac{5}{2^4} + \cdots.$$

This series converges by the Ratio test, but to what does it converge? One way to answer this question is to consider the Cauchy product of $\sum_{n=0}^{\infty} x^n$ with itself. Since $\sum_{n=0}^{\infty} x^n$ converges absolutely for $|x| < 1$, the Cauchy product

$$\left(\sum_{n=0}^{\infty} x^n \right) \left(\sum_{n=0}^{\infty} x^n \right) = \sum_{n=0}^{\infty} \left(\sum_{k=0}^{n} x^k x^{n-k} \right)$$
$$= \sum_{n=0}^{\infty} \left(\sum_{k=0}^{n} x^n \right)$$
$$= \sum_{n=0}^{\infty} (n+1) x^n$$

converges for $|x| < 1$ by Theorem 7.14. (It actually converges absolutely for

$|x| < 1$ by Exercise 2 or by the Ratio test.) Hence,

$$\sum_{n=0}^{\infty}(n+1)x^n = \left(\frac{1}{1-x}\right)^2$$

for $|x| < 1$. Letting $x = 1/2$, $\sum_{n=0}^{\infty}(n+1)/2^n = 4$.

Although the following would be more appropriate in Chapter 9, it is hard not to consider this from another viewpoint. For $|x| < 1$,

$$\sum_{n=0}^{\infty}(n+1)x^n = \sum_{n=0}^{\infty}nx^n + \sum_{n=0}^{\infty}x^n$$

$$= x\sum_{n=1}^{\infty}nx^{n-1} + \frac{1}{1-x}$$

$$= x\sum_{m=0}^{\infty}(m+1)x^m + \frac{1}{1-x},$$

where we made the substitution $m = n - 1$. Replacing m by n and transposing,

$$(1-x)\sum_{n=0}^{\infty}(n+1)x^n = \frac{1}{1-x}$$

and so

$$\sum_{n=0}^{\infty}(n+1)x^n = \left(\frac{1}{1-x}\right)^2$$

for $|x| < 1$.

Exercises

1. Show that the Cauchy product of the alternating harmonic series $\sum_{n=0}^{\infty} (-1)^n/(n+1)$ with itself converges. [*Hint:* Use

$$\frac{1}{(k+1)(n-k+1)} = \frac{1}{n+2}\left(\frac{1}{k+1} + \frac{1}{n-k+1}\right)$$

to obtain

$$\sum_{n=0}^{\infty} c_n = 2\sum_{n=0}^{\infty} \frac{(-1)^n}{n+2}\left(1 + \frac{1}{2} + \frac{1}{3} + \cdots + \frac{1}{n+1}\right).$$

Now use the Alternating Series test to obtain convergence. To get the nth term going to 0, use upper and lower sums for $f(x) = 1/x$ to obtain $\ln(n+2) \leq 1 + \frac{1}{2} + \frac{1}{3} + \cdots + 1/(n+1) \leq 1 + \ln(n+1)$.]

2. Show that the Cauchy product of two absolutely convergent series converges absolutely.

3. Find the sum of

$$\sum_{n=1}^{\infty} \frac{n}{3^n}.$$

4. Find the sum of

$$\sum_{n=0}^{\infty} (-1)^n \frac{n+1}{2^n}.$$

5. (a) Show that if $\sum_{n=0}^{\infty} a_n$ converges and $\sum_{n=0}^{\infty} b_n$ converges absolutely, then $\sum_{n=0}^{\infty} a_n b_n$ (sometimes called the *inner product* series) converges absolutely.

   (b) Show, by example, that if both series are conditionally convergent, then the inner product series may diverge.

Sequences and Series of Functions

In Section 8.1 we explain the terminology and the main problem in terms of sequences of functions; in Section 8.2 we examine how this new concept relates to continuity, integration, and differentiation. We consider series of functions in Section 8.3, where we construct a continuous, nowhere differentiable function on $\mathbb{R}$. In Section 8.4 we prove the classical Weierstrass Approximation Theorem.

8.1 Function Sequences

Definition 8.1 Let $D \subset \mathbb{R}$ and, for each n in $\mathbb{N}$, let f_n be a function from D into $\mathbb{R}$. Then $(f_n)_{n \in \mathbb{N}}$ or $(f_n)_{n=1}^{\infty}$ is a *sequence of functions* (or a *function sequence*) on D. For each x in D, if the sequence of real numbers $(f_n(x))_{n \in \mathbb{N}}$ converges, then the sequence of functions $(f_n)_{n \in \mathbb{N}}$ *converges pointwise* on D (or is *pointwise convergent on D*). If $(f_n)_{n \in \mathbb{N}}$ is pointwise convergent on D, then $F : D \to \mathbb{R}$, defined by

$$F(x) = \lim_{n \to \infty} f_n(x)$$

for each x in D, is the *limit function* (or simply the *limit*) of the function sequence $(f_n)_{n \in \mathbb{N}}$, and $(f_n)_{n \in \mathbb{N}}$ *converges pointwise to F on D*.

Notation When $(f_n)_{n \in \mathbb{N}}$ converges pointwise to F on D, we write $f_n \to F$ pointwise on D, or just $f_n \to F$. Thus, the arrow without further qualification indicates pointwise convergence on the appropriate domain.

Example 8.1 Let $f_n : [0, 1] \to \mathbb{R}$ be given by $f_n(x) = x^n$ for each n in $\mathbb{N}$ (Figure 8.1). Then $f_n \to F$ pointwise on $[0, 1]$, where

$$F(x) = \begin{cases} 0 & \text{if } 0 \le x < 1 \\ 1 & \text{if } x = 1. \end{cases}$$

To see this, fix x and let $n \to \infty$. For instance, if $0 < x < 1$, then $\lim_{n \to \infty} x^n = 0$ by Example 3.5.

Thus, we have a sequence of continuous functions whose limit function is not continuous. Also, note that the word "continuous" may be replaced by "differentiable" in the preceding sentence.

Example 8.2 Let $f_n : \mathbb{R} \to \mathbb{R}$ be given by $f_n(x) = (\sin nx)/\sqrt{n}$ for each n in $\mathbb{N}$. Since $|(\sin nx)/\sqrt{n}| \le 1/\sqrt{n}$, $f_n \to F$ pointwise on $\mathbb{R}$ where $F(x) = 0$ for all x in $\mathbb{R}$. Since $f_n'(x) = \sqrt{n} \cos nx$ for each n and x, each f_n and F are differentiable, but $(f_n')_{n \in \mathbb{N}}$ does not converge pointwise to F'. For instance, $f_n'(0) = \sqrt{n} \to \infty$ and $F'(0) = 0$.

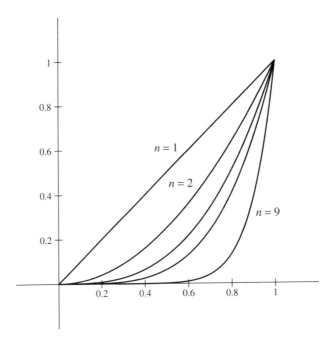

Figure 8.1

Example 8.3   Let $f_n : [0, 1] \to \mathbb{R}$ be given by $f_n(x) = nx(1 - x^2)^n$ for each n in $\mathbb{N}$ (Figure 8.2).

We first show that $f_n \to F$ pointwise on $[0, 1]$ where $F(x) = 0$ for all x in $[0, 1]$. Fix x in $(0, 1)$. Then

$$\lim_{n \to \infty} f_n(x) = x \lim_{n \to \infty} n(1 - x^2)^n \qquad (\infty \cdot 0 \text{ form})$$

$$= x \lim_{n \to \infty} \frac{n}{(1 - x^2)^{-n}} \qquad (\tfrac{\infty}{\infty} \text{ form})$$

$$= x \lim_{n \to \infty} \frac{1}{-(1 - x^2)^{-n} \ln(1 - x^2)} \qquad \begin{array}{l}\text{(by L'Hôpital, note that} \\ \text{the derivative is with} \\ \text{respect to } n)\end{array}$$

$$= \frac{-x}{\ln(1 - x^2)} \lim_{n \to \infty} (1 - x^2)^n$$

$$= 0.$$

For n in $\mathbb{N}$,

$$\int_0^1 f_n = n \int_0^1 x(1 - x^2)^n dx$$

$$= \frac{n}{2} \int_0^1 y^n dy \qquad (\text{by the substitution } y = 1 - x^2)$$

$$= \frac{1}{2}\left(\frac{n}{n + 1}\right).$$

Thus, $\lim_{n \to \infty} \int_0^1 f_n = \frac{1}{2} \neq 0 = \int_0^1 F = \int_0^1 \left(\lim_{n \to \infty} f_n\right)$. In other words, we cannot interchange the two operations of limit and integral in this example.

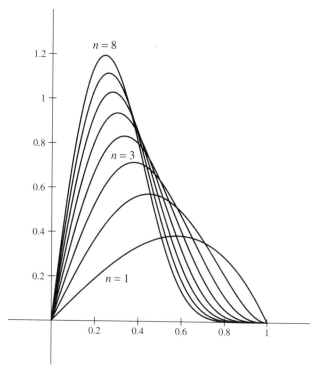

Figure 8.2

Example 8.4   This is a standard example showing that the limit of a sequence of Riemann integrable functions need not be Riemann integrable.

Enumerate the rational numbers in $[0, 1]$ as a sequence of distinct points $(r_n)_{n=1}^{\infty}$. For each n in $\mathbb{N}$, define $f_n : [0, 1] \to \mathbb{R}$ by

$$f_n(x) = \begin{cases} 1 & \text{if } x = r_k \text{ for some } k \le n \\ 0 & \text{otherwise.} \end{cases}$$

Because each f_n agrees with a continuous function except at a finite number of points, each f_n is Riemann integrable on $[0, 1]$ by Theorem 6.5. Let $F : [0, 1] \to \mathbb{R}$ be given by

$$F(x) = \begin{cases} 1 & \text{if } x \in \mathbb{Q} \\ 0 & \text{if } x \notin \mathbb{Q}. \end{cases}$$

Then $f_n \to F$ pointwise on $[0, 1]$ and F is not Riemann integrable on $[0, 1]$ by Example 6.1.

Main Problem   If each f_n has a certain property (such as continuity, differentiability, integrability) and $f_n \to F$, will F have this property? From Examples 8.1 and 8.4 the answer is clearly no. Moreover, if each f_n and F are differentiable or integrable, is $\lim_{n\to\infty} f_n' = F'$ or is $\lim_{n\to\infty} \int_a^b f_n = \int_a^b F$? From Examples 8.2 and 8.3 the answer again is no. Obviously, we need a stronger form of convergence.

Before defining this stronger form of convergence, we consider our problem from another viewpoint in the context of continuity. Given $f_n \to F$ pointwise on D with c an accumulation point in D and each f_n continuous at c, then

$$F \text{ is continuous at } c \Leftrightarrow \lim_{x \to c} F(x) = F(c)$$

$$\Leftrightarrow \lim_{x \to c}[\lim_{n \to \infty} f_n(x)] = \lim_{n \to \infty} f_n(c)$$

$$\Leftrightarrow \lim_{x \to c}[\lim_{n \to \infty} f_n(x)] = \lim_{n \to \infty}[\lim_{x \to c} f_n(x)].$$

Thus the question of F being continuous at c is equivalent to the question of when one can interchange the order of the limits.

Remark   We restate Definition 8.1. For each n in $\mathbb{N}$, given functions f_n and F from D into $\mathbb{R}$, we have that $f_n \to F$ pointwise on D

if and only if for each x in D, $\lim_{n \to \infty} f_n(x) = F(x)$

if and only if for each x in D and for each $\varepsilon > 0$, there is an n_0 (depending on both x and ε) in $\mathbb{N}$ such that if $n \geq n_0$, then $|f_n(x) - F(x)| < \varepsilon$.

Intuitively, referring to Example 8.1 and Figure 8.1 with $\varepsilon = \frac{1}{4}$, the closer x is to 1, the larger n_0 has to be to ensure that $x^n \leq x^{n_0} < \varepsilon$ for $n \geq n_0$.

Definition 8.2   Given a sequence of functions $(f_n)_{n \in \mathbb{N}}$ on D and a function F on D, the sequence $(f_n)_{n \in \mathbb{N}}$ *converges uniformly* (or is *uniformly convergent*) to F on D if for each $\varepsilon > 0$ there is an n_0 (depending only on ε) in $\mathbb{N}$ such that if $n \geq n_0$, then $|f_n(x) - F(x)| < \varepsilon$ for all x in D.

If $(f_n)_{n \in \mathbb{N}}$ converges uniformly to F on D, we will sometimes write $f_n \to F$ uniformly on D. Clearly, uniform convergence implies pointwise convergence. As Figure 8.3 illustrates with $D = [a, b]$, for $f_n \to F$ uniformly on D, given $\varepsilon > 0$, the entire graph of f_n must lie within the 2ε band centered on the graph of F for all sufficiently large n.

Example 8.5   Let f_n and F be as in Example 8.1. For $0 < \delta < 1$, we show that $f_n \to F$ uniformly on $[0, \delta]$. Let $\varepsilon > 0$. Choose n_0 in $\mathbb{N}$ such that $\delta^{n_0} < \varepsilon$. If $n \geq n_0$, $|f_n(x) - F(x)| = x^n \leq x^{n_0} \leq \delta^{n_0} < \varepsilon$ for all x in $[0, \delta]$.

Next, we show that $(f_n)_{n \in \mathbb{N}}$ does not converge to F uniformly on $[0, 1]$. Suppose that $f_n \to F$ uniformly on $[0, 1]$ and let $\varepsilon = \frac{1}{2}$. By Definition 8.2, there is an n_0 in $\mathbb{N}$ such that $x^n < \frac{1}{2}$ for all x in $[0, 1)$ and for all $n \geq n_0$. But if $(\frac{1}{2})^{1/n_0} < x < 1$, then $x^{n_0} > \frac{1}{2}$, which is a contradiction. So $(f_n)_{n \in \mathbb{N}}$ does not converge uniformly to F on $[0, 1]$.

Example 8.6   Let f_n and F be as in Example 8.2. Then $f_n \to F$ uniformly on $\mathbb{R}$ because if $\varepsilon > 0$, we can choose n_0 in $\mathbb{N}$ with $1/n_0 < \varepsilon^2$. Then for all x in $\mathbb{R}$ and for all $n \geq n_0$, $|f_n(x) - F(x)| \leq 1/\sqrt{n} \leq 1/\sqrt{n_0} < \varepsilon$. Thus, even uniform convergence does not guarantee that the limit of the derivatives will be the derivative of the limit.

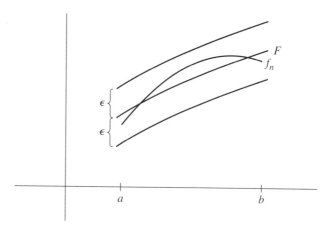

Figure 8.3

Example 8.7   Let $f_n(x) = nx/(1 + n^2x^2)$ for n in $\mathbb{N}$ (Figure 8.4).

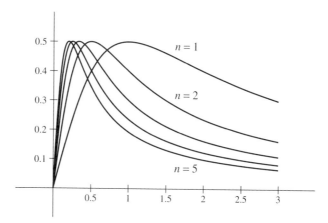

Figure 8.4

Then $(f_n)_{n \in \mathbb{N}}$ converges pointwise to the zero function on $(0, \infty)$. Let $\delta > 0$. Then $x \geq \delta$ implies that

$$f_n(x) \leq \frac{nx}{n^2x^2} = \frac{1}{nx} \leq \frac{1}{n\delta}.$$

Given $\varepsilon > 0$, choose n_0 in $\mathbb{N}$ with $1/n_0 < \varepsilon\delta$. Then $x \geq \delta$ and $n \geq n_0$ imply $f_n(x) \leq 1/(n_0\delta) < \varepsilon$. Therefore, $(f_n)_{n \in \mathbb{N}}$ converges uniformly on $[\delta, \infty)$ for all $\delta > 0$.

We now show that $(f_n)_{n \in \mathbb{N}}$ does not converge uniformly on $(0, \infty)$. The key is to notice that $f_n(1/n) = \frac{1}{2}$ for each n. Thus, given $\varepsilon = \frac{1}{4}$ and n_0 in $\mathbb{N}$, $f_n(1/n) > \varepsilon$ for all $n \geq n_0$ and so $(f_n)_{n \in \mathbb{N}}$ cannot converge uniformly on $(0, \infty)$. The reader should compare the idea here with that of Exercise 5.

Our next result will be useful in the following sections.

Proposition 8.1 (Cauchy condition) Let $(f_n)_{n \in \mathbb{N}}$ be a function sequence on D. Then $(f_n)_{n \in \mathbb{N}}$ converges uniformly on D if and only if for each $\varepsilon > 0$ there is an n_0 in $\mathbb{N}$ such that if $m \geq n_0$ and $n \geq n_0$, then $|f_m(x) - f_n(x)| < \varepsilon$ for all x in D.

Proof   Suppose that $f_n \to F$ uniformly on D and let $\varepsilon > 0$. By Definition 8.2, choose n_0 in $\mathbb{N}$ such that if $n \geq n_0$, then $|f_n(x) - F(x)| < \varepsilon/2$ for all x in D. Then $m \geq n_0$, $n \geq n_0$, and x in D imply that

$$|f_m(x) - f_n(x)| \leq |f_m(x) - F(x)| + |F(x) - f_n(x)| < \varepsilon.$$

Conversely, assume the statement after the if and only if. Then for each x in D, $(f_n(x))_{n \in \mathbb{N}}$ is a Cauchy sequence in $\mathbb{R}$ and hence converges to a real number, which we call $F(x)$. Thus, $f_n \to F$ pointwise on D. To show that the convergence is uniform, let $\varepsilon > 0$ and choose n_0 in $\mathbb{N}$ such that $m \geq n_0$, $n \geq n_0$, and x in D imply that $|f_n(x) - f_m(x)| < \varepsilon/2$. Since $f_m(x) \to F(x)$ as $m \to \infty$, holding $n \geq n_0$ and x in D fixed and letting $m \to \infty$ gives $|f_n(x) - F(x)| \leq \varepsilon/2$. Hence, $n \geq n_0$ implies $|f_n(x) - F(x)| < \varepsilon$ for all x in D, and so $f_n \to F$ uniformly on D.   ∎

Applying Proposition 8.1 to the function sequence in Example 8.4 is one way to tell that the convergence is not uniform, because if $n < m$, then $|f_m(r_m) - f_n(r_m)| = 1$.

Exercises

1. Let $f_n(x) = x/n$ for x in $[0, \infty)$. Show that $(f_n)_{n \in \mathbb{N}}$ converges uniformly on $[0, b]$ for all $b > 0$ but not on $[0, \infty)$.

2. Let $f_n(x) = x^n/(1 + x^n)$ for x in $[0, 1]$. Show that $(f_n)_{n \in \mathbb{N}}$ converges uniformly on $[0, b]$ for all $0 < b < 1$ but not on $[0, 1]$.

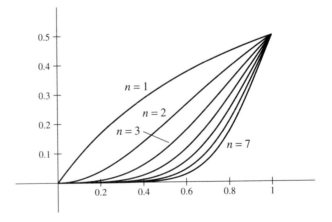

3. Let $f_n(x) = nx(1 - x)^n$ for x in $[0, 1]$. Show that $(f_n)_{n \in \mathbb{N}}$ converges pointwise but not uniformly to the zero function on $[0, 1]$.

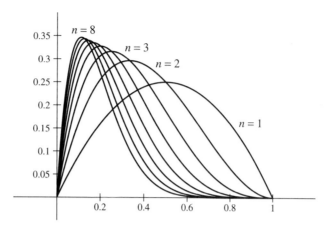

4. A function sequence $(f_n)_{n \in \mathbb{N}}$ is *uniformly bounded* on D if there is an $M > 0$ such that $|f_n(x)| \leq M$ for all x in D and all n in $\mathbb{N}$. Show that if $(f_n)_{n \in \mathbb{N}}$ is uniformly convergent on D and each f_n is bounded on D, then $(f_n)_{n \in \mathbb{N}}$ is uniformly bounded on D. Use this to conclude that the function sequence in Example 8.3 is not uniformly convergent.

5. Suppose that $f_n \to F$ pointwise on D and let $M_n = \sup_{x \in D} |f_n(x) - F(x)|$ for each n in $\mathbb{N}$. Show that $f_n \to F$ uniformly on D if and only if $\lim_{n \to \infty} M_n = 0$.

6. Let $f_n(x) = 1/(nx + 1)$ and $g_n(x) = x/(nx + 1)$ for x in $(0, 1)$. Show that $(f_n)_{n \in \mathbb{N}}$ does not converge uniformly on $(0, 1)$ but $(g_n)_{n \in \mathbb{N}}$ does converge uniformly on $(0, 1)$.

7. Let $f_n(x) = x/(1 + nx^2)$ for x in $\mathbb{R}$ (see graph below). Show that $(f_n)_{n \in \mathbb{N}}$ converges uniformly to the zero function on $\mathbb{R}$. [*Hint:* By calculus, the M_n defined in Exercise 5 is $1/(2\sqrt{n})$.] Then show that

$$f_n'(x) \underset{n}{\to} \begin{cases} 0 & \text{if} \quad x \neq 0 \\ 1 & \text{if} \quad x = 0. \end{cases}$$

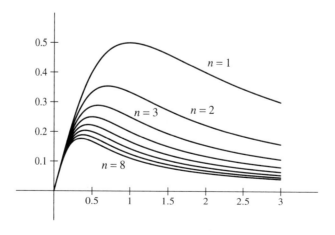

8. Let $f_n(x) = x + (1/n)$ for x in $\mathbb{R}$.

    (a) Show that $(f_n)_{n \in \mathbb{N}}$ converges uniformly to $F(x) = x$ on $\mathbb{R}$.

    (b) Show that $(f_n^2)_{n \in \mathbb{N}}$ does not converge uniformly on $\mathbb{R}$.

9. Let $(f_n)_{n \in \mathbb{N}}$ and $(g_n)_{n \in \mathbb{N}}$ converge uniformly on D.

    (a) Show that $(f_n \pm g_n)_{n \in \mathbb{N}}$ converges uniformly on D.

    (b) If each f_n and each g_n are bounded, show that $(f_n g_n)_{n \in \mathbb{N}}$ converges uniformly on D. [*Hint:* First use Exercise 4.]

10. Use Proposition 8.1 to show that if $(f_n)_{n \in \mathbb{N}}$ converges uniformly on D_1 and on D_2, then $(f_n)_{n \in \mathbb{N}}$ converges uniformly on $D_1 \cup D_2$.

11. Let $f_n, F : D \to [c, d]$ for each n in $\mathbb{N}$. If $f_n \to F$ uniformly on D and g is a continuous function on $[c, d]$, show that $g \circ f_n \to g \circ F$ uniformly on D. [*Hint:* First use the uniform continuity of g on $[c, d]$.]

8.2 Preservation Theorems

In this section we ask which properties of the function sequence are preserved (or carried over to the limit function) under uniform convergence.

Uniform Convergence and Continuity

Theorem 8.1 If $f_n \to F$ uniformly on D and each f_n is continuous on D, then F is continuous on D.

Proof Fix c in D and let $\varepsilon > 0$. To show that F is continuous at c, we want a $\delta > 0$ such that if x is in D and $|x - c| < \delta$, then $|F(x) - F(c)| < \varepsilon$.

    Since $f_n \to F$ uniformly on D, there is an n_0 in $\mathbb{N}$ such that $|f_n(x) - F(x)| < \varepsilon/3$ for all x in D and all $n \geq n_0$. Fix an $n \geq n_0$. Since f_n is continuous at c, there is a $\delta > 0$ such that if x is in D and $|x - c| < \delta$, then $|f_n(x) - f_n(c)| < \varepsilon/3$. Let x be in D with $|x - c| < \delta$. Then

$$|F(x) - F(c)| \leq |F(x) - f_n(x)| + |f_n(x) - f_n(c)| + |f_n(c) - F(c)|$$
$$< \frac{\varepsilon}{3} + \frac{\varepsilon}{3} + \frac{\varepsilon}{3}$$
$$= \varepsilon,$$

and so F is continuous at c. ∎

    From the material in Section 8.1 following the paragraph entitled Main Problem, it follows that if each f_n is continuous at an accumulation point c in D, then uniform convergence implies that

$$\lim_{x \to c}[\lim_{n \to \infty} f_n(x)] = \lim_{n \to \infty}[\lim_{x \to c} f_n(x)].$$

Also note that Theorem 8.1 is an easy way to show that the convergence in Example 8.1 is not uniform.

Remark   Theorem 8.1 shows that uniform convergence is sufficient to guarantee that the limit function is continuous when each member of the function sequence is continuous. By Example 8.3 or Exercise 1 in Section 8.1, uniform convergence is not necessary. The next theorem gives a partial converse to Theorem 8.1.

Theorem 8.2   Let $f_n \to F$ pointwise on $[a, b]$, and let each f_n and F be continuous on $[a, b]$. If $f_n(x) \geq f_{n+1}(x)$ for all n in $\mathbb{N}$ and all x in $[a, b]$, then $f_n \to F$ uniformly on $[a, b]$.

Proof   First let $F(x) = 0$ for all x in $[a, b]$. Then the monotonicity condition implies that $f_n(x) \geq 0$ for all n in $\mathbb{N}$ and all x in $[a, b]$. Suppose that $(f_n)_{n \in \mathbb{N}}$ does not converge uniformly to F on $[a, b]$. By Definition 8.2, there is an $\varepsilon > 0$ such that for all n_0 in $\mathbb{N}$ there exist an $n \geq n_0$ and an x in $[a, b]$ with $f_n(x) \geq \varepsilon$. Letting $n_0 = 1$, there exist an $n_1 \geq n_0$ and an x_1 in $[a, b]$ with $f_{n_1}(x_1) \geq \varepsilon$. Letting $n_0 = n_1 + 1$, there exist an $n_2 \geq n_0 > n_1$ and an x_2 in $[a, b]$ with $f_{n_2}(x_2) \geq \varepsilon$. Continuing, we obtain a strictly increasing sequence of positive integers $(n_i)_{i \in \mathbb{N}}$ and a sequence $(x_i)_{i \in \mathbb{N}}$ in $[a, b]$ with $f_{n_i}(x_i) \geq \varepsilon$ for all i in $\mathbb{N}$.

By the Bolzano-Weierstrass Theorem for sequences (Theorem 3.10) and Theorem 3.3, there exist a subsequence $\left(x_{i_k}\right)_{k=1}^{\infty}$ of $(x_i)_{i \in \mathbb{N}}$ and an x in $[a, b]$ with $x_{i_k} \underset{k}{\to} x$. Fix an m in $\mathbb{N}$. For $k \geq m$,

$$f_{n_{i_m}}(x_{i_k}) \geq f_{n_{i_k}}(x_{i_k}) \geq \varepsilon.$$

Since $f_{n_{i_m}}$ is continuous on $[a, b]$, $f_{n_{i_m}}(x_{i_k}) \underset{k}{\to} f_{n_{i_m}}(x)$. Hence, $f_{n_{i_m}}(x) \geq \varepsilon$, and this holds for all m in $\mathbb{N}$. However, since $f_n(x) \underset{n}{\to} 0$, there is an N_1 in $\mathbb{N}$ such that $n \geq N_1$ implies that $f_n(x) < \varepsilon$. In particular, for $m \geq N_1$, $f_{n_{i_m}}(x) < \varepsilon$, which is a contradiction. Therefore, $f_n \to 0$ uniformly on $[a, b]$.

Next let F be arbitrary. Setting $g_n(t) = f_n(t) - F(t)$, we have that each g_n is continuous on $[a, b]$, $(g_n)_{n \in \mathbb{N}}$ converges pointwise to the zero function on $[a, b]$, and $g_n \geq g_{n+1}$ on $[a, b]$ for each n in $\mathbb{N}$. By the previous case, $g_n \to 0$ uniformly on $[a, b]$, and hence $f_n \to F$ uniformly on $[a, b]$.   ∎

The function sequence $(f_n)_{n \in \mathbb{N}}$ on $(0, 1)$ given in Exercise 6 in Section 8.1 has the property that $f_n \geq f_{n+1}$ on $(0, 1)$, but the convergence is not uniform. Hence, an arbitrary domain will not work in Theorem 8.2. However, Theorem 8.2 remains valid if we change the monotonicity condition to $f_n(x) \leq f_{n+1}(x)$ for all n in $\mathbb{N}$ and all x in $[a, b]$, because then $(1 - f_n)_{n \in \mathbb{N}}$ and $1 - F$ satisfy the hypothesis of Theorem 8.2, and so $1 - f_n \to 1 - F$ uniformly on $[a, b]$. Hence, $f_n \to F$ uniformly on $[a, b]$.

Example 8.8   Let

$$f_n(x) = \begin{cases} \dfrac{1}{n} & \text{if } x \in \mathbb{Q} \\ 0 & \text{if } x \in \mathbb{R} \setminus \mathbb{Q}. \end{cases}$$

Then each f_n is discontinuous at every point of $\mathbb{R}$. Since $|f_n(x)| \leq 1/n$ for all x in $\mathbb{R}$ and all n in $\mathbb{N}$, $(f_n)_{n \in \mathbb{N}}$ converges uniformly to the zero function on $\mathbb{R}$, which is of course continuous. To quote Gelbaum and Olmsted, p. 76, "uniform convergence preserves good behavior, not bad behavior."

Uniform Convergence and Integration

Theorem 8.3   If $f_n \to F$ uniformly on $[a, b]$ and each f_n is in $\mathcal{R}[a, b]$, then F is in $\mathcal{R}[a, b]$ and $\int_a^b F = \lim_{n \to \infty} \int_a^b f_n$.

Proof   To show that F is in $\mathcal{R}[a, b]$, we use Theorem 6.2. First note that F is bounded since each f_n is bounded. Let $\varepsilon > 0$ and note that we want a partition P of $[a, b]$ such that $U(P, F) - L(P, F) < \varepsilon$. For any n we have

$$U(P, F) - L(P, F) \leq |U(P, F) - U(P, f_n)| +$$
$$|U(P, f_n) - L(P, f_n)| +$$
$$|L(P, f_n) - L(P, F)|. \tag{1}$$

Since $f_n \to F$ uniformly on $[a, b]$, there is an n in $\mathbb{N}$ such that $|f_n(x) - F(x)| < \varepsilon/[3(b - a)]$ for all x in $[a, b]$. Since f_n is in $\mathcal{R}[a, b]$, there is a partition $P = \{x_i\}_{i=0}^k$ of $[a, b]$ such that $U(P, f_n) - L(P, f_n) < \varepsilon/3$. Now we need to consider the first and third summands on the right of (1).

Letting $M_i = \sup\{F(x) : x_{i-1} \leq x \leq x_i\}$ and $M_i^n = \sup\{f_n(x) : x_{i-1} \leq x \leq x_i\}$ for $i = 1, \ldots, k$, we have

$$|U(P, F) - U(P, f_n)| = \left| \sum_{i=1}^k (M_i - M_i^n)\Delta x_i \right|$$
$$\leq \sum_{i=1}^k |M_i - M_i^n| \Delta x_i$$
$$\leq \sum_{i=1}^k \frac{\varepsilon}{3(b - a)} \Delta x_i \qquad \text{(by our choice of } f_n\text{)}$$
$$= \frac{\varepsilon}{3}.$$

Similarly, $|L(P, f_n) - L(P, F)| \leq \varepsilon/3$. Combining these into (1), we have that $U(P, F) - L(P, F) < \varepsilon$, and so F is in $\mathcal{R}[a, b]$.

To show that $\int_a^b F = \lim_{n \to \infty} \int_a^b f_n$, let $\varepsilon > 0$ and choose n_0 in $\mathbb{N}$ such that $n \geq n_0$ implies that $|f_n(x) - F(x)| < \varepsilon/[2(b - a)]$ for all x in $[a, b]$. Then for $n \geq n_0$,

$$\left| \int_a^b F - \int_a^b f_n \right| = \left| \int_a^b (F - f_n) \right|$$
$$\leq \int_a^b |F - f_n| \qquad \text{(Corollary 6.1)}$$
$$\leq \int_a^b \frac{\varepsilon}{2(b - a)} \qquad \text{(Proposition 6.4)}$$
$$< \varepsilon.$$

Hence, $\lim_{n \to \infty} \int_a^b f_n = \int_a^b F$.   ∎

Theorem 8.3 is an easy way to show that the convergence in Example 8.3 is not uniform. Also, Theorem 8.3 shows that uniform convergence is sufficient to guarantee that the limit function is integrable when each member of the function sequence is integrable and that the integral of the limit is the limit of the integrals. However, uniform convergence is not necessary. In Example 8.1 the convergence is not uniform, but the limit function is integrable and $\int_0^1 f_n = 1/(n+1) \to 0 = \int_0^1 F$.

Remark   The reader may be wondering about uniform convergence and improper integrals. Usually one considers these individually. As an example, let

$$f_n(x) = \begin{cases} \dfrac{1}{n} & \text{if} \quad 0 \le x \le n \\ 0 & \text{if} \quad x > n. \end{cases}$$

Then $(f_n)_{n \in \mathbb{N}}$ converges uniformly to the zero function F on $[0, \infty)$. However, $\int_0^\infty f_n = 1$ for each n, and so $\int_0^\infty F \ne \lim_{n \to \infty} \int_0^\infty f_n$.

Uniform Convergence and Differentiation

In Example 8.2 and Exercise 7 in Section 8.1, we have $f_n \to F$ uniformly but $\left(f_n'\right)_{n \in \mathbb{N}}$ does not converge to F' (even pointwise). Exercise 6 provides us with an example of a sequence of differentiable functions that converges uniformly to a function that is not differentiable at a certain point. Thus the analogue of Theorems 8.1 and 8.3 for differentiation requires stronger hypotheses.

Theorem 8.4   Suppose that $(f_n)_{n \in \mathbb{N}}$ is a sequence of differentiable functions defined on a bounded interval I and that $(f_n(x_0))_{n \in \mathbb{N}}$ converges for some point x_0 in I. If $\left(f_n'\right)_{n \in \mathbb{N}}$ converges uniformly on I, then $(f_n)_{n \in \mathbb{N}}$ converges uniformly on I to a differentiable function F and $F'(x) = \lim_{n \to \infty} f_n'(x)$ for all x in I.

Proof   Let a and b be the endpoints of I with $a < b$, and let $\varepsilon > 0$. Choose n_0 in $\mathbb{N}$ such that if $n \ge n_0$ and $m \ge n_0$, then

$$|f_n(x_0) - f_m(x_0)| < \frac{\varepsilon}{2}$$

[since $(f_n(x_0))_{n \in \mathbb{N}}$ is Cauchy] and

$$\left|f_n'(x) - f_m'(x)\right| < \frac{\varepsilon}{2(b-a)}$$

for all x in I (by Proposition 8.1). Since $f_n - f_m$ is differentiable on I, by the Mean Value Theorem (Theorem 5.5), for $x \ne t$ in I, we have

$$|f_n(x) - f_m(x) - [f_n(t) - f_m(t)]| = \left|f_n'(x_1) - f_m'(x_1)\right| |x - t|$$

$$\text{(for some } x_1 \text{ between } x \text{ and } t\text{)}$$

$$\le \frac{\varepsilon}{2(b-a)} |x - t|$$

$$\le \frac{\varepsilon}{2} \tag{2}$$

whenever n and $m \ge n_0$. Note that (2) holds if $x = t$. Hence,

$$|f_n(x) - f_m(x)| \le$$

$$|f_n(x) - f_m(x) - [f_n(x_0) - f_m(x_0)]| + |f_n(x_0) - f_m(x_0)| < \varepsilon$$

for all x in I whenever n and $m \ge n_0$. By the Cauchy condition (Proposition 8.1), $(f_n)_{n \in \mathbb{N}}$ converges uniformly on I to a function F.

Fix c in I. We want to show that F is differentiable at c and $F'(c) = \lim\limits_{n \to \infty} f_n'(c)$. Define a new function sequence $(g_n)_{n \in \mathbb{N}}$ on I by

$$g_n(x) = \begin{cases} \dfrac{f_n(x) - f_n(c)}{x - c} & \text{if} \quad x \ne c \\ f_n'(c) & \text{if} \quad x = c. \end{cases}$$

Since $g_n(c) = f_n'(c)$, $(g_n(c))_{n \in \mathbb{N}}$ converges. For $x \ne c$,

$$g_n(x) - g_m(x) = \frac{f_n(x) - f_m(x) - [f_n(c) - f_m(c)]}{x - c}$$

and so $(g_n)_{n \in \mathbb{N}}$ converges uniformly on $I \setminus \{c\}$. By Exercise 10 in Section 8.1, $(g_n)_{n \in \mathbb{N}}$ converges uniformly on I to a function g.

Since f_n is differentiable at c, $\lim\limits_{x \to c} g_n(x) = g_n(c)$ and so each g_n is continuous at c. By the proof of Theorem 8.1, g is continuous at c. Thus

$$\begin{aligned} g(c) &= \lim_{x \to c} g(x) \\ &= \lim_{x \to c} \left[\lim_{n \to \infty} g_n(x) \right] \\ &= \lim_{x \to c} \left[\lim_{n \to \infty} \frac{f_n(x) - f_n(c)}{x - c} \right] \\ &= \lim_{x \to c} \frac{F(x) - F(c)}{x - c} \end{aligned}$$

since $f_n \to F$ on I. Therefore, F is differentiable at c and

$$F'(c) = g(c) = \lim_{n \to \infty} g_n(c) = \lim_{n \to \infty} f_n'(c).$$

Since c is an arbitrary point of I, this completes the proof.   ∎

If $I = [a, b]$ and each f_n' is assumed to be continuous on $[a, b]$, then a simpler proof of Theorem 8.4 can be based on the Fundamental Theorem of Calculus. We ask for this proof in Exercise 10.

Example 8.9   The purpose of this example is to show that Theorem 8.4 is not true if I is not a bounded interval. Let

$$f_n(x) = \frac{\ln(1 + nx^2)}{2n}$$

for x in $[0, \infty)$. By L'Hôpital's rule, $(f_n)_{n \in \mathbb{N}}$ converges pointwise to the zero function on $[0, \infty)$. Also, $f_n(0) = 0$ for each n in $\mathbb{N}$. Since

$$f_n'(x) = \frac{x}{1 + nx^2},$$

$(f_n')_{n \in \mathbb{N}}$ converges uniformly to the zero function on $[0, \infty)$ by Exercise 7 in Section 8.1.

However, $(f_n)_{n \in \mathbb{N}}$ does not converge uniformly on $[0, \infty)$ because, in the notation of Exercise 5 in Section 8.1,

$$M_n = \sup_{x \in [0, \infty)} |f_n(x) - 0| = \infty$$

for each n in $\mathbb{N}$.

Exercises

1. Use Theorem 8.1 to show that $f_n(x) = e^{-nx}$ does not converge uniformly on $[0, \infty)$.

2. Use Theorem 8.1 to show that $f_n(x) = \sin^n x$ does not converge uniformly on $[0, \pi/2]$.

3. For each n in $\mathbb{N}$, define f_n on $[0, 2]$ by

$$f_n(x) = \begin{cases} 0 & \text{if } x = 0 \text{ or } x \geq \dfrac{2}{n} \\ n & \text{if } x = \dfrac{1}{n} \\ \text{linear} & \text{otherwise.} \end{cases}$$

   Show that $(f_n)_{n \in \mathbb{N}}$ converges pointwise but not uniformly to the zero function on $[0, 2]$. [*Hint:* Use Theorem 8.3.]

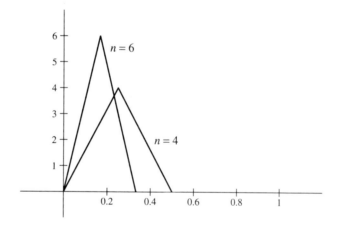

4. Use Theorem 8.3 to show that $f_n(x) = n^2 x(1 - x^2)^n$ converges pointwise but not uniformly to the zero function on $[0, 1]$. The graph is similar to Figure 8.2.

5. Let $f_n(x) = x^n/n$ on $[0, 1]$. Show that $(f_n)_{n \in \mathbb{N}}$ converges uniformly to a differentiable function on $[0, 1]$, but $(f_n')_{n \in \mathbb{N}}$ does not converge uniformly on $[0, 1]$.

6. Show that $f_n(x) = \sqrt{x^2 + (1/n^2)}$ converges uniformly to $|x|$ on $\mathbb{R}$. [*Hint:* Rationalize to show that $\left| f_n(x) - \sqrt{x^2} \right| \leq 1/n$ for each n.]

7. Let $f_n \to F$ uniformly on D with each f_n continuous on D. If x is in D and $(x_n)_{n \in \mathbb{N}}$ is a sequence in D with $x_n \to x$, show that $\lim_{n \to \infty} f_n(x_n) = F(x)$.

8. If $f_n \to F$ pointwise on D and each f_n is monotone increasing on D, show that F is monotone increasing on D. A similar result holds for monotone decreasing.

9. Suppose that $f_n \to F$ pointwise on $[a, b]$, each f_n is monotone on $[a, b]$, and F is continuous on $[a, b]$. Show that $f_n \to F$ uniformly on $[a, b]$. [*Hint:* F is uniformly continuous on $[a, b]$.]

10. Use the Fundamental Theorem of Calculus to provide a proof of Theorem 8.4 under the additional assumption that each f_n' is continuous on $I = [a, b]$. [*Hint:* For x in $[a, b]$, $\int_{x_0}^x f_n' = f_n(x) - f_n(x_0)$. If $f_n' \to g$ uniformly on $[a, b]$, then Theorem 8.3 implies that $\lim_{n \to \infty} [f_n(x) - f_n(x_0)] = \int_{x_0}^x g$. It follows that $f_n \to F$ pointwise on $[a, b]$, where $F(x) = \lim_{n \to \infty} f_n(x_0) + \int_{x_0}^x g$. By Theorem 6.12, $F'(x) = g(x)$ on $[a, b]$. Now show that $f_n \to F$ uniformly on $[a, b]$.]

8.3 Series of Functions

We define convergence of a series of functions in terms of convergence of the corresponding partial sums. This allows us to obtain easily the series analogues of earlier theorems in this chapter for function sequences. We then obtain the Weierstrass M-test, which gives sufficient conditions for a series of functions to converge uniformly; and we use this test to construct a continuous function on $\mathbb{R}$ that is nowhere differentiable.

Definition 8.3   Let $(f_n)_{n \in \mathbb{N}}$ be a sequence of functions defined on D. Then $\sum_{n=1}^{\infty} f_n$ is a *series of functions* (or a *function series*) on D. For each n in $\mathbb{N}$ and x in D, let

$$s_n(x) = \sum_{k=1}^{n} f_k(x) = f_1(x) + f_2(x) + \cdots + f_n(x).$$

Then $(s_n)_{n \in \mathbb{N}}$ is the function sequence of *partial sums* of the series of functions $\sum_{n=1}^{\infty} f_n$. The function series $\sum_{n=1}^{\infty} f_n$ *converges pointwise* (or is *pointwise convergent*) to a function F on D, denoted by

$$F = \sum_{n=1}^{\infty} f_n \text{ or } F(x) = \sum_{n=1}^{\infty} f_n(x)$$

for all x in D, if the sequence of partial sums $(s_n)_{n \in \mathbb{N}}$ converges pointwise to F on D. The function series $\sum_{n=1}^{\infty} f_n$ *converges uniformly* (or is *uniformly convergent*) on D if the sequence of partial sums $(s_n)_{n \in \mathbb{N}}$ converges uniformly on D.

Remark   In the notation of Definition 8.3, when $\sum_{n=1}^{\infty} f_n$ converges pointwise to F, then F is the limit of the sequence of partial sums and F is the *sum* of the series $\sum\limits_{n=1}^{\infty} f_n$. If $\sum\limits_{n=1}^{\infty} f_n$ converges uniformly to F on D, we will sometimes write this as $\sum\limits_{n=1}^{\infty} f_n = F$ uniformly on D. Clearly, uniform convergence implies pointwise convergence.

Example 8.10   For each n in $\mathbb{N}$, define $f_n : \mathbb{R} \to \mathbb{R}$ by

$$f_n(x) = \frac{x^2}{(1 + x^2)^{n-1}}.$$

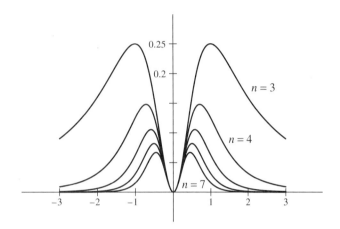

For $x \neq 0$,

$$\sum_{n=1}^{\infty} \frac{x^2}{(1+x^2)^{n-1}} = x^2 \left[1 + \frac{1}{1+x^2} + \left(\frac{1}{1+x^2} \right)^2 + \cdots \right]$$

$$= x^2 \cdot \frac{1}{1 - [1/(1+x^2)]}$$

$$= 1 + x^2.$$

Hence,

$$F(x) = \sum_{n=1}^{\infty} f_n(x) = \begin{cases} 0 & if \quad x = 0 \\ 1 + x^2 & if \quad x \neq 0. \end{cases}$$

Thus, $\sum\limits_{n=1}^{\infty} f_n$ converges pointwise to F on $\mathbb{R}$. That this convergence is not uniform is most easily seen from Theorem 8.5 below.

We next state the analogues of Proposition 8.1 and Theorems 8.1, 8.3, and 8.4 for function series. Their validity follows by applying the hypothesis to the sequence of partial sums $(s_n)_{n\in\mathbb{N}}$ and using the corresponding results for function sequences.

Proposition 8.2   (Cauchy condition) The function series $\sum\limits_{n=1}^{\infty} f_n$ converges uniformly on D if and only if for each $\varepsilon > 0$ there is an n_0 in $\mathbb{N}$ such that if $n > m \geq n_0$, then $\left| \sum\limits_{k=m+1}^{n} f_k(x) \right| < \varepsilon$ for all x in D.

Proof   Note that $s_n(x) - s_m(x) = \sum\limits_{k=m+1}^{n} f_k(x)$ for $n > m$ and apply Proposition 8.1. ∎

Theorem 8.5   If $\sum\limits_{n=1}^{\infty} f_n = F$ uniformly on D and each f_n is continuous on D, then F is continuous on D.

Proof   Each s_n is continuous on D and $s_n \to F$ uniformly on D. Now apply Theorem 8.1. ∎

If each f_n is continuous at an accumulation point c in D and $\sum\limits_{n=1}^{\infty} f_n = F$ uniformly on D, then Theorem 8.5 implies that

$$\lim_{x \to c} \sum_{n=1}^{\infty} f_n(x) = \lim_{x \to c} F(x)$$
$$= F(c)$$
$$= \sum_{n=1}^{\infty} f_n(c)$$
$$= \sum_{n=1}^{\infty} \lim_{x \to c} f_n(x).$$

Thus, in this situation we can interchange the two operations of limit and sum.

Theorem 8.6   If $\sum\limits_{n=1}^{\infty} f_n = F$ uniformly on $[a, b]$ and each f_n is in $\mathcal{R}[a, b]$, then F is in $\mathcal{R}[a, b]$ and $\int_a^b F = \int_a^b \sum\limits_{n=1}^{\infty} f_n = \sum\limits_{n=1}^{\infty} \int_a^b f_n$. (Thus, the function series may be integrated term by term.)

Proof   Apply Theorem 8.3 to the sequence of partial sums $(s_n)_{n \in \mathbb{N}}$. ∎

Theorem 8.7   Suppose that $(f_n)_{n \in \mathbb{N}}$ is a sequence of differentiable functions defined on a bounded interval I and that $\sum\limits_{n=1}^{\infty} f_n(x_0)$ converges for some point x_0 in I. If $\sum\limits_{n=1}^{\infty} f_n'$ converges uniformly on I, then $\sum\limits_{n=1}^{\infty} f_n$ converges uniformly on I to a differentiable function F and $F'(x) = \sum\limits_{n=1}^{\infty} f_n'(x)$ for all x in I.

Proof   Note that $(s_n(x_0))_{n \in \mathbb{N}}$ converges and that $(s_n')_{n \in \mathbb{N}}$ converges uniformly on I. Now apply Theorem 8.4. ∎

The following theorem gives us a useful but different type of result.

Theorem 8.8   (Weierstrass M-test) Let $(f_n)_{n \in \mathbb{N}}$ be a sequence of functions on D, and let $(M_n)_{n \in \mathbb{N}}$ be a sequence of nonnegative numbers satisfying $|f_n(x)| \leq M_n$ for all n in $\mathbb{N}$ and all x in D. If $\sum\limits_{n=1}^{\infty} M_n$ converges, then $\sum\limits_{n=1}^{\infty} f_n$ converges uniformly on D.

Proof   We use the Cauchy condition. Let $\varepsilon > 0$. Since $\sum\limits_{n=1}^{\infty} M_n$ converges, by Theorem 7.1, there is an n_0 in $\mathbb{N}$ such that if $n > m \geq n_0$, then $\sum\limits_{k=m+1}^{n} M_k < \varepsilon$.

Then $n > m \geq n_0$ implies $\left| \sum\limits_{k=m+1}^{n} f_k(x) \right| \leq \sum\limits_{k=m+1}^{n} |f_k(x)| \leq \sum\limits_{k=m+1}^{n} M_k < \varepsilon$ for all x in D. By Proposition 8.2, $\sum\limits_{n=1}^{\infty} f_n$ converges uniformly on D. ∎

Example 8.11 $\sum\limits_{n=1}^{\infty} (\sin nx)/n^2$ converges uniformly on $\mathbb{R}$ since $\left|(\sin nx)/n^2\right|$
$\leq 1/n^2$ and $\sum\limits_{n=1}^{\infty} 1/n^2$ converges. Note that the Weierstrass M-test does not tell
us anything about the convergence of $\sum\limits_{n=1}^{\infty} (\sin nx)/n$ since $\sum\limits_{n=1}^{\infty} 1/n$ diverges. See
Exercise 5.

Remark   The proof of the Weierstrass M-test actually shows that $\sum\limits_{n=1}^{\infty} |f_n|$ con-
verges uniformly, which is a stronger conclusion than that given in the statement
of Theorem 8.8. Example 8.10 is an example of an absolutely convergent series that does not converge uniformly. If a function series is going to converge
uniformly but not absolutely, then the Weierstrass M-test will not apply.

Example 8.12 $\sum\limits_{n=1}^{\infty} (-1)^n[(x^2+n)/n^2]$ converges uniformly on $[0, 1]$ but not
absolutely. Since

$$\lim_{n\to\infty} \frac{(x^2+n)/n^2}{1/n} = 1,$$

the Limit Comparison test (Theorem 7.5) shows that $\sum\limits_{n=1}^{\infty} (-1)^n[(x^2+n)/n^2]$ is
not absolutely convergent at any x in $\mathbb{R}$.

By the Alternating Series test (Theorem 7.6), $\sum\limits_{n=1}^{\infty} (-1)^n[(x^2+n)/n^2]$ converges for each x in $\mathbb{R}$. Let $F(x) = \sum\limits_{n=1}^{\infty} (-1)^n[(x^2+n)/n^2]$ for each x in $[0, 1]$
and let $s_n(x)$ be the nth partial sum of this series. By Theorem 7.6,

$$|F(x) - s_n(x)| \leq \frac{x^2 + (n+1)}{(n+1)^2} \leq \frac{1 + (n+1)}{(n+1)^2} \underset{n}{\to} 0$$

independent of x. Therefore, $(s_n)_{n\in\mathbb{N}}$ and hence $\sum\limits_{n=1}^{\infty} (-1)^n[(x^2+n)/n^2]$ converge
uniformly on $[0, 1]$. A slight modification of the argument above shows that
$\sum\limits_{n=1}^{\infty} (-1)^n[(x^2+n)/n^2]$ converges uniformly on any bounded interval.

We end this section with an example of a continuous function on $\mathbb{R}$ that
is nowhere differentiable. The original example is due to Weierstrass and was
first published by one of his students in 1874.

Example 8.13   This example is taken from p. 141 of Rudin.   Define
$\varphi : [0, 2] \to \mathbb{R}$ by

$$\varphi(x) = \begin{cases} x & \text{if } 0 \leq x \leq 1 \\ 2 - x & \text{if } 1 \leq x \leq 2. \end{cases}$$

Extend φ to $\mathbb{R}$ by $\varphi(x + 2) = \varphi(x)$ for all x in $\mathbb{R}$ (and draw φ). Then φ is
continuous on $\mathbb{R}$, φ has period 2, and $0 \leq \varphi(x) \leq 1$ for all x in $\mathbb{R}$.
Define $F : \mathbb{R} \to \mathbb{R}$ by

$$F(x) = \sum_{n=0}^{\infty} \left(\frac{3}{4}\right)^n \varphi(4^n x). \tag{3}$$

[It is convenient here to start with $n = 0$ rather than $n = 1$. Since $F(x) =$

$\sum_{n=1}^{\infty} \left(\frac{3}{4}\right)^{n-1} \varphi(4^{n-1}x)$, all of our theorems still apply.] Since $\left|\left(\frac{3}{4}\right)^{n} \varphi(4^{n}x)\right| \leq$ $\left(\frac{3}{4}\right)^{n}$, the series in (3) converges uniformly on $\mathbb{R}$ by the Weierstrass M-test. By Theorem 8.5, F is continuous on $\mathbb{R}$.

To show that F is not differentiable at any point of $\mathbb{R}$, we use Proposition 5.3. Fix c in $\mathbb{R}$ and m in $\mathbb{N}$. Choose an integer k such that $k \leq 4^{m}c < k+1$ and set

$$a_m = 4^{-m}k \quad \text{and} \quad b_m = 4^{-m}(k+1).$$

For $n = 0, 1, 2, 3, \dots$, consider $\delta_n = 4^{n}b_m - 4^{n}a_m = 4^{n-m}$. If $n > m$, δ_n is an even integer; if $n = m$, $\delta_n = 1$, $4^{n}a_m$ and $4^{n}b_m$ are consecutive integers, and so $\varphi(4^{n}b_m) - \varphi(4^{n}a_m) = \pm 1$ by definition of φ; if $n < m$, then no integer lies between $4^{n}a_m$ and $4^{n}b_m$. Therefore,

$$\left|\varphi(4^{n}b_m) - \varphi(4^{n}a_m)\right| = \begin{cases} 0 & \text{if } n > m \\ 4^{n-m} & \text{if } n \leq m. \end{cases} \tag{4}$$

By (3) and (4),

$$F(b_m) - F(a_m) = \sum_{n=0}^{\infty} \left(\frac{3}{4}\right)^{n} \left[\varphi(4^{n}b_m) - \varphi(4^{n}a_m)\right]$$

$$= \sum_{n=0}^{m} \left(\frac{3}{4}\right)^{n} \left[\varphi(4^{n}b_m) - \varphi(4^{n}a_m)\right]$$

$$= \pm \left(\frac{3}{4}\right)^{m} - \sum_{n=0}^{m-1} \left(\frac{3}{4}\right)^{n} \left[\varphi(4^{n}a_m) - \varphi(4^{n}b_m)\right].$$

Therefore,

$$|F(b_m) - F(a_m)| \geq \left(\frac{3}{4}\right)^{m} - \left|\sum_{n=0}^{m-1} \left(\frac{3}{4}\right)^{n} \left[\varphi(4^{n}a_m) - \varphi(4^{n}b_m)\right]\right|$$

$$\geq \left(\frac{3}{4}\right)^{m} - \sum_{n=0}^{m-1} \left(\frac{3}{4}\right)^{n} 4^{n-m}.$$

Since

$$\sum_{n=0}^{m-1} \left(\frac{3}{4}\right)^{n} 4^{n-m} = \frac{1}{4^{m}} \left(1 + 3 + 3^{2} + \cdots + 3^{m-1}\right)$$

$$= \frac{1}{4^{m}} \cdot \frac{3^{m}-1}{3-1}$$

$$< \frac{1}{2} \left(\frac{3}{4}\right)^{m},$$

$|F(b_m) - F(a_m)| > \frac{1}{2} \left(\frac{3}{4}\right)^{m}$. Therefore,

$$\left|\frac{F(b_m) - F(a_m)}{b_m - a_m}\right| > \frac{\frac{1}{2}\left(\frac{3}{4}\right)^{m}}{4^{-m}} = \frac{1}{2}(3^{m})$$

and so
$$\lim_{m \to \infty} \frac{F(b_m) - F(a_m)}{b_m - a_m}$$
does not exist in $\mathbb{R}$.

Since $b_m - a_m = 4^{-m} \to 0$, both $a_m \to c$ and $b_m \to c$ as $m \to \infty$. If $a_m < c < b_m$ for all m in $\mathbb{N}$, then Proposition 5.3 implies that F is not differentiable at c. For a_m equal to c, $4^m c$ must be an integer. This implies that either c is an integer or a rational number whose denominator is a power of 2. In this case there is an m_0 in $\mathbb{N}$ such that $a_m = c$ for all $m \geq m_0$ and so Definition 5.1 implies that F is not differentiable at c.

Letting $f_n(x) = \left(\frac{3}{4}\right)^n \varphi(4^n x)$ and $s_n(x) = \sum_{k=0}^{n} f_k(x)$ for $n = 0, 1, 2, \ldots$, the graphs of s_n given below (intuitively) indicate that $F = \lim_{n \to \infty} s_n$ has a "sawtooth" at every point.

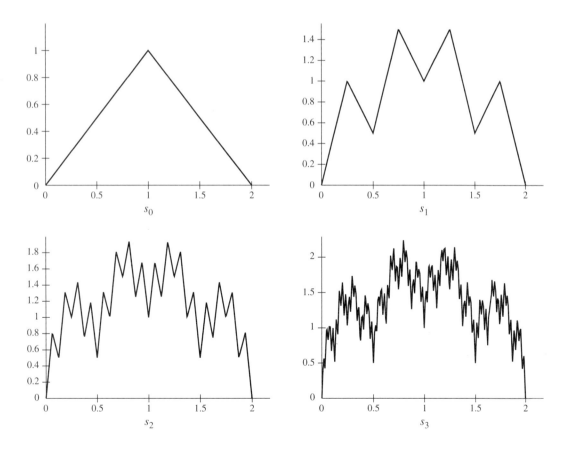

≡ Exercises ≡

1. Use the Weierstrass M-test to show that $\sum\limits_{n=1}^{\infty} f_n$ converges uniformly on D.

   (a) $f_n(x) = \dfrac{\cos nx}{n^2}$, $D = \mathbb{R}$.

   (b) $f_n(x) = x^n$, $D = \left[-\dfrac{1}{2}, \dfrac{1}{2}\right]$.

   (c) $f_n(x) = \dfrac{1}{n^2 + x^2}$, $D = \mathbb{R}$.

   (d) $f_n(x) = \left(\dfrac{\tan x}{3}\right)^n$, $D = \left[-\dfrac{\pi}{4}, \dfrac{\pi}{4}\right]$.

2. Show that $\sum\limits_{n=0}^{\infty} x(1-x)^n$ converges pointwise but not uniformly on $[0, 1]$.

3. Show that if $\sum\limits_{n=1}^{\infty} a_n$ converges absolutely, then $\sum\limits_{n=1}^{\infty} a_n \sin nx$ and $\sum\limits_{n=1}^{\infty} a_n \cos nx$ converge uniformly on $\mathbb{R}$.

4. Prove Dirichlet's test for uniform convergence: if $(f_n)_{n\in\mathbb{N}}$ and $(g_n)_{n\in\mathbb{N}}$ are sequences of functions on D satisfying

   (a) $\sum\limits_{n=1}^{\infty} f_n$ has uniformly bounded partial sums,

   (b) $g_n \to 0$ uniformly on D, and

   (c) $g_{n+1}(x) \le g_n(x)$ for all x in D and all n in $\mathbb{N}$,

   then $\sum\limits_{n=1}^{\infty} f_n g_n$ converges uniformly on D. [*Hint:* Use Lemma 7.1 and Proposition 8.2.]

5. Use Exercise 4 to show that $\sum\limits_{n=1}^{\infty} (\sin nx)/n$ converges uniformly on $[\pi/2, 3\pi/2]$.

   $$\left[\text{\textit{Hint:} For } x \neq 0, \pm 2\pi, \ldots .\right.$$
   $$\left.\sum_{k=1}^{n} \sin kx = \frac{\cos \frac{1}{2}x - \cos(n + \frac{1}{2})x}{2 \sin \frac{1}{2}x}.\right]$$

6. Show that $\sum\limits_{n=1}^{\infty} (-1)^{n+1}/(n+x^2)$ converges uniformly but not absolutely on $\mathbb{R}$.

7. Let $(g_n)_{n\in\mathbb{N}}$ be a sequence of functions on D such that $g_{n+1}(x) \le g_n(x)$ for all x in D and all n in $\mathbb{N}$. If $g_n \to 0$ uniformly on D, show that $\sum\limits_{n=1}^{\infty} (-1)^n g_n$ converges uniformly on D.

8. Show that the Riemann zeta function given by $\zeta(x) = \sum\limits_{n=1}^{\infty} 1/n^x$ converges uniformly on $[a, \infty)$ for all $a > 1$. Conclude that $\zeta'(x) = -\sum\limits_{n=1}^{\infty} (\ln n)/n^x$ for $x > 1$.

9. This exercise provides another example of a continuous, nowhere differentiable function on $\mathbb{R}$. Let $\varphi(x) = |x|$ on $[-1, 1]$ and extend φ to $\mathbb{R}$ by $\varphi(x+2) = \varphi(x)$ for all x in $\mathbb{R}$. Let $F(x) = \sum\limits_{n=0}^{\infty} \left(\frac{3}{4}\right)^n \varphi(4^n x)$. Follow the development of Example 8.13 to show that F is continuous on $\mathbb{R}$ but nowhere differentiable.

8.4 Weierstrass Approximation Theorem

In this section we prove the classical Weierstrass Approximation Theorem, Theorem 8.10 below, which states that on a closed interval we can uniformly approximate a continuous function by a polynomial. The proof we give also appears in Fridy. The reader can find two totally different proofs in Apostol, p. 481, and in Rudin, p. 146.

> ***Definition 8.4***   Given $f : [0, 1] \to \mathbb{R}$ and n in $\mathbb{N}$, the *nth Bernstein polynomial for f*, denoted by B_n, is given by
>
> $$B_n(x) = \sum_{k=0}^{n} f\left(\frac{k}{n}\right)\binom{n}{k} x^k (1 - x)^{n-k}$$
>
> where $\binom{n}{k} = \dfrac{n!}{k!(n-k)!}.$

The polynomial B_n has degree at most n, and its coefficients depend on the values of f at $0, 1/n, 2/n, 3/n, \ldots, 1$.

Example 8.14   For $f(x) = \sqrt{x}$,

$$B_1(x) = f(0)(1 - x) + f(1)x = x,$$

$$B_2(x) = f(0)(1 - x)^2 + f\left(\frac{1}{2}\right) 2x(1 - x) + f(1)x^2$$

$$= \sqrt{2}x(1 - x) + x^2,$$

$$B_3(x) = f(0)(1 - x)^3 + f\left(\frac{1}{3}\right) 3x(1 - x)^2 +$$

$$f\left(\frac{2}{3}\right) 3x^2(1 - x) + f(1)x^3$$

$$= \sqrt{3}\, x(1 - x)^2 + \sqrt{6}\, x^2(1 - x) + x^3.$$

In anticipation of the proof to be given below, we now develop certain results. Fix y in $[0, 1]$. For any x in $[0, 1]$ with $x + y \neq 0$ and any $n \geq 2$, by the Binomial Theorem,

$$\sum_{k=0}^{n} \binom{n}{k} x^k y^{n-k} = (x + y)^n. \tag{5}$$

Differentiating both sides of (5) with respect to x, we obtain

$$\sum_{k=1}^{n} \binom{n}{k} k x^{k-1} y^{n-k} = n (x + y)^{n-1},$$

which when multiplied by x/n gives

$$\sum_{k=0}^{n} \frac{k}{n} \binom{n}{k} x^k y^{n-k} = x (x + y)^{n-1}. \tag{6}$$

Differentiating both sides of (6) with respect to x, we obtain

$$\sum_{k=1}^{n} \frac{k^2}{n} \binom{n}{k} x^{k-1} y^{n-k} = (x+y)^{n-1} + x(n-1)(x+y)^{n-2},$$

which when multiplied by x/n gives

$$\sum_{k=0}^{n} \frac{k^2}{n^2} \binom{n}{k} x^k y^{n-k} = \frac{x}{n}(x+y)^{n-1} + \left(1 - \frac{1}{n}\right) x^2 (x+y)^{n-2}. \qquad (7)$$

Note that (5), (6), and (7) also hold if $n = 1$.

For any n in $\mathbb{N}$ and x in $[0, 1]$, replacing y by $1 - x$ in (5), (6), and (7), we obtain

$$\sum_{k=0}^{n} \binom{n}{k} x^k (1-x)^{n-k} = 1, \qquad (5')$$

$$\sum_{k=0}^{n} \frac{k}{n} \binom{n}{k} x^k (1-x)^{n-k} = x, \qquad (6')$$

and

$$\sum_{k=0}^{n} \frac{k^2}{n^2} \binom{n}{k} x^k (1-x)^{n-k} = \frac{x}{n} + \left(1 - \frac{1}{n}\right) x^2. \qquad (7')$$

Using (5'), (6'), and (7'), for any n in $\mathbb{N}$ and x in $[0, 1]$, we obtain

$$\sum_{k=0}^{n} \left(\frac{k}{n} - x\right)^2 \binom{n}{k} x^k (1-x)^{n-k}$$

$$= \sum_{k=0}^{n} \left(\frac{k^2}{n^2} - 2x\frac{k}{n} + x^2\right) \binom{n}{k} x^k (1-x)^{n-k}$$

$$= \frac{x}{n} + \left(1 - \frac{1}{n}\right) x^2 - 2x^2 + x^2$$

$$= \frac{x(1-x)}{n}. \qquad (8)$$

Theorem 8.9 (Bernstein Approximation Theorem) If f is continuous on $[0, 1]$, then the sequence of Bernstein polynomials converges uniformly to f on $[0, 1]$.

Proof   Let $\varepsilon > 0$. Since f is uniformly continuous on $[0, 1]$, there is a $\delta > 0$ such that if x and y are in $[0, 1]$ with $|x - y| < \delta$, then $|f(x) - f(y)| < \varepsilon/2$. Multiplying (5') by $f(x)$ we obtain

$$f(x) - B_n(x) = \sum_{k=0}^{n} \left[f(x) - f\left(\frac{k}{n}\right)\right] \binom{n}{k} x^k (1-x)^{n-k} \qquad (9)$$

for all x in $[0, 1]$ and all n in $\mathbb{N}$.

For each k in $\{0, 1, 2, \ldots, n\}$ and x in $[0, 1]$, let

$$\varphi_k(x) = \left[f(x) - f\left(\frac{k}{n}\right)\right] \binom{n}{k} x^k (1-x)^{n-k},$$

let $S = \{k \in \{0, 1, 2, \ldots, n\} : |(k/n) - x| < 1/n^{1/4}\}$, and let $T = \{0, 1, 2, \ldots, n\} \setminus S$. Then (9) can be rewritten as

$$f(x) - B_n(x) = \sum_{k \in S} \varphi_k(x) + \sum_{k \in T} \varphi_k(x). \tag{10}$$

We first consider $\sum\limits_{k \in S} \varphi_k(x)$. Choose n_1 in $\mathbb{N}$ such that $1/n_1^{1/4} < \delta$. If $n \geq n_1$ and $k \in S$, then $|(k/n) - x| < \delta$ and so $|f(k/n) - f(x)| < \varepsilon/2$. Hence, $n \geq n_1$ implies

$$\left| \sum_{k \in S} \varphi_k(x) \right| \leq \sum_{k \in S} \left| f(x) - f\left(\frac{k}{n}\right) \right| \binom{n}{k} x^k (1 - x)^{n-k}$$

$$< \frac{\varepsilon}{2} \sum_{k=0}^{n} \binom{n}{k} x^k (1 - x)^{n-k}$$

$$= \frac{\varepsilon}{2}.$$

We now consider $\sum\limits_{k \in T} \varphi_k(x)$. Choose $M > 0$ such that $|f(x)| \leq M$ for all x in $[0, 1]$. If k is in T, then $|(k/n) - x| \geq 1/n^{1/4}$ and so $[(k/n) - x]^2 \geq 1/n^{1/2}$. Hence,

$$\left| \sum_{k \in T} \varphi_k(x) \right| \leq \sum_{k \in T} \left| f(x) - f\left(\frac{k}{n}\right) \right| \binom{n}{k} x^k (1 - x)^{n-k}$$

$$\leq 2M \sum_{k \in T} \binom{n}{k} x^k (1 - x)^{n-k}$$

$$\leq 2M \sum_{k \in T} n^{1/2} \left(\frac{k}{n} - x\right)^2 \binom{n}{k} x^k (1 - x)^{n-k}$$

$$\leq 2M n^{1/2} \sum_{k=0}^{n} \left(\frac{k}{n} - x\right)^2 \binom{n}{k} x^k (1 - x)^{n-k}$$

$$= \left(2M n^{1/2}\right) \frac{x(1 - x)}{n} \qquad \text{[by (8)]}$$

$$\leq \frac{2M}{\sqrt{n}}.$$

Choose n_2 in $\mathbb{N}$ such that $1/\sqrt{n_2} < \varepsilon/(4M)$. Then $n \geq n_2$ implies $\left| \sum\limits_{k \in T} \varphi_k(x) \right| < \varepsilon/2$.

Let $n_0 = \max\{n_1, n_2\}$. Then $n \geq n_0$ implies, by (10), that $|f(x) - B_n(x)| < \varepsilon$ for all x in $[0, 1]$. Since n_0 was chosen independently of x, $B_n \to f$ uniformly on $[0, 1]$. ∎

Theorem 8.10   (Weierstrass Approximation Theorem) If f is continuous on $[a, b]$, then there exists a sequence of polynomials that converges uniformly to f on $[a, b]$.

Proof   Define g on $[0, 1]$ by $g(t) = f(a + (b-a)t)$. Then g is continuous on $[0, 1]$ and so, by Theorem 8.9, there is a sequence of polynomials $(B_n)_{n \in \mathbb{N}}$ that converges to g uniformly on $[0, 1]$. That is, given $\varepsilon > 0$, $|g(t) - B_n(t)| < \varepsilon$ for all t in $[0, 1]$ and all sufficiently large n. If x is in $[a, b]$, then $t = (x-a)/(b-a)$ is in $[0, 1]$ and $g(t) = f(x)$. Hence,

$$|f(x) - B_n((x-a)/(b-a))| < \varepsilon$$

for all x in $[a, b]$ and all sufficiently large n. Since each $B_n((x-a)/(b-a))$ is a polynomial in x, the conclusion follows.   ∎

The proof of Theorem 8.9 indicates how large n must be in order for B_n to approximate f on $[0, 1]$ and hence, by the proof of Theorem 8.10, for $B_n((x-a)/(b-a))$ to approximate f on $[a, b]$. For a fixed $\varepsilon > 0$, once δ and M (as in the proof of Theorem 8.9) have been found, we choose n such that $1/n^{1/4} < \delta$ and $1/\sqrt{n} < \varepsilon/(4M)$. For example, if $f(x) = \sqrt{x}$ on $[0, 1]$, then $\delta = \varepsilon^2/8$ (this is a good exercise for the reader) and $M = 1$. Hence, we must choose $n > \max\{4096/\varepsilon^8, 16/\varepsilon^2\}$. If $\varepsilon = \frac{1}{10}$, then $n > 4096(10^8)$.

As in Exercise 4, however, it is often necessary to know only that there is a polynomial that uniformly approximates f on $[a, b]$.

☰ Exercises ☰

1. Find B_1, B_2, and B_3 for
   (a) $f(x) = 1 - x$;
   (b) $f(x) = x^2$;
   (c) $f(x) = e^x$.

2. By considering $f(x) = 1/x$ on $(0, 1)$, show that Theorem 8.10 would not hold if $[a, b]$ were replaced by (a, b).

3. Let $f(x) = |x|$ on $[a, b]$, where $a < 0 < b$. Show that there is a sequence of polynomials $(Q_n)_{n \in \mathbb{N}}$ such that $Q_n(0) = 0$ for each n and $Q_n \to f$ uniformly on $[a, b]$.

4. Suppose that f is continuous on $[0, 1]$ and

$$\int_0^1 f(x)x^n dx = 0$$

for $n = 0, 1, 2, 3, \ldots$. Show that $f(x) = 0$ on $[0, 1]$. [*Hint:* First show that $\int_0^1 f(x)P(x)dx = 0$ for all polynomials P. Then use Theorem 8.10 to show that $\int_0^1 f^2 = 0$.]

9 Power Series

In Section 9.1 we show that every power series has a radius of convergence R, where $0 \le R \le \infty$. When $R > 0$, the power series has a nondegenerate interval of convergence. First we show that a power series converges absolutely in the interior of its interval of convergence. Then, relating a power series to the function series of Chapter 8, we show that a power series converges uniformly on every closed subinterval contained in the interior of its interval of convergence. This uniform convergence allows us to differentiate and integrate a power series term by term on an appropriate interval.

In Section 9.2 we show that the power series representation of a function is unique, and therefore this power series must be the Taylor series for the function.

9.1 Convergence

Definition 9.1 Given a sequence $(a_n)_{n=0}^{\infty}$ of real numbers and a real number c, an infinite series of the form

$$\sum_{n=0}^{\infty} a_n (x - c)^n = a_0 + a_1 (x - c) + a_2 (x - c)^2 + \cdots \qquad (1)$$

is a *power series in* $(x - c)$, or a *power series centered at* c. The numbers a_n are called the *coefficients* of the power series.

The immediate question is for which real numbers x does the power series (1) converge? Clearly, this series always converges at $x = c$. The familiar geometric series $\sum\limits_{n=0}^{\infty} x^n = 1 + x + x^2 + x^3 + \cdots$, which converges for $|x| < 1$ and diverges for $|x| \ge 1$ (Example 7.3), is a power series with $c = 0$ and each $a_n = 1$. The power series $\sum\limits_{n=0}^{\infty} x^n/n!$ converges for all real numbers x by Exercise 1 in Section 7.2, and the power series $\sum\limits_{n=0}^{\infty} n! x^n$ converges only for $x = 0$ by the Ratio test. In order to give a definitive answer to our immediate question, we first extend the Root test of Theorem 7.8.

Theorem 9.1   (This is also called the Root test.) Given a series $\sum\limits_{n=0}^{\infty} a_n$ of real numbers, let

$$L = \limsup_{n \to \infty} \sqrt[n]{|a_n|}.$$

Then

1. the series $\sum\limits_{n=0}^{\infty} a_n$ converges absolutely if $L < 1$;

2. the series $\sum\limits_{n=0}^{\infty} a_n$ diverges if $L > 1$;

3. the test is inconclusive if $L = 1$.

Proof

1. First choose a real number r such that $L < r < 1$. If $\sqrt[n]{|a_n|} > r$ for infinitely many n, then $\left(\sqrt[n]{|a_n|}\right)_{n=0}^{\infty}$ would have a subsequential limit greater than or equal to r (see the first paragraph of Section 3.8). Since L is the largest subsequential limit of $\left(\sqrt[n]{|a_n|}\right)_{n=0}^{\infty}$, there is an n_0 in $\mathbb{N}$ with $\sqrt[n]{|a_n|} \leq r$ for all $n \geq n_0$ or, equivalently, $|a_n| \leq r^n$ for all $n \geq n_0$. Hence, $\sum\limits_{n=0}^{\infty} |a_n|$ converges by the Comparison test (Theorem 7.4).

2. Since $L > 1$, $|a_n| \geq 1$ for infinitely many n and so $\lim\limits_{n \to \infty} a_n \neq 0$.

3. Use the series in the proof of part 3 of Theorem 7.8.   ∎

Absolute Convergence

Theorem 9.2   Given the power series $\sum\limits_{n=0}^{\infty} a_n(x - c)^n$, let $L = \limsup\limits_{n \to \infty} \sqrt[n]{|a_n|}$ and $R = 1/L$ (where $R = 0$ if $L = \infty$ and $R = \infty$ if $L = 0$). Then $\sum\limits_{n=0}^{\infty} a_n(x - c)^n$ converges absolutely if $|x - c| < R$ and diverges if $|x - c| > R$. For $|x - c| = R$, the power series may or may not converge.

Proof   Since $\limsup\limits_{n \to \infty} \sqrt[n]{|a_n(x - c)^n|} = L\,|x - c| = |x - c|/R$, the result follows from Theorem 9.1.   ∎

Definition 9.2   The extended real number R defined in Theorem 9.2 is called the *radius of convergence* of the power series $\sum\limits_{n=0}^{\infty} a_n(x - c)^n$. For $R > 0$, the set of real numbers x such that $\sum\limits_{n=0}^{\infty} a_n(x - c)^n$ converges is called the *interval of convergence* of the power series $\sum\limits_{n=0}^{\infty} a_n(x - c)^n$.

Example 9.1   From the paragraph following Definition 9.1, $\sum\limits_{n=0}^{\infty} x^n$ has $R = 1$ and converges absolutely on its interval of convergence $(-1, 1)$; $\sum\limits_{n=0}^{\infty} x^n/n!$ has $R = \infty$ and converges absolutely on its interval of convergence $(-\infty, \infty)$; while $\sum\limits_{n=0}^{\infty} n!x^n$ has $R = 0$ and converges only at $x = 0$.

Example 9.2   Our purpose here is to provide an example in which $\lim\limits_{n\to\infty} \sqrt[n]{|a_n|}$ does not exist but Theorem 9.2 still applies. Consider the power series

$$\sum_{n=0}^{\infty} a_n x^n = 1 + 2x + x^2 + 2^3 x^3 + x^4 + 2^5 x^5 + \cdots,$$

where $a_n = 1$ if n is even and $a_n = 2^n$ if n is odd. Then

$$\sqrt[n]{|a_n|} = \begin{cases} 1 & \text{if } n \text{ is even} \\ 2 & \text{if } n \text{ is odd.} \end{cases}$$

Since $\limsup\limits_{n\to\infty} \sqrt[n]{|a_n|} = 2$, $R = \frac{1}{2}$.

Given a power series $\sum\limits_{n=0}^{\infty} a_n(x-c)^n$, if $\lambda = \lim\limits_{n\to\infty} |a_{n+1}/a_n|$ exists in $[0, \infty]$, then the radius of convergence $R = 1/\lambda$ (see Exercise 1). Thus, when applicable, we can use the Ratio test to obtain the radius of convergence. Note that the Ratio test is not applicable in Example 9.2.

Example 9.3   Since $\lim\limits_{n\to\infty} n^2/(n+1)^2 = 1$, the Ratio test implies that $\sum\limits_{n=1}^{\infty} x^n/n^2$ has $R = 1$. Since $\sum\limits_{n=1}^{\infty} 1/n^2$ converges, $\sum\limits_{n=1}^{\infty} x^n/n^2$ converges absolutely on its interval of convergence $[-1, 1]$. Similarly, $\sum\limits_{n=1}^{\infty} x^n/n$ has $R = 1$, but $\sum\limits_{n=1}^{\infty} 1/n$ is the divergent harmonic series and $\sum\limits_{n=1}^{\infty} (-1)^n/n$ is conditionally convergent by the Alternating Series test. Thus $\sum\limits_{n=1}^{\infty} x^n/n$ has interval of convergence $[-1, 1)$ but converges absolutely only on $(-1, 1)$.

Remark   At this point we know that the power series $\sum\limits_{n=0}^{\infty} a_n(x-c)^n$ with radius of convergence $R > 0$ converges absolutely on the interval $(c - R, c + R)$, where $(c - R, c + R) = (-\infty, \infty)$ if $R = \infty$. When R is real, anything may happen at the endpoints of the interval of convergence. Thus, the endpoints need to be checked separately for each power series.

Uniform Convergence

Given the power series (1) $\sum\limits_{n=0}^{\infty} a_n(x - c)^n$, let $f_n(x) = a_n(x - c)^n$ for each $n = 0, 1, 2, 3, \ldots$. Then the power series (1) is just the function series $\sum\limits_{n=0}^{\infty} f_n$ of Section 8.3. Although each f_n is defined on $\mathbb{R}$, the reader should think of the common domain of each f_n as the interval of convergence of the power series (1), since outside of this interval the series $\sum\limits_{n=0}^{\infty} f_n$ diverges.

Theorem 9.3   Suppose that the power series (1) $\sum\limits_{n=0}^{\infty} a_n(x - c)^n$ has radius of convergence $R > 0$. If $0 < r < R$, then this power series converges uniformly on $[c - r, c + r]$. Thus, a power series converges uniformly on every closed subinterval of $(c - R, c + R)$.

Proof   For x in $[c - r, c + r]$, $|f_n(x)| = |a_n(x - c)^n| \leq |a_n| r^n$. Since $\sum\limits_{n=0}^{\infty} a_n(x-c)^n$ converges absolutely at $x = c+r$, $\sum\limits_{n=0}^{\infty} |a_n| r^n$ converges. Letting

$M_n = |a_n| r^n$ for each $n = 0, 1, 2, \ldots$, the Weierstrass M-test (Theorem 8.8) implies that $\sum_{n=0}^{\infty} a_n(x - c)^n$ converges uniformly on $[c - r, c + r]$.

The last statement of the theorem follows by noting that a closed subinterval of $(c - R, c + R)$ is contained in an interval of the form $[c - r, c + r]$ where $0 < r < R$. ∎

Corollary 9.1   Assume the hypothesis of Theorem 9.3, and let $F(x) = \sum_{n=0}^{\infty} a_n(x - c)^n$ for x in $(c - R, c + R)$. Then

1. F is continuous on $(c - R, c + R)$;

2. F is differentiable on $(c - R, c + R)$ and $F'(x) = \sum_{n=1}^{\infty} na_n(x - c)^{n-1}$ for x in $(c - R, c + R)$;

3. F is Riemann integrable on every closed subinterval $[s, t]$ of $(c - R, c + R)$ and
$$\int_s^t F = \sum_{n=0}^{\infty} a_n \int_s^t (x - c)^n dx.$$

(Note that parts 2 and 3 imply that term by term differentiation and integration are valid on the appropriate interval.)

Proof

2. Since $\lim_{n \to \infty} \sqrt[n]{n} = 1$ [Exercise 9(c) in Section 3.1], $\limsup_{n \to \infty} \sqrt[n]{n |a_n|} = \limsup_{n \to \infty} \sqrt[n]{|a_n|}$, and so the power series $\sum_{n=1}^{\infty} na_n(x - c)^{n-1}$ also has radius of convergence R. For $0 < r < R$, Theorem 9.3 implies that $\sum_{n=1}^{\infty} na_n(x-c)^{n-1}$ converges uniformly on $[c - r, c + r]$. Using Theorem 8.7, $F'(x) = \sum_{n=1}^{\infty} na_n(x - c)^{n-1}$ on $[c - r, c + r]$. Given x with $|x - c| < R$, choose r such that $|x - c| < r < R$. Then x is in $[c - r, c + r]$, so part 2 holds on $(c - R, c + R)$.

1. This follows since differentiability implies continuity (Theorem 5.1). It also follows from Theorem 8.5.

3. This follows from Theorem 8.6 since the power series is uniformly convergent on every closed subinterval of $(c - R, c + R)$. ∎

Example 9.4   Consider the geometric series $\sum_{n=0}^{\infty} x^n = 1/(1 - x)$ for $|x| < 1$.

Differentiating, we obtain $\sum_{n=1}^{\infty} nx^{n-1} = 1/(1-x)^2$ for $|x| < 1$. In the last series, first let $m = n-1$ and then replace m with n to obtain $\sum_{n=0}^{\infty} (n+1)x^n = 1/(1-x)^2$, which was derived in Example 7.18 by two different techniques.

Example 9.5   Again consider the geometric series $\sum_{n=0}^{\infty} t^n = 1/(1 - t)$ for $|t| < 1$. Replacing t by $-t$, we have $\sum_{n=0}^{\infty} (-1)^n t^n = 1/(1 + t)$ for $|t| < 1$. Integrating both sides of the last equation from 0 to x, where $|x| < 1$, we obtain

$$\sum_{n=0}^{\infty}(-1)^n \frac{t^{n+1}}{n+1}\bigg|_0^x = \ln(1+t)|_0^x$$

or

$$\sum_{n=0}^{\infty}(-1)^n \frac{x^{n+1}}{n+1} = \ln(1+x)$$

or

$$\sum_{n=1}^{\infty}(-1)^{n+1} \frac{x^n}{n} = \ln(1+x). \tag{2}$$

By Corollary 9.1, equation (2) is valid for $|x| < 1$. If we put $x = 1$ on the left side of (2), we obtain the convergent alternating harmonic series $\sum_{n=1}^{\infty}(-1)^{n+1}/n = 1 - \frac{1}{2} + \frac{1}{3} - \frac{1}{4} + \cdots$. Of course, we obtain $\ln 2$ if we put $x = 1$ on the right side of (2). That

$$\ln 2 = 1 - \frac{1}{2} + \frac{1}{3} - \frac{1}{4} + \cdots$$

follows from the next theorem.

Theorem 9.4 (Abel's Limit Theorem) Assuming that the power series $\sum_{n=0}^{\infty} a_n(x-c)^n$ has radius of convergence R, where $0 < R < \infty$, let $F(x) = \sum_{n=0}^{\infty} a_n(x-c)^n$ for x in $(c-R, c+R)$. If this power series converges at $x = c+R$, then F is continuous from the left at $c + R$; that is,

$$\lim_{x \to c+R^-} F(x) = \sum_{n=0}^{\infty} a_n R^n.$$

(For the corresponding result at $x = c - R$, see Exercise 9.)

Proof   First note that the substitution $y = x - c$ replaces $\sum_{n=0}^{\infty} a_n(x-c)^n$ with $\sum_{n=0}^{\infty} a_n y^n$, valid for $|y| < R$. Letting $y = Rz$ replaces $\sum_{n=0}^{\infty} a_n y^n$ with $\sum_{n=0}^{\infty} (a_n R^n) z^n$. Since

$$\limsup_{n\to\infty} \sqrt[n]{|a_n R^n|} = R \limsup_{n\to\infty} \sqrt[n]{|a_n|} = R \cdot \frac{1}{R} = 1,$$

the power series $\sum_{n=0}^{\infty} (a_n R^n) z^n$ has radius of convergence 1. If we can prove that this series is continuous from the left at $z = 1$, then F will be continuous from the left at $x = c + R$.

Thus, it suffices to consider $F(x) = \sum_{n=0}^{\infty} a_n x^n$ with radius of convergence 1 and with the additional assumption that $\sum_{n=0}^{\infty} a_n$ converges. We must show that

$$\lim_{x \to 1^-} F(x) = \sum_{n=0}^{\infty} a_n.$$

Let $s_{-1} = 0$, $s_n = a_0 + a_1 + \cdots + a_n$ for $n = 0, 1, 2, \ldots$, and $s = \sum\limits_{n=0}^{\infty} a_n$. For m in $\mathbb{N}$,

$$\sum_{n=0}^{m} a_n x^n = \sum_{n=0}^{m} (s_n - s_{n-1}) x^n$$

$$= \sum_{n=0}^{m-1} s_n x^n + s_m x^m - \sum_{n=1}^{m} s_{n-1} x^n$$

$$= (1 - x) \sum_{n=0}^{m-1} s_n x^n + s_m x^m.$$

For $|x| < 1$, $\lim\limits_{m \to \infty} s_m x^m = s \cdot 0 = 0$, and so

$$F(x) = (1 - x) \sum_{n=0}^{\infty} s_n x^n$$

for $|x| < 1$.

Let $\varepsilon > 0$. Since $s_n \to s$, choose n_0 in $\mathbb{N}$ such that $|s_n - s| < \varepsilon/2$ for all $n \geq n_0$. For $0 < x < 1$,

$$|F(x) - s| = \left| (1 - x) \sum_{n=0}^{\infty} s_n x^n - s \right|$$

$$= \left| (1 - x) \sum_{n=0}^{\infty} (s_n - s) x^n \right| \qquad \left(\text{since } \sum_{n=0}^{\infty} x_n = \frac{1}{1-x} \right)$$

$$\leq (1 - x) \sum_{n=0}^{n_0-1} |s_n - s| x^n + (1 - x) \sum_{n=n_0}^{\infty} |s_n - s| x^n$$

$$\leq (1 - x) \sum_{n=0}^{n_0-1} |s_n - s| + \frac{\varepsilon}{2}.$$

Letting $M = \max\{|s_n - s| : n = 0, 1, 2, \ldots, n_0 - 1\}$, we have that

$$|F(x) - s| \leq (1 - x) M n_0 + \frac{\varepsilon}{2}$$

for $0 < x < 1$. If $M \leq \varepsilon/(2n_0)$, then $|F(x) - s| < \varepsilon$ for all x in $(0, 1)$. If $M > \varepsilon/(2n_0)$, let $\delta = \varepsilon/(2Mn_0)$. Then $1 - \delta < x < 1$ implies $0 < 1 - x < \delta$ and $|F(x) - s| < \varepsilon$. Therefore, $\lim\limits_{x \to 1^-} F(x) = s$. ∎

In Exercise 10 we ask the reader to show that the converse of Theorem 9.4 is false. As an application of Theorem 9.4, we prove the result about the Cauchy product of two series mentioned after Theorem 7.14.

Corollary 9.2 If $\sum\limits_{n=0}^{\infty} a_n$, $\sum\limits_{n=0}^{\infty} b_n$, and their Cauchy product $\sum\limits_{n=0}^{\infty} c_n$ converge to A, B, and C, respectively, then $C = AB$.

Proof Let $f(x) = \sum\limits_{n=0}^{\infty} a_n x^n$, $g(x) = \sum\limits_{n=0}^{\infty} b_n x^n$, and $h(x) = \sum\limits_{n=0}^{\infty} c_n x^n$. Since each of these power series converges at $x = 1$, each is absolutely convergent for $|x| < 1$. By Theorem 7.14, $f(x)g(x) = h(x)$ for $|x| < 1$. By Theorem 9.4,

$$AB = \lim_{x \to 1^-} f(x)g(x) = \lim_{x \to 1^-} h(x) = C. \qquad \blacksquare$$

Exercises

1. Given the power series $\sum\limits_{n=0}^{\infty} a_n(x - c)^n$, show that if $\lambda = \lim\limits_{n \to \infty} |a_{n+1}/a_n|$ exists in $[0, \infty]$, then $R = 1/\lambda$ (where $R = 0$ if $\lambda = \infty$ and $R = \infty$ if $\lambda = 0$).

2. Find the radius and interval of convergence for each of the following power series.

   (a) $\sum\limits_{n=1}^{\infty} \dfrac{2^n x^n}{n^2}$ (b) $\sum\limits_{n=2}^{\infty} \dfrac{x^n}{(\ln n)^n}$

   (c) $\sum\limits_{n=1}^{\infty} \dfrac{(x - 4)^n}{n}$ (d) $\sum\limits_{n=1}^{\infty} \dfrac{n!(x - 2)^n}{n^n}$

3. Let $\sum\limits_{n=0}^{\infty} a_n x^n$ be a power series in which each a_n is an integer and infinitely many of the a_n's are nonzero. Show that $R \leq 1$. [*Hint:* Suppose that $R > 1$.]

4. (a) This is also called the Ratio test. Given a series $\sum\limits_{n=0}^{\infty} a_n$ where each a_n is nonzero, show that

   1. $\sum\limits_{n=0}^{\infty} a_n$ converges absolutely if $\limsup\limits_{n \to \infty} |a_{n+1}/a_n| < 1$;

   2. $\sum\limits_{n=0}^{\infty} a_n$ diverges if $\liminf\limits_{n \to \infty} |a_{n+1}/a_n| > 1$;

   3. the test is inconclusive if $\liminf\limits_{n \to \infty} |a_{n+1}/a_n| \leq 1 \leq \limsup\limits_{n \to \infty} |a_{n+1}/a_n|$.

   (b) Show that the test given in part (a) fails for the power series of Example 9.2.

5. (a) Show that $\sum\limits_{n=1}^{\infty} n/2^n = 2$. [*Hint:* Differentiate $\sum\limits_{n=0}^{\infty} x^n$ and then multiply by x.]

   (b) Show that $\sum\limits_{n=1}^{\infty} n^2/2^n = 6$. [*Hint:* Differentiate $\sum\limits_{n=1}^{\infty} nx^n$ and then multiply by x.]

6. Show that $\sum\limits_{n=1}^{\infty} 1/(n2^n) = \ln 2$. [*Hint:* Integrate $\sum\limits_{n=0}^{\infty} t^n$ from 0 to x.]

7. Show that $\pi/4 = 1 - \frac{1}{3} + \frac{1}{5} - \frac{1}{7} + \cdots = \sum\limits_{n=0}^{\infty} (-1)^n/(2n + 1)$. [*Hint:* Manipulate $\sum\limits_{n=0}^{\infty} t^n$ to obtain $\sum\limits_{n=0}^{\infty} (-1)^n t^{2n} = 1/(1 + t^2)$. Integrate from 0 to x to obtain a power series for $\arctan x$, valid for $|x| < 1$. Now use Theorem 9.4.]

8. Prove that if $\sum\limits_{n=0}^{\infty} a_n$ converges, then $\sum\limits_{n=0}^{\infty} a_n x^n$ converges uniformly on $[0, 1]$.

Use this result to give another proof of Theorem 9.4.

9. Assume that $\sum\limits_{n=0}^{\infty} a_n(x - c)^n$ has radius of convergence R, where $0 < R < \infty$, and let $F(x) = \sum\limits_{n=0}^{\infty} a_n(x - c)^n$ for x in $(c - R, c + R)$. Show that if this power series converges at $x = c - R$, then F is continuous from the right at $c - R$; that is, show that $\lim\limits_{x \to c-R+} F(x) = \sum\limits_{n=0}^{\infty} a_n(-R)^n$. Also, the power series is uniformly convergent on $[c - R, c]$.

10. Show that the converse of Theorem 9.4 is false. [*Hint:* Consider $\sum\limits_{n=0}^{\infty} (-1)^n x^n$.]

9.2 Taylor Series

A power series that converges on a nondegenerate interval represents a function on this interval. We show that such a function has derivatives of all orders in this interval and that such a power series representation is unique. From the uniqueness it follows that the power series must be the Taylor series for the function. We next reverse our thinking and ask: Given a function, necessarily having derivatives of all orders, when is this function represented by its Taylor series? We give necessary and sufficient conditions for this to happen, and we consider various examples.

> **Definition 9.3** Suppose that $\sum\limits_{n=0}^{\infty} a_n(x - c)^n$ has radius of convergence $R > 0$, and let $F(x) = \sum\limits_{n=0}^{\infty} a_n(x - c)^n$ for each x in $(c - R, c + R)$. Then F is *represented* by the power series $\sum\limits_{n=0}^{\infty} a_n(x - c)^n$ in $(c - R, c + R)$, and $\sum\limits_{n=0}^{\infty} a_n(x - c)^n$ is a *power series expansion of F centered at c.*

The power series expansion may also include the endpoints $c \pm R$. Our first goal is to show that the power series expansion of a function is unique. Recall that for a function F, $F^{(0)} = F$ and $F^{(k)}$ is the kth derivative of F.

Theorem 9.5 If F is represented by the power series $\sum\limits_{n=0}^{\infty} a_n(x - c)^n$ in $(c - R, c + R)$, then F has derivatives of all orders in $(c - R, c + R)$ and

$$F^{(k)}(x) = \sum_{n=k}^{\infty} \frac{n!}{(n - k)!} a_n(x - c)^{n-k} \tag{3}$$

for each x in $(c - R, c + R)$ and for each $k = 0, 1, 2, \ldots$. In particular, $F^{(k)}(c) = k! a_k$.

Proof   From part 2 of Corollary 9.1, we can differentiate F term by term to obtain

$$F'(x) = \sum_{n=1}^{\infty} n a_n (x - c)^{n-1}$$

$$= \sum_{n=1}^{\infty} \frac{n!}{(n-1)!} a_n (x - c)^{n-1}$$

for x in $(c - R, c + R)$. From the proof of part 2 of Corollary 9.1, this last series also has radius of convergence R, and so we can differentiate it term by term to obtain

$$F''(x) = \sum_{n=2}^{\infty} \frac{n!}{(n-1)!} (n-1) a_n (x - c)^{n-2}$$

$$= \sum_{n=2}^{\infty} \frac{n!}{(n-2)!} a_n (x - c)^{n-2}$$

for x in $(c - R, c + R)$. Equation (3) follows by induction and by part 2 of Corollary 9.1. Putting $x = c$ in (3), we obtain $F^{(k)}(c) = k! a_k$.   ∎

Corollary 9.3   The power series expansion of a function is unique.

Proof   Let F be represented by the power series $\sum_{n=0}^{\infty} a_n(x-c)^n$ in $(c-R, c+R)$. From Theorem 9.5 on page 198, each $a_n = F^{(n)}(c)/n!$. Therefore $F(x) = \sum_{n=0}^{\infty} [F^{(n)}(c)/n!](x - c)^n$.   ∎

Remark   Suppose that $\sum_{n=0}^{\infty} a_n x^n = \sum_{n=0}^{\infty} b_n x^n$ on $(-R, R)$ for some $R > 0$. Letting $F(x) = \sum_{n=0}^{\infty} a_n x^n$ for x in $(-R, R)$, we have that $a_n = F^{(n)}(0)/n! = b_n$ for $n = 0, 1, 2, \ldots$. Equating $a_n = b_n$ for each n is the underlying principle for finding power series solutions of differential equations.

Definition 9.4   Let f have derivatives of all orders on an interval I, and let c be an interior point of I. The *Taylor series for f centered at c* is

$$\sum_{n=0}^{\infty} \frac{f^{(n)}(c)}{n!} (x - c)^n = f(c) + \frac{f'(c)}{1!}(x - c) + \frac{f''(c)}{2!}(x - c)^2 + \cdots.$$

When $c = 0$, this series is also called the *Maclaurin series for f*.

For a function f to be represented by a power series, Theorem 9.5 implies that f must have derivatives of all orders, and Corollary 9.3 implies that this power series must be the Taylor series for f. When a function is represented by its Taylor series centered at c, the function is called *analytic* at c.

Example 9.6 From equation (2) in Example 9.5,

$$\ln(1 + x) = \sum_{n=1}^{\infty} \frac{(-1)^{n+1}x^n}{n} = x - \frac{x^2}{2} + \frac{x^3}{3} - \frac{x^4}{4} + \cdots \tag{4}$$

valid for $-1 < x \leq 1$. Thus, $\ln(1 + x)$ is represented by its Maclaurin series $\sum_{n=1}^{\infty} (-1)^{n+1}x^n/n$ on the interval $(-1, 1]$. Replacing x with $x - 1$ in (4) gives

$$\ln x = \sum_{n=1}^{\infty} \frac{(-1)^{n+1}(x - 1)^n}{n}$$

$$= (x - 1) - \frac{(x - 1)^2}{2} + \frac{(x - 1)^3}{3} - \frac{(x - 1)^4}{4} + \cdots$$

valid for $-1 < x - 1 \leq 1$ or, equivalently, $0 < x \leq 2$. Thus, $\ln x$ is represented by its Taylor series $\sum_{n=1}^{\infty} (-1)^{n+1}(x - 1)^n/n$ centered at 1 on the interval $(0, 2]$.

Our second goal is to determine which functions are represented by their Taylor series. We first point out that a Taylor series need not converge at any point except where it is centered; an example can be found in Gelbaum and Olmsted, p. 68. An easier example is the following.

Example 9.7 Let

$$f(x) = \begin{cases} e^{-(1/x^2)} & \text{if } x \neq 0 \\ 0 & \text{if } x = 0. \end{cases}$$

Then $f^{(n)}(0) = 0$ for $n = 0, 1, 2, \ldots$ (see Exercise 1). Thus, the Maclaurin series for f converges everywhere to the constant function 0, and so f is not represented by its Maclaurin series.

In order to achieve our second goal, we now recall Section 5.3 on Taylor's Theorem. Let f have derivatives of all orders on an interval I, and let c be an interior point of I. As in Section 5.3, letting

$$P_n(x) = \sum_{k=0}^{n} \frac{f^{(k)}(c)}{k!}(x - c)^k$$

denote the nth Taylor polynomial of f at c and letting $R_n(x)$ denote the remainder term, we have

$$f(x) = P_n(x) + R_n(x)$$

for all x in I. From Taylor's Theorem (Theorem 5.6),

$$R_n(x) = \frac{f^{(n+1)}(t)}{(n + 1)!}(x - c)^{n+1}$$

for some t between c and x [at $x = c$, $P_n(c) = f(c)$ for all n and $R_n(c) = 0$ for all t].

Theorem 9.6 Let f have derivatives of all orders on an interval I, and let c be an interior point of I. Then f is represented by its Taylor series on I if and only if $\lim_{n \to \infty} R_n(x) = 0$ for all x in I.

Proof   First note that the Taylor polynomials are the partial sums of the Taylor series. Hence, for all x in I,

$$f(x) = \sum_{n=0}^{\infty} \frac{f^{(n)}(c)}{n!} (x - c)^n$$

$$\Leftrightarrow f(x) = \lim_{n \to \infty} P_n(x)$$

$$\Leftrightarrow \lim_{n \to \infty} [f(x) - P_n(x)] = 0$$

$$\Leftrightarrow \lim_{n \to \infty} R_n(x) = 0.$$ ∎

The following table lists some standard Maclaurin series that are convergent to the given functions on the indicated intervals.

Table 1

Function	Interval
(1) $\dfrac{1}{1 - x} = \sum\limits_{n=0}^{\infty} x^n = 1 + x + x^2 + x^3 + \cdots$	$-1 < x < 1$
(2) $e^x = \sum\limits_{n=0}^{\infty} \dfrac{x^n}{n!} = 1 + x + \dfrac{x^2}{2!} + \dfrac{x^3}{3!} + \cdots$	$-\infty < x < \infty$
(3) $\sin x = \sum\limits_{n=0}^{\infty} \dfrac{(-1)^n x^{2n+1}}{(2n+1)!} = x - \dfrac{x^3}{3!} + \dfrac{x^5}{5!} - \dfrac{x^7}{7!} + \cdots$	$-\infty < x < \infty$
(4) $\cos x = \sum\limits_{n=0}^{\infty} \dfrac{(-1)^n x^{2n}}{(2n)!} = 1 - \dfrac{x^2}{2!} + \dfrac{x^4}{4!} - \dfrac{x^6}{6!} + \cdots$	$-\infty < x < \infty$
(5) $\ln(1 + x) = \sum\limits_{n=0}^{\infty} \dfrac{(-1)^n x^{n+1}}{n + 1} = x - \dfrac{x^2}{2} + \dfrac{x^3}{3} - \dfrac{x^4}{4} + \cdots$	$-1 < x \le 1$
(6) $\arctan x = \sum\limits_{n=0}^{\infty} \dfrac{(-1)^n x^{2n+1}}{2n + 1} = x - \dfrac{x^3}{3} + \dfrac{x^5}{5} - \dfrac{x^7}{7} + \cdots$	$-1 \le x \le 1$
(7) $\sinh x = \sum\limits_{n=0}^{\infty} \dfrac{x^{2n+1}}{(2n+1)!} = x + \dfrac{x^3}{3!} + \dfrac{x^5}{5!} + \dfrac{x^7}{7!} + \cdots$	$-\infty < x < \infty$
(8) $\cosh x = \sum\limits_{n=0}^{\infty} \dfrac{x^{2n}}{(2n)!} = 1 + \dfrac{x^2}{2!} + \dfrac{x^4}{4!} + \dfrac{x^6}{6!} + \cdots$	$-\infty < x < \infty$
(9) $(1 + x)^p = 1 + \sum\limits_{n=1}^{\infty} \dfrac{p(p-1)(p-2) \cdot \,\cdots\, \cdot (p - (n-1))}{n!} x^n$	
$\qquad = 1 + px + \dfrac{p(p-1)}{2!} x^2 + \dfrac{p(p-1)(p-2)}{3!} x^3 + \cdots$	$-1 < x < 1$

The series in (9) is the *binomial series* that we investigate in Example 9.9 below. The indicated interval is valid for any p. What happens at ± 1 depends on the value of p, and this is examined in Example 9.9 and Exercise 10. Of course, if p is a nonnegative integer, then the binomial series terminates to the usual binomial expansion.

Note that (5) of Table 1 follows from Example 9.5, and the reader should have obtained (6) of Table 1 in Exercise 7 in Section 9.1.

Example 9.8 We first consider (2) of Table 1. In Example 5.6, for $f(x) = e^x$, we found that

$$P_n(x) = 1 + x + \frac{x^2}{2!} + \cdots + \frac{x^n}{n!}$$

and

$$R_n(x) = e^t \frac{x^{n+1}}{(n+1)!}$$

for some t between 0 and x. Since $\sum_{n=0}^{\infty} x^n/n!$ converges absolutely on $\mathbb{R}$ (Example 9.1), $\lim_{n \to \infty} x^n/n! = 0$ for each x in $\mathbb{R}$. Therefore, $\lim_{n \to \infty} R_n(x) = 0$ for each x in $\mathbb{R}$. By Theorem 9.6, e^x is represented by its Maclaurin series, $\sum_{n=0}^{\infty} x^n/n!$, on $\mathbb{R}$.

We now consider (8) of Table 1. Replacing x by $-x$ in (2), we have

$$e^{-x} = \sum_{n=0}^{\infty} \frac{(-1)^n x^n}{n!} = 1 - x + \frac{x^2}{2!} - \frac{x^3}{3!} + \cdots$$

valid for all x in $\mathbb{R}$. Hence,

$$\cosh x = \frac{1}{2}(e^x + e^{-x})$$

$$= \frac{1}{2} \left(\left[1 + x + \frac{x^2}{2!} + \frac{x^3}{3!} + \cdots \right] + \left[1 - x + \frac{x^2}{2!} - \frac{x^3}{3!} + \cdots \right] \right)$$

$$= 1 + \frac{x^2}{2!} + \frac{x^4}{4!} + \frac{x^6}{6!} + \cdots$$

is also valid on $\mathbb{R}$.

Example 9.9 We consider (9) of Table 1. Most of the following is taken from Hewitt and Stromberg. Let $\binom{p}{0} = 1$ and let

$$\binom{p}{n} = \frac{p(p-1)(p-2) \cdot \cdots \cdot (p-(n-1))}{n!}$$

for n in $\mathbb{N}$. Then the series in (9) can be written as $\sum_{n=0}^{\infty} \binom{p}{n} x^n$. If p is a nonnegative integer, then all but finitely many of the numbers $\binom{p}{n}$ are 0, and so $\sum_{n=0}^{\infty} \binom{p}{n} x^n$

converges for all x. For p not a nonnegative integer,

$$\left| \frac{\binom{p}{n+1}}{\binom{p}{n}} \right| = \frac{|p-n|}{n+1}$$

$$= \frac{n-p}{n+1} \qquad \text{(for } n > p\text{)}$$

$$\underset{n}{\to} 1,$$

and so $\sum\limits_{n=0}^{\infty} \binom{p}{n} x^n$ has radius of convergence 1 by the Ratio test. That is, $\sum\limits_{n=0}^{\infty} \binom{p}{n} x^n$ converges absolutely for $|x| < 1$ and diverges for $|x| > 1$ for all p in $\mathbb{R}$, p not a nonnegative integer.

We next show for all real numbers p that $(1 + x)^p = \sum\limits_{n=0}^{\infty} \binom{p}{n} x^n$ for x in $(-1, 1)$. Although the reader may wish to try this using Theorem 9.6, we proceed in a completely different manner. For x in $(-1, 1)$ and p in $\mathbb{R}$, let $f_p(x) = \sum\limits_{n=0}^{\infty} \binom{p}{n} x^n$. By part 2 of Corollary 9.1,

$$f_p'(x) = \sum_{n=1}^{\infty} n \binom{p}{n} x^{n-1} = \sum_{n=0}^{\infty} (n+1) \binom{p}{n+1} x^n$$

for $|x| < 1$. Since $(n+1)\binom{p}{n+1} = p\binom{p-1}{n}$,

$$f_p'(x) = p \sum_{n=0}^{\infty} \binom{p-1}{n} x^n = p f_{p-1}(x) \tag{5}$$

for $|x| < 1$. Also for $|x| < 1$,

$$(1+x) f_{p-1}(x) = (1+x) \sum_{n=0}^{\infty} \binom{p-1}{n} x^n$$

$$= \sum_{n=0}^{\infty} \binom{p-1}{n} x^n + \sum_{n=0}^{\infty} \binom{p-1}{n} x^{n+1}$$

$$= 1 + \sum_{n=1}^{\infty} \binom{p-1}{n} x^n + \sum_{n=1}^{\infty} \binom{p-1}{n-1} x^n$$

$$= 1 + \sum_{n=1}^{\infty} \left[\binom{p-1}{n} + \binom{p-1}{n-1} \right] x^n$$

$$= 1 + \sum_{n=1}^{\infty} \binom{p}{n} x^n$$

$$= f_p(x).$$

Combining this with (5), we have

$$(1 + x)f_p'(x) = pf_p(x) \tag{6}$$

for all x in $(-1, 1)$. One could now solve the differential equation (6) directly using the initial condition $f_p(0) = 1$, or one could observe that

$$\frac{d}{dx}\left[f_p(x)(1 + x)^{-p}\right] = (1 + x)^{-p-1}\left[(1 + x)f_p'(x) - pf_p(x)\right] = 0$$

by (6). Hence, $f_p(x)/(1+x)^p$ is a constant function on $(-1, 1)$. Letting $x = 0$, we see that the constant value of this function is 1; that is, $f_p(x) = (1 + x)^p$ for all x in $(-1, 1)$.

Let $p > 0$ and not an integer. We now show that $\sum_{n=0}^{\infty} \left|\binom{p}{n}\right|$ converges. Let $a_n = \left|\binom{p}{n}\right|$ for $n = 0, 1, 2, \ldots$. Then $a_{n+1}/a_n = (n - p)/(n + 1)$ whenever $n \geq [p] + 1$, where $[p]$ denotes the largest integer less than or equal to p. For $n \geq [p] + 1$,

$$(n + 1)a_{n+1} = na_n - pa_n$$

and so

$$na_n - (n + 1)a_{n+1} = pa_n > 0. \tag{7}$$

Therefore, $(na_n)_{n=[p]+1}^{\infty}$ is a decreasing sequence, and so it has a limit. Let $\lim_{n \to \infty} na_n = \gamma \geq 0$. Since

$$\sum_{n=0}^{m} [na_n - (n + 1)a_{n+1}] = -(m + 1)a_{m+1} \underset{m}{\to} -\gamma,$$

the series $\sum_{n=0}^{\infty} [na_n - (n + 1)a_{n+1}]$ converges. By (7), $a_n = (1/p)[na_n - (n + 1)a_{n+1}]$ for $n \geq [p] + 1$, and so the series $\sum_{n=0}^{\infty} a_n = \sum_{n=0}^{\infty} \left|\binom{p}{n}\right|$ converges.

Since $\sum_{n=0}^{\infty} \binom{p}{n}$ terminates if p is a nonnegative integer, we have that $\sum_{n=0}^{\infty} \left|\binom{p}{n}\right|$ converges for all $p \geq 0$. Since $\left|\binom{p}{n}x^n\right| \leq \left|\binom{p}{n}\right|$ for $|x| \leq 1$, $\sum_{n=0}^{\infty} \binom{p}{n}x^n$ converges absolutely on $[-1, 1]$ for all $p \geq 0$ by the Comparison test. By Theorem 9.4 and Exercise 9 in Section 9.1, $\sum_{n=0}^{\infty} \binom{p}{n}x^n = (1 + x)^p$ for all $p \geq 0$ and for all x in $[-1, 1]$.

Exercises

1. Let
$$f(x) = \begin{cases} e^{-(1/x^2)} & \text{if} \quad x \neq 0 \\ 0 & \text{if} \quad x = 0. \end{cases}$$
   Show that $f^{(n)}(0) = 0$ for $n = 0, 1, 2, 3, \ldots$.

2. Derive (7) of Table 1.

3. (a) Find the Maclaurin series and its interval of convergence for $f(x) = 2^x$.
   (b) Find the Taylor series for e^x centered at 1. [*Hint:* Use (2) of Table 1 for both part (a) and part (b).]

4. Suppose that f has derivatives of all orders on an interval I, and let c be an interior point of I. If there is an $M > 0$ such that $\left| f^{(n)}(x) \right| \leq M^n$ for all $n = 0, 1, 2, 3, \ldots$ and for all x in I, show that $f(x) = \sum_{n=0}^{\infty} [f^{(n)}(c)/n!](x - c)^n$ for all x in I.

5. (a) Derive (3) and (4) of Table 1.
   (b) Replace x by $x - \pi/2$ in (4) of Table 1 to obtain the Taylor series for $\sin x$ centered at $\pi/2$. What is its interval of convergence?

6. Assume that (2) of Table 1 holds when x is a complex number. Replace x by ix, where $i = \sqrt{-1}$ and x is real, to show that $e^{ix} = \cos x + i \sin x$. Note that this implies $1 + e^{\pi i} = 0$.

7. Find the Maclaurin series and its interval of convergence for
$$\frac{1}{2} \ln \left(\frac{1 + x}{1 - x} \right).$$

8. Use (9) of Table 1 to obtain a series expansion of $(3 + x)^{2/3}$. What is the interval of convergence?

9. Use Raabe's test (Theorem 7.9) to show that the binomial series is absolutely convergent at $x = 1$ when $p > 0$. Conclude from this that the binomial series is absolutely convergent on $[-1, 1]$ when $p > 0$.

10. (a) If $p \leq -1$, show that the interval of convergence of the binomial series is $(-1, 1)$.
    (b) If $-1 < p < 0$, show that the binomial series is conditionally convergent at 1 and divergent at -1. Hence, its interval of convergence is $(-1, 1]$.

11. Derive the Maclaurin series and its interval of convergence for $\arcsin x$. Use this to obtain
$$\frac{\pi}{6} = \frac{1}{2} + \frac{1}{2} \cdot \frac{1}{3 \cdot 2^3} + \frac{1 \cdot 3}{2 \cdot 4} \cdot \frac{1}{5 \cdot 2^5} + \frac{1 \cdot 3 \cdot 5}{2 \cdot 4 \cdot 6} \cdot \frac{1}{7 \cdot 2^7} + \cdots.$$
    [*Hint:* Use (9) of Table 1 to obtain a series expansion of $(1 - t^2)^p$. Letting $p = -\frac{1}{2}$, integrate this from 0 to x.]

The Riemann-Stieltjes Integral

The Riemann-Stieltjes integral on an interval $[a, b]$ involves an integrand f and an integrator α, and is usually denoted by $\int_a^b f \, d\alpha$ or $\int_a^b f(x)d\alpha(x)$. When $\alpha(x) = x$, this is the usual Riemann integral discussed in Chapter 6. When α is the greatest integer function, any finite sum or countably infinite sum can be expressed as a Riemann-Stieltjes integral (albeit improper when the sum is infinite). We establish this in Section 10.1. Thus, Riemann-Stieltjes integrals make possible the simultaneous treatment of continuous and discrete random variables in probability theory.

To facilitate the development of this chapter, the first three sections involve a monotone increasing integrator. This allows for a development parallel to that in Chapter 6; however, there are differences. In Section 10.4 we introduce functions of bounded variation, and we develop the connection between functions of bounded variation and monotone functions. We then consider in Section 10.5 the Riemann-Stieltjes integral when the integrator is a function of bounded variation.

10.1 Monotone Increasing Integrators

Our setting is as follows: for a and b real with $a < b$, f is a bounded function and α is a monotone increasing function from $[a, b]$ into $\mathbb{R}$.

Recall from Chapter 6 that a partition P of $[a, b]$ is a finite set of points $\{x_0, x_1, x_2, \ldots, x_n\}$ where n is in $\mathbb{N}$ with $a = x_0 < x_1 < x_2 < \cdots < x_n = b$, and that $\mathcal{P}$ or $\mathcal{P}[a, b]$ denotes the set of partitions of $[a, b]$. As in Chapter 6, for $P = \{x_i\}_{i=0}^n$ in $\mathcal{P}$, we set

$$M_i = \sup\{f(x) : x_{i-1} \leq x \leq x_i\}$$

and

$$m_i = \inf\{f(x) : x_{i-1} \leq x \leq x_i\}.$$

We also set $\Delta\alpha_i = \alpha(x_i) - \alpha(x_{i-1})$ for each $i = 1, 2, \ldots, n$. Since α is monotone increasing on $[a, b]$, each $\Delta\alpha_i \geq 0$. This last observation is what makes many of the proofs in this section virtually the same as the corresponding proofs in Chapter 6; basically, $\Delta\alpha_i$ replaces Δx_i in Chapter 6.

Definition 10.1 For $P = \{x_i\}_{i=0}^n$ in $\mathcal{P}$, set

$$U(P, f, \alpha) = \sum_{i=1}^n M_i \Delta\alpha_i$$

(the *upper sum* of f with respect to α and P) and

$$L(P, f, \alpha) = \sum_{i=1}^{n} m_i \Delta\alpha_i$$

(the *lower sum* of f with respect to α and P). Set

$$\overline{\int_a^b} f \, d\alpha = \inf\{U(P, f, \alpha) : P \in \mathcal{P}\}$$

(the *upper Riemann-Stieltjes integral* of f with respect to α over $[a, b]$) and

$$\underline{\int_a^b} f \, d\alpha = \sup\{L(P, f, \alpha) : P \in \mathcal{P}\}$$

(the *lower Riemann-Stieltjes integral* of f with respect to α over $[a, b]$).

Remark   Since f is bounded, there exist real numbers m and M such that $m \le f(x) \le M$ for all x in $[a, b]$. For any $P = \{x_i\}_{i=0}^{n}$ in $\mathcal{P}$, since $\sum_{i=1}^{n} \Delta\alpha_i = [\alpha(x_1) - \alpha(x_0)] + [\alpha(x_2) - \alpha(x_1)] + \cdots + [\alpha(x_n) - \alpha(x_{n-1})] = \alpha(b) - \alpha(a)$, $m[\alpha(b) - \alpha(a)] \le L(P, f, \alpha) \le U(P, f, \alpha) \le M[\alpha(b) - \alpha(a)]$. Therefore, $\overline{\int_a^b} f \, d\alpha$ and $\underline{\int_a^b} f \, d\alpha$ always exist in $\mathbb{R}$.

Note that the Remark above is the same as the corresponding Remark after Definition 6.2 with $x_i - x_{i-1}$ replaced by $\alpha(x_i) - \alpha(x_{i-1})$. With the same type of modification, the proof of Proposition 6.1 carries over to our setting in this section.

Theorem 10.1

1. If P^* is a refinement of P (see Definition 6.3), then $L(P, f, \alpha) \le L(P^*, f, \alpha)$ and $U(P^*, f, \alpha) \le U(P, f, \alpha)$.

2. If P_1 and P_2 are in $\mathcal{P}$, then $L(P_1, f, \alpha) \le U(P_2, f, \alpha)$.

3. $\underline{\int_a^b} f \, d\alpha \le \overline{\int_a^b} f \, d\alpha$.

Proof   See the proofs of Propositions 6.1 and 6.2 and Theorem 6.1.   ∎

Definition 10.2   The bounded function f is *Riemann-Stieltjes integrable* with respect to the monotone increasing function α on $[a, b]$ if $\underline{\int_a^b} f \, d\alpha = \overline{\int_a^b} f \, d\alpha$, and this common value is the *Riemann-Stieltjes integral* (or simply the *Stieltjes integral*) of f with respect to α over $[a, b]$.

Notation and Terminology   Recall that $\mathcal{R}$ or $\mathcal{R}[a, b]$ denotes the set of Riemann integrable functions on $[a, b]$. Let $\mathcal{R}(\alpha)$ or $\mathcal{R}_a^b(\alpha)$ denote the set of functions that are Riemann-Stieltjes integrable with respect to α on $[a, b]$. For

f in $\mathcal{R}(\alpha)$, we denote the Riemann-Stieltjes integral of f with respect to α on $[a, b]$ by $\int_a^b f\, d\alpha$ or $\int_a^b f(x)\, d\alpha(x)$. Also, f is called the *integrand* and α is called the *integrator*; of course, $[a, b]$ is called the *interval of integration*.

Note that when $\alpha(x) = x$, the Riemann-Stieltjes integral becomes the Riemann integral. In Example 10.3 and Exercise 8, we will see that every finite or countably infinite sum is also a Riemann-Stieltjes integral.

Example 10.1   Let
$$f(x) = \begin{cases} 1 & \text{if } x \in \mathbb{Q} \\ 0 & \text{if } x \notin \mathbb{Q}. \end{cases}$$
Let α be a monotone increasing function on $[a, b]$ with $a < b$. For $P = \{x_i\}_{i=0}^n$ in $\mathcal{P}$, $L(P, f, \alpha) = \sum_{i=1}^n m_i \Delta\alpha_i = 0$ and $U(P, f, \alpha) = \sum_{i=1}^n M_i \Delta\alpha_i = \sum_{i=1}^n \Delta\alpha_i = \alpha(b) - \alpha(a)$. Thus $\underline{\int_a^b} f\, d\alpha = 0$ and $\overline{\int_a^b} f\, d\alpha = \alpha(b) - \alpha(a)$, and so f is in $\mathcal{R}(\alpha)$ on $[a, b]$ if and only if α is a constant function on $[a, b]$.

Existence

The following theorem is the analogue of Theorem 6.2.

Theorem 10.2   The function f is in $\mathcal{R}(\alpha)$ on $[a, b]$ if and only if for every $\varepsilon > 0$ there is a partition P of $[a, b]$ such that $U(P, f, \alpha) - L(P, f, \alpha) < \varepsilon$.

Proof   Modify the proof of Theorem 6.2 appropriately. For example, replace $L(P, f)$ by $L(P, f, \alpha)$. ∎

Example 10.2   Let $a < c < b$ and define $\alpha : [a, b] \to \mathbb{R}$ by
$$\alpha(x) = \begin{cases} k_1 & \text{if } a \le x < c \\ k_2 & \text{if } c \le x \le b, \end{cases}$$
where $k_1 < k_2$ (see Figure 10.1).

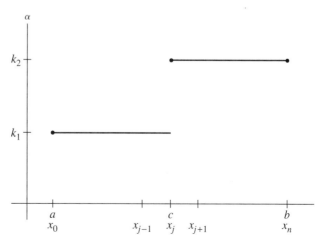

Figure 10.1

Let $P = \{x_i\}_{i=0}^n$ be any partition of $[a, b]$ with $x_j = c$, where $1 \leq j \leq n - 1$. For any bounded function f on $[a, b]$, since $\Delta\alpha_i = 0$ for all i except $i = j$,

$$U(P, f, \alpha) = \sum_{i=1}^n M_i \Delta\alpha_i = M_j \Delta\alpha_j = M_j(k_2 - k_1)$$

and

$$L(P, f, \alpha) = \sum_{i=1}^n m_i \Delta\alpha_i = m_j(k_2 - k_1).$$

We now consider different functions f. If f is constant on $[a, c]$ (and arbitrary but bounded on $(c, b]$), then $M_j = m_j = f(c)$, and so f is in $\mathcal{R}(\alpha)$ on $[a, b]$ by Theorem 10.2. From part 1 of Theorem 10.1, it follows that $\int_a^b f \, d\alpha = f(c)(k_2 - k_1)$.

Next, suppose that

$$f(x) = \begin{cases} L_1 & \text{if} \quad a \leq x < c \\ L_2 & \text{if} \quad c \leq x \leq b, \end{cases}$$

where $L_1 < L_2$. For the partition P defined at the beginning of this example,

$$U(P, f, \alpha) - L(P, f, \alpha) = (M_j - m_j)(k_2 - k_1) = (L_2 - L_1)(k_2 - k_1).$$

From part 1 of Theorem 10.1, it follows that Theorem 10.2 cannot be satisfied for any $\varepsilon < (L_2 - L_1)(k_2 - k_1)$, and so f is not in $\mathcal{R}(\alpha)$ on $[a, b]$. The key point here is that both f and α are discontinuous from the left at c, which, in the case of f, means that $f(c-) = \lim_{x \to c^-} f(x) \neq f(c)$.

Now let us consider an arbitrary bounded function f on $[a, b]$. Then f is in $\mathcal{R}(\alpha)$ on $[a, b]$ if and only if for all $\varepsilon > 0$ there is a partition $P = \{x_i\}_{i=0}^n$ of $[a, b]$ with $x_j = c$ in P such that $M_j - m_j < \varepsilon/(k_2 - k_1)$. Since

$$M_j = \sup\{f(x) : x_{j-1} \leq x \leq c\} \geq f(c)$$

and

$$m_j = \inf\{f(x) : x_{j-1} \leq x \leq c\} \leq f(c),$$

$M_j - m_j$ can be made arbitrarily small if and only if both M_j and m_j can be made arbitrarily close to $f(c)$. Thus, f is in $\mathcal{R}(\alpha)$ on $[a, b]$ if and only if $f(c-) = f(c)$ or, equivalently, if and only if f is continuous from the left at c. When f is continuous from the left at c, $\int_a^b f \, d\alpha = f(c)(k_2 - k_1)$.

Basic Properties

Example 10.2 shows that we cannot obtain an analogue of Theorem 6.5 for Riemann-Stieltjes integrals. In Section 10.3 we show that the analogue of Theorem 6.3 for Riemann-Stieltjes sums (to be defined in Section 10.3) does not always hold. Consequently, we now alter our development from that of Chapter 6.

Proposition 10.1   Below α, α_1, and α_2 are monotone increasing functions on $[a, b]$.

1. If f_1 and f_2 are in $\mathcal{R}(\alpha)$ on $[a, b]$ and c is in $\mathbb{R}$, then $f_1 \pm f_2$ and cf_1 are in $\mathcal{R}(\alpha)$ on $[a, b]$; moreover,

$$\int_a^b (f_1 \pm f_2)\, d\alpha = \int_a^b f_1\, d\alpha \pm \int_a^b f_2\, d\alpha$$

and

$$\int_a^b cf_1\, d\alpha = c \int_a^b f_1\, d\alpha.$$

2. If f is in $\mathcal{R}(\alpha_1)$ and $\mathcal{R}(\alpha_2)$ on $[a, b]$ and $c > 0$, then f is in $\mathcal{R}(\alpha_1 + \alpha_2)$ and f is in $\mathcal{R}(c\alpha_1)$ on $[a, b]$, and

$$\int_a^b f\, d(\alpha_1 + \alpha_2) = \int_a^b f\, d\alpha_1 + \int_a^b f\, d\alpha_2$$

and

$$\int_a^b f\, d(c\alpha_1) = c \int_a^b f\, d\alpha_1.$$

3. If f_1 and f_2 are in $\mathcal{R}(\alpha)$ on $[a, b]$ and $f_1(x) \le f_2(x)$ for all x in $[a, b]$, then $\int_a^b f_1\, d\alpha \le \int_a^b f_2\, d\alpha$.

4. If f is in $\mathcal{R}(\alpha)$ on $[a, b]$ and $a < c < b$, then f is in $\mathcal{R}(\alpha)$ on $[a, c]$ and f is in $\mathcal{R}(\alpha)$ on $[c, b]$, and $\int_a^b f\, d\alpha = \int_a^c f\, d\alpha + \int_c^b f\, d\alpha$.

Proof

1. This proof necessarily differs from the proof of Proposition 6.3 (the analogous result for Riemann integrals) since we used Riemann sums in the proof of Proposition 6.3. However, the proof is similar to the proof of part 2, and so we leave this proof as Exercise 7.

2. Let $\alpha = \alpha_1 + \alpha_2$ and note that α is monotone increasing on $[a, b]$. For $P = \{x_i\}_{i=0}^n$ in $\mathcal{P}[a, b]$,

$$\Delta\alpha_i = \alpha(x_i) - \alpha(x_{i-1}) = [\alpha_1(x_i) - \alpha_1(x_{i-1})] + [\alpha_2(x_i) - \alpha_2(x_{i-1})]$$

$$= (\Delta\alpha_1)_i + (\Delta\alpha_2)_i,$$

and so

$$U(P, f, \alpha) = U(P, f, \alpha_1) + U(P, f, \alpha_2)$$

and $\qquad\qquad\qquad\qquad\qquad\qquad\qquad\qquad\qquad\qquad\qquad$ (1)

$$L(P, f, \alpha) = L(P, f, \alpha_1) + L(P, f, \alpha_2).$$

Let $\varepsilon > 0$. By Theorem 10.2 there exist P_1 and P_2 in $\mathcal{P}[a, b]$ such that

$$U(P_i, f, \alpha_i) - L(P_i, f, \alpha_i) < \frac{\varepsilon}{2} \qquad\qquad\qquad (2)$$

for $i = 1$ and 2. Letting $P = P_1 \cup P_2$ be a common refinement of P_1 and P_2, part 1 of Theorem 10.1 implies that (2) holds with P_i replaced by P for $i = 1$ and 2. Then (1) implies that $U(P, f, \alpha) - L(P, f, \alpha) < \varepsilon$, and so f is in $\mathcal{R}(\alpha)$ on $[a, b]$ by Theorem 10.2.

With this same P, since $L(P, f, \alpha_i) \leq \int_a^b f\, d\alpha_i \leq U(P, f, \alpha_i)$, we have that $U(P, f, \alpha_i) - \int_a^b f\, d\alpha_i < \varepsilon/2$ for $i = 1$ and 2. Then

$$\int_a^b f\, d\alpha \leq U(P, f, \alpha) = U(P, f, \alpha_1) + U(P, f, \alpha_2)$$

$$< \int_a^b f\, d\alpha_1 + \int_a^b f\, d\alpha_2 + \varepsilon.$$

Since ε is arbitrary,

$$\int_a^b f\, d\alpha \leq \int_a^b f\, d\alpha_1 + \int_a^b f\, d\alpha_2. \tag{3}$$

Either use the fact that $\int_a^b f\, d\alpha_i - L(P, f, \alpha_i) < \varepsilon/2$ for $i = 1$ and 2 with an argument similar to the one given above, or replace f by $-f$ in (3) and use the second equality in part 1 to obtain that $\int_a^b f\, d\alpha \geq \int_a^b f\, d\alpha_1 + \int_a^b f\, d\alpha_2$.

Next, let α be monotone increasing on $[a, b]$, let f be in $\mathcal{R}(\alpha)$ on $[a, b]$, and let $c > 0$ (this keeps $c\alpha$ monotone increasing). For any P in $\mathcal{P}[a, b]$, $U(P, f, c\alpha) = cU(P, f, \alpha)$ and $L(P, f, c\alpha) = cL(P, f, \alpha)$. Given $\varepsilon > 0$, choose P in $\mathcal{P}[a, b]$ with $U(P, f, \alpha) - L(P, f, \alpha) < \varepsilon/c$ by Theorem 10.2. Then $U(P, f, c\alpha) - L(P, f, c\alpha) < \varepsilon$, and so f is in $\mathcal{R}(c\alpha)$ on $[a, b]$.

For any P in $\mathcal{P}[a, b]$,

$$\int_a^b f\, d(c\alpha) \leq U(P, f, c\alpha) = cU(P, f, \alpha),$$

and so

$$\int_a^b f\, d(c\alpha) \leq c \inf\{U(P, f, \alpha) : P \in \mathcal{P}\} = c \int_a^b f\, d\alpha.$$

Replacing f by $-f$ and using part 1, we obtain the reverse inequality, and so $\int_a^b f\, d(c\alpha) = c \int_a^b f\, d\alpha$.

3. and 4. These proofs are analogous to the proofs of Propositions 6.4 and 6.5, respectively. ∎

Example 10.3   The purpose of this example is to realize any finite sum as a Riemann-Stieltjes integral. First, let $\alpha(x) = [x]$ denote the greatest integer function—that is, $[x]$ is the largest integer less than or equal to x. By Example 10.2 it follows that for k an integer, $\int_k^{k+1} f\, d[x] = f(k+1)$ whenever the bounded function f is continuous from the left at $k + 1$. Hence, for n in $\mathbb{N}$,

$$\int_0^n f\, d[x] = \int_0^1 f\, d[x] + \int_1^2 f\, d[x] + \cdots + \int_{n-1}^n f\, d[x]$$

$$= f(1) + f(2) + \cdots + f(n)$$

whenever the bounded function f is continuous from the left at $1, 2, \ldots, n$.

Given a finite sum $\sum\limits_{k=1}^n a_k$, define $f : [0, n] \to \mathbb{R}$ by

$$f(x) = \begin{cases} a_k & \text{if } k - 1 < x \leq k \text{ for } k = 1, 2, \ldots, n \\ 0 & \text{if } x = 0. \end{cases}$$

Then $\sum\limits_{k=1}^n a_k = \sum\limits_{k=1}^n f(k) = \int_0^n f\, d[x]$.

For a specific example, let n be in $\mathbb{N}$, let $0 < p < 1$, let $q = 1 - p$, and let $a_k = \binom{n}{k} p^k q^{n-k}$ for $k = 0, 1, 2, \ldots, n$. Letting

$$f(x) = \begin{cases} a_k & \text{if} \quad k - 1 < x \leq k \text{ for } k = 0, 1, 2, \ldots, n \\ 0 & \text{if} \quad x = -1 \end{cases}$$

[and using an argument similar to that given above for $\int_{-1}^{0} f \, d \, [x] = f(0)$], we have $\int_{-1}^{n} f \, d \, [x] = \sum\limits_{k=0}^{n} \binom{n}{k} p^k q^{n-k} = (p + q)^n = 1^n = 1$. Those readers who have had a course in probability will recognize a_k as the probability of k successes in n trials for a binomial random variable.

Similar to Definition 6.8, we have the following.

Definition 10.3   For f in $\mathcal{R}(\alpha)$ on $[a, b]$, we set $\int_{b}^{a} f \, d\alpha = - \int_{a}^{b} f \, d\alpha$; and for any function f defined at a, we set $\int_{a}^{a} f \, d\alpha = 0$.

Similar to the Remark following Proposition 6.5, if $a < b < c$ and f is in $\mathcal{R}(\alpha)$ on $[a, c]$, then the equation in part 4 of Proposition 10.1 still holds.

We end this section with results analogous to Theorem 6.4 and Corollary 6.1.

Theorem 10.3   Let f be in $\mathcal{R}(\alpha)$ on $[a, b]$ with the range of f contained in the closed interval $[m, M]$. Let $\varphi : [m, M] \to \mathbb{R}$ be continuous on $[m, M]$. Then $\varphi \circ f$ is in $\mathcal{R}(\alpha)$ on $[a, b]$.

Proof   In the proof of Theorem 6.4, replace $b - a$ by $\alpha(b) - \alpha(a)$. Of course, $U(P, f)$ and $L(P, f)$ become $U(P, f, \alpha)$ and $L(P, f, \alpha)$, respectively. The inequalities in the proof remain the same when Δx_i is replaced by $\Delta \alpha_i$ since each $\Delta \alpha_i$ is nonnegative.   ∎

Corollary 10.1   If f and g are in $\mathcal{R}(\alpha)$ on $[a, b]$, then

1. fg is in $\mathcal{R}(\alpha)$ on $[a, b]$;

2. $|f|$ is in $\mathcal{R}(\alpha)$ on $[a, b]$ and $\left| \int_{a}^{b} f \, d\alpha \right| \leq \int_{a}^{b} |f| \, d\alpha$;

3. $\max\{f, g\}$ and $\min\{f, g\}$ are in $\mathcal{R}(\alpha)$ on $[a, b]$.

Proof   See the proof of Corollary 6.1.   ∎

═ Exercises ═

In the following exercises f is bounded and α is monotone increasing on $[a, b]$.

1. Let $f(x) = c$ on $[a, b]$. Show that f is in $\mathcal{R}(\alpha)$ on $[a, b]$ and that $\int_{a}^{b} f \, d\alpha = c[\alpha(b) - \alpha(a)]$.

2. Let $\alpha(x) = c$ on $[a, b]$. Show that f is in $\mathcal{R}(\alpha)$ on $[a, b]$ and that $\int_{a}^{b} f \, d\alpha = 0$.

3. Let
$$f(x) = \begin{cases} 1 & \text{if} \quad 0 < x \le 1 \\ 2 & \text{if} \quad x = 0 \end{cases} \quad \text{and} \quad \alpha(x) = \begin{cases} 0 & \text{if} \quad 0 < x \le 1 \\ -1 & \text{if} \quad x = 0. \end{cases}$$
Show that f is not in $\mathcal{R}(\alpha)$ on $[0, 1]$. Note that f and α are both discontinuous from the right at 0.

4. Let
$$g(x) = \begin{cases} 1 & \text{if} \quad 0 \le x < 1 \\ 2 & \text{if} \quad x = 1 \end{cases}$$
and let α be defined as in Exercise 3. Show that g is in $\mathcal{R}(\alpha)$ on $[0, 1]$ and find $\int_0^1 g \, d\alpha$.

5. Let $a < c < b$ and let
$$\alpha(x) = \begin{cases} k_1 & \text{if} \quad a \le x \le c \\ k_2 & \text{if} \quad c < x \le b, \end{cases}$$
where $k_1 < k_2$.

(a) If
$$f(x) = \begin{cases} L_1 & \text{if} \quad a \le x \le c \\ L_2 & \text{if} \quad c < x \le b, \end{cases}$$
where $L_1 < L_2$, show that f is not in $\mathcal{R}(\alpha)$ on $[a, b]$.

(b) Show that an arbitrary bounded function f is in $\mathcal{R}(\alpha)$ on $[a, b]$ if and only if f is continuous from the right at c, and then $\int_a^b f \, d\alpha = f(c)(k_2 - k_1)$. [*Hint:* See Example 10.2.]

6. Suppose that α is continuous at a point c in $[a, b]$. If
$$f(x) = \begin{cases} 0 & \text{if} \quad x \ne c \\ 1 & \text{if} \quad x = c, \end{cases}$$
show that f is in $\mathcal{R}(\alpha)$ on $[a, b]$ and that $\int_a^b f \, d\alpha = 0$.

7. Prove part 1 of Proposition 10.1.

8. Analogous to Definition 6.10, for a in $\mathbb{R}$ we define the *improper Riemann-Stieltjes integral* of f with respect to α on $[a, \infty)$ by
$$\int_a^\infty f \, d\alpha = \lim_{t \to \infty} \int_a^t f \, d\alpha,$$
where f is assumed to be in $\mathcal{R}(\alpha)$ on $[a, t]$ for all $t \ge a$. The improper integral *converges* if this limit is a real number; otherwise, it *diverges*.

   Show that any convergent infinite series can be realized as a convergent improper Riemann-Stieltjes integral. [*Hint:* By defining f appropriately, $\sum_{k=1}^\infty a_k = \int_0^\infty f \, d[x]$.] Also note that the improper integral diverges if the series diverges.

9. (a) Let $0 < p < 1$, let $q = 1 - p$, and let
$$f(x) = \begin{cases} pq^{k-1} & \text{if} \quad k - 1 < x \le k \text{ for } k = 1, 2, 3, \dots \\ 0 & \text{if} \quad x = 0. \end{cases}$$
Show that $\int_0^\infty f \, d[x] = 1$. ($pq^{k-1}$ is the probability of the first success occurring on trial k for a geometric random variable.)

(b) Let $0 < \lambda < \infty$ and let

$$f(x) = \begin{cases} \dfrac{\lambda^k e^{-\lambda}}{k!} & \text{if} \quad k - 1 < x \le k \text{ for } k = 0, 1, 2, \dots \\ 0 & \text{if} \quad x = -1. \end{cases}$$

Show that $\int_{-1}^{\infty} f\, d[x] = 1$. ($\lambda^k e^{-\lambda}/k!$ is the probability of k occurrences of a "rare" event for a Poisson random variable.)

10. Prove the analogue of Theorem 8.3 for Riemann-Stieltjes integrals: if the sequence of functions $(f_n)_{n \in \mathbb{N}}$ converges uniformly to F on $[a, b]$ and if each f_n is in $\mathcal{R}(\alpha)$ on $[a, b]$, then F is in $\mathcal{R}(\alpha)$ on $[a, b]$ and

$$\int_a^b F\, d\alpha = \lim_{n \to \infty} \int_a^b f_n\, d\alpha.$$

[*Hint:* In the proof of Theorem 8.3, replace $b - a$ by $\alpha(b) - \alpha(a)$.]

11. Prove the analogue of Exercise 10 in Section 6.3 for Riemann-Stieltjes integrals: if f and g are in $\mathcal{R}(\alpha)$ on $[a, b]$, then

$$\left| \int_a^b fg\, d\alpha \right| \le \left[\left(\int_a^b f^2 d\alpha \right) \left(\int_a^b g^2 d\alpha \right) \right]^{1/2}.$$

10.2 Families of Integrable Functions

In this section, where α is still monotone increasing, we show that a continuous integrand and a monotone integrand (with the additional assumption that α is continuous) are Riemann-Stieltjes integrable. In Theorem 10.7 we derive a sufficient condition for a function not to be Riemann-Stieltjes integrable, and in Theorem 10.9 we derive the integration by parts formula.

Continuous Functions

Theorem 10.4   If f is continuous on $[a, b]$, then f is in $\mathcal{R}(\alpha)$ on $[a, b]$.

Proof   By Exercise 2 in Section 10.1 we can assume that $\alpha(a) < \alpha(b)$. In the proof of Theorem 6.7, replace $b - a$ with $\alpha(b) - \alpha(a)$, and of course Δx_i with $\Delta \alpha_i$.   ∎

Analogous to Theorem 6.8, we have the following.

Theorem 10.5   (First Mean Value Theorem) If f is continuous on $[a, b]$, then there is a c in $[a, b]$ such that

$$\int_a^b f\, d\alpha = f(c)\,[\alpha(b) - \alpha(a)].$$

Proof   If α is a constant function, any c in $[a, b]$ will do. Otherwise, as in the proof of Theorem 6.8,

$$m\,[\alpha(b) - \alpha(a)] \le \int_a^b f\, d\alpha \le M\,[\alpha(b) - \alpha(a)],$$

where m and M are, respectively, the absolute minimum and maximum values of f on $[a, b]$. Now apply the Intermediate Value Theorem to f. ∎

Example 10.4 If

$$\alpha(x) = \begin{cases} 0 & \text{if} \quad a \leq x < b \\ 1 & \text{if} \quad x = b \end{cases}$$

and f is continuous, then, as in Example 10.2, $\int_a^b f \, d\alpha = f(b)[\alpha(b) - \alpha(a)]$. Thus, it may not be possible to choose the c guaranteed by Theorem 10.5 in (a, b).

The following theorem is similar to Theorem 6.12.

Theorem 10.6 Suppose that f is in $\mathcal{R}(\alpha)$ on $[a, b]$, and define F on $[a, b]$ by $F(x) = \int_a^x f \, d\alpha$ for each x in $[a, b]$.

1. Every point of continuity of α is a point of continuity of F.

2. If c is in $[a, b]$ with f continuous at c and α differentiable at c, then F is differentiable at c and $F'(c) = f(c)\alpha'(c)$.

Proof

1. Let $M > 0$ with $|f(x)| \leq M$ for all x in $[a, b]$. For x and x_0 in $[a, b]$,

$$|F(x) - F(x_0)| = \left| \int_{x_0}^x f \, d\alpha \right| \leq M \, |\alpha(x) - \alpha(x_0)|$$

by part 2 of Corollary 10.1 and Exercise 1 in Section 10.1. Hence, if α is continuous at x_0, then so is F.

2. For $x \neq c$, by the proof of Theorem 10.5, there is a constant K_x such that

$$F(x) - F(c) = \int_c^x f \, d\alpha = K_x \, [\alpha(x) - \alpha(c)],$$

where $\inf f(t) \leq K_x \leq \sup f(t)$ with the inf and sup being taken over all t between c and x inclusive. Therefore,

$$\begin{aligned} F'(c) &= \lim_{x \to c} \frac{F(x) - F(c)}{x - c} \\ &= \left(\lim_{x \to c} K_x \right) \cdot \lim_{x \to c} \left(\frac{\alpha(x) - \alpha(c)}{x - c} \right) \\ &= f(c)\alpha'(c) \end{aligned}$$

since f is continuous at c and α is differentiable at c. ∎

Because of Exercise 2 in Section 10.1 and Example 10.2, it should be clear that we cannot obtain an analogue of Corollary 6.2. A new type of result in our present setting is Theorem 10.7 below. In anticipation of its proof, let us note that for $a < c \leq b$ and a function $g : [a, b] \to \mathbb{R}$, we have that g is continuous from the left at c if and only if for all $\varepsilon > 0$ there is a $\delta > 0$ such that if x is in $[a, b]$ and $c - \delta < x < c$, then $|g(x) - g(c)| < \varepsilon$. Thus, g is discontinuous from the left at c if and only if there exists an $\varepsilon > 0$ such that for all $\delta > 0$ there is an x' in $[a, b]$ with $c - \delta < x' < c$ and $|g(x') - g(c)| \geq \varepsilon$.

Theorem 10.7 Let $a < c \leq b$ and assume that both f and α are discontinuous from the left at c. Then $\int_a^b f \, d\alpha$ cannot exist. A similar result holds if f and α are both discontinuous from the right at c, where $a \leq c < b$.

Proof Let $P = \{x_i\}_{i=0}^n$ be in $\mathcal{P}[a, b]$, with $x_j = c$ for some j in $\{1, 2, \ldots, n\}$. Then

$$U(P, f, \alpha) - L(P, f, \alpha) = \sum_{i=1}^n (M_i - m_i) \Delta \alpha_i$$

$$\geq (M_j - m_j) \Delta \alpha_j \qquad \text{(since each summand is nonnegative)}$$

$$= (M_j - m_j) \big[\alpha(c) - \alpha(x_{j-1})\big].$$

Since f is discontinuous from the left at c, there is an $\varepsilon_f > 0$ such that $M_j - m_j \geq \varepsilon_f$. Since α is discontinuous from the left at c, there is an $\varepsilon_\alpha > 0$ such that the point x_{j-1} may be chosen so that $\alpha(c) - \alpha(x_{j-1}) \geq \varepsilon_\alpha$. Then $U(P, f, \alpha) - L(P, f, \alpha) \geq \varepsilon_f \varepsilon_\alpha$. From part 1 of Theorem 10.1, it follows that Theorem 10.2 cannot be satisfied, and so f is not in $\mathcal{R}(\alpha)$ on $[a, b]$.

The argument when both f and α are discontinuous from the right at c is asked for in Exercise 4. ∎

The reader should compare Theorem 10.7 with Example 10.2 and Exercises 3 and 5 in Section 10.1.

Monotone Functions

Example 10.2 shows that a monotone function is not necessarily Riemann-Stieltjes integrable. The following theorem is the analogue of Theorem 6.9.

Theorem 10.8 If f is monotone on $[a, b]$ and α is continuous on $[a, b]$, then f is in $\mathcal{R}(\alpha)$ on $[a, b]$. [Of course, α is still monotone increasing on $[a, b]$.]

Proof First suppose that f is monotone increasing on $[a, b]$. If $f(a) = f(b)$, then f is a constant function on $[a, b]$ and hence in $\mathcal{R}(\alpha)$ on $[a, b]$; so assume that $f(a) < f(b)$. Let $\varepsilon > 0$. Since α is uniformly continuous on $[a, b]$, there is a $\delta > 0$ such that if x and y are in $[a, b]$ with $|x - y| < \delta$, then $|\alpha(x) - \alpha(y)| < \varepsilon / [f(b) - f(a)]$. Let $P = \{x_i\}_{i=0}^n$ be in $\mathcal{P}[a, b]$ with $\|P\| < \delta$. Then

$$U(P, f, \alpha) - L(P, f, \alpha) = \sum_{i=1}^n (M_i - m_i) \Delta \alpha_i$$

$$= \sum_{i=1}^n \big[f(x_i) - f(x_{i-1})\big]\big[\alpha(x_i) - \alpha(x_{i-1})\big]$$

$$< \frac{\varepsilon}{f(b) - f(a)} \sum_{i=1}^n \big[f(x_i) - f(x_{i-1})\big]$$

$$= \frac{\varepsilon}{f(b) - f(a)} [f(b) - f(a)]$$

$$= \varepsilon,$$

and so f is in $\mathcal{R}(\alpha)$ on $[a, b]$ by Theorem 10.2.

If f is monotone decreasing on $[a, b]$, then $-f$ is monotone increasing on $[a, b]$, and so $f = -(-f)$ is in $\mathcal{R}(\alpha)$ on $[a, b]$ by part 1 of Proposition 10.1. ■

The analogue of Theorem 6.10, the Second Mean Value Theorem for Riemann integrals, is given in Exercise 6. In Theorem 10.9 below, we establish a remarkable connection between $\int_a^b f \, d\alpha$ and $\int_a^b \alpha \, df$ under the appropriate hypothesis. First, we need a lemma.

Lemma 10.1   If f and α are both monotone increasing on $[a, b]$ and P is in $\mathcal{P}[a, b]$, then

$$U(P, f, \alpha) = f(b)\alpha(b) - f(a)\alpha(a) - L(P, \alpha, f)$$

and

$$L(P, f, \alpha) = f(b)\alpha(b) - f(a)\alpha(a) - U(P, \alpha, f).$$

Proof   Let $P = \{x_i\}_{i=0}^n$ be in $\mathcal{P}[a, b]$. Since f is monotone increasing on $[a, b]$,

$$U(P, f, \alpha) = \sum_{i=1}^n f(x_i)\Delta\alpha_i$$

$$= f(x_1)\left[\alpha(x_1) - \alpha(x_0)\right] + f(x_2)\left[\alpha(x_2) - \alpha(x_1)\right] + \cdots +$$
$$f(x_{n-1})\left[\alpha(x_{n-1}) - \alpha(x_{n-2})\right] + f(x_n)\left[\alpha(x_n) - \alpha(x_{n-1})\right]$$

$$= f(x_n)\alpha(x_n) + \alpha(x_{n-1})\left[f(x_{n-1}) - f(x_n)\right] + \cdots +$$
$$\alpha(x_1)\left[f(x_1) - f(x_2)\right] - f(x_1)\alpha(x_0)$$

$$= f(x_n)\alpha(x_n) - f(x_0)\alpha(x_0) + \alpha(x_0)\left[f(x_0) - f(x_1)\right] +$$
$$\sum_{i=1}^{n-1}\alpha(x_i)\left[f(x_i) - f(x_{i+1})\right]$$

$$= f(b)\alpha(b) - f(a)\alpha(a) - \sum_{i=0}^{n-1}\alpha(x_i)\left[f(x_{i+1}) - f(x_i)\right]$$

$$= f(b)\alpha(b) - f(a)\alpha(a) - L(P, \alpha, f)$$

since α is monotone increasing.

The second equation follows by interchanging f and α in the first equation.   ■

Theorem 10.9   Suppose that f and α are both monotone increasing on $[a, b]$. Then f is in $\mathcal{R}(\alpha)$ on $[a, b]$ if and only if α is in $\mathcal{R}(f)$ on $[a, b]$, and in this case

$$\int_a^b f \, d\alpha = f(b)\alpha(b) - f(a)\alpha(a) - \int_a^b \alpha \, df.$$

(This is the *integration by parts formula*.)

Proof   Let $A = f(b)\alpha(b) - f(a)\alpha(a)$ and rewrite the result of Lemma 10.1 as

$$U(P, f, \alpha) = A - L(P, \alpha, f)$$

and

$$L(P, f, \alpha) = A - U(P, \alpha, f),$$

where P is in $\mathcal{P}[a, b]$. Then

$$U(P, f, \alpha) - L(P, f, \alpha) = U(P, \alpha, f) - L(P, \alpha, f),$$

and so f is in $\mathcal{R}(\alpha)$ on $[a, b]$ if and only if α is in $\mathcal{R}(f)$ on $[a, b]$ by Theorem 10.2.

Let f be in $\mathcal{R}(\alpha)$ on $[a, b]$, and let $\varepsilon > 0$. Since $\int_a^b \alpha \, df = \sup\{L(P, \alpha, f) : P \in \mathcal{P}[a, b]\}$, there exists a partition P in $\mathcal{P}[a, b]$ such that

$$L(P, \alpha, f) > \int_a^b \alpha \, df - \varepsilon.$$

Then

$$\int_a^b \alpha \, df < L(P, \alpha, f) + \varepsilon$$
$$= A - U(P, f, \alpha) + \varepsilon,$$

and so

$$U(P, f, \alpha) < A - \int_a^b \alpha \, df + \varepsilon.$$

Since ε is arbitrary,

$$\int_a^b f \, d\alpha \le A - \int_a^b \alpha \, df.$$

The reverse inequality is obtained similarly. ∎

Example 10.5   By Example 10.3,

$$\int_0^n x \, d[x] = \sum_{k=1}^n k = \frac{n(n + 1)}{2}.$$

By Theorem 10.9,

$$\int_0^n x \, d[x] = n[n] - 0[0] - \int_0^n [x] \, dx$$

$$= n^2 - \sum_{k=0}^{n-1} \int_k^{k+1} [x] \, dx$$

$$= n^2 - \sum_{k=0}^{n-1} k$$

$$= n^2 - \frac{(n - 1)(n)}{2}$$

$$= \frac{n(n + 1)}{2}.$$

≡ Exercises ≡

In the following exercises α is monotone increasing on $[a, b]$.

1. If f is continuous and nonnegative on $[a, b]$ with $\alpha(a) < \alpha(b)$, show that there is a c in $[a, b]$ such that

$$f(c) = \left[\frac{1}{\alpha(b) - \alpha(a)} \int_a^b f^2 \, d\alpha \right]^{1/2}.$$

   (This is the analogue of Exercise 8 in Section 6.4.)

2. If f is continuous on $[a, b]$ and g is in $\mathcal{R}(\alpha)$ on $[a, b]$ with g nonnegative, show that there is a c in $[a, b]$ such that $\int_a^b fg \, d\alpha = f(c) \int_a^b g \, d\alpha$. (This is the analogue of Exercise 9 in Section 6.4.)

3. Let f be continuous on $[a, b]$. Let β be another monotone increasing function on $[a, b]$ such that $\beta = \alpha$ on $[a, b]$ except at the interior point c of (a, b). Show that $\int_a^b f \, d\alpha = \int_a^b f \, d\beta$. Also show by example that the result may fail if c is an endpoint of $[a, b]$.

4. Complete the proof of Theorem 10.7 when f and α are both discontinuous from the right at c, where $a \le c < b$.

5. Let α be continuous on $[a, b]$ and suppose that $\int_a^b f \, d\alpha = 0$ for all monotone functions f on $[a, b]$. Show that α is a constant function on $[a, b]$.

6. Let f be monotone and α continuous on $[a, b]$. Show that there is a c in $[a, b]$ such that

$$\int_a^b f \, d\alpha = f(a) \left[\alpha(c) - \alpha(a) \right] + f(b) \left[\alpha(b) - \alpha(c) \right].$$

   [*Hint:* For f monotone increasing, use Theorems 10.9 and 10.5].

7. By considering different functions f, show that $\int_0^2 f \, d[x]$ sometimes satisfies the equation in Exercise 6 and sometimes does not.

8. Use Theorem 10.9 to calculate $\int_0^n [x] \, d(x^2)$ where n is in $\mathbb{N}$.

10.3 Riemann-Stieltjes Sums

The first part of this section is analogous to Section 6.2. However, we do not obtain an exact analogue of Theorem 6.3, which states that a bounded function is Riemann integrable on $[a, b]$ if and only if the limit of the Riemann sums exists in $\mathbb{R}$. In the second part of this section, we derive some useful properties of the Riemann-Stieltjes integral.

In this section we still have that f is a bounded function and α is a monotone increasing function on $[a, b]$ with $a < b$.

Sums

Definition 10.4   Let $P = \{x_i\}_{i=0}^n$ be in $\mathcal{P}[a, b]$, and for each $i = 1, 2, \ldots, n$ choose points $t_1, t_2, \ldots, t_n$ with $x_{i-1} \le t_i \le x_i$. A *Riemann-Stieltjes sum* for f with respect to α and P, denoted by $S(P, f, \alpha)$, is given by

$$S(P, f, \alpha) = \sum_{i=1}^n f(t_i) \Delta \alpha_i.$$

The definition above corresponds to Definition 6.6. The Remark following Definition 6.6, with Riemann sums replaced by Riemann-Stieltjes sums, is still valid in our present setting; and we always have

$$L(P, f, \alpha) \le S(P, f, \alpha) \le U(P, f, \alpha).$$

Recall from Definition 6.5 that for $P = \{x_i\}_{i=0}^n$ in $\mathcal{P}[a, b]$, the mesh of P is given by $\|P\| = \max\{\Delta x_i : i = 1, 2, \ldots, n\}$. Analogous to Definition 6.7, we have the following.

Definition 10.5   Let I be in $\mathbb{R}$. Then $\lim_{\|P\| \to 0} S(P, f, \alpha) = I$ if for every $\varepsilon > 0$ there is a $\delta > 0$ such that if $P = \{x_i\}_{i=0}^n$ is in $\mathcal{P}[a, b]$ with $\|P\| < \delta$, then $|S(P, f, \alpha) - I| < \varepsilon$ for all possible choices of the t_i's in $[x_{i-1}, x_i]$.

Theorem 10.10   (1) If $\lim_{\|P\| \to 0} S(P, f, \alpha)$ exists in $\mathbb{R}$, then f is in $\mathcal{R}(\alpha)$ on $[a, b]$ and

$$\int_a^b f \, d\alpha = \lim_{\|P\| \to 0} S(P, f, \alpha). \tag{4}$$

If (2) f is continuous on $[a, b]$ or (3) f is in $\mathcal{R}(\alpha)$ and α is continuous on $[a, b]$, then (4) holds.

Proof

1. If α is a constant function, then every $S(P, f, \alpha)$ is zero and part 1 follows from Exercise 2 in Section 10.1. Assume $\alpha(a) < \alpha(b)$, $\lim_{\|P\| \to 0} S(P, f, \alpha) = I$ is in $\mathbb{R}$, and let $\varepsilon > 0$. By Definition 10.5, there is a $\delta > 0$ such that if P is in $\mathcal{P}$ with $\|P\| < \delta$, then $|S(P, f, \alpha) - I| < \varepsilon/4$ for all possible choices of the t_i's. Fix one $P = \{x_i\}_{i=0}^n$ in $\mathcal{P}$ with $\|P\| < \delta$. For each $i = 1, 2, \ldots, n$, since $M_i = \sup\{f(x) : x \in [x_{i-1}, x_i]\}$, there is a t_i in $[x_{i-1}, x_i]$ with

$$f(t_i) > M_i - \frac{\varepsilon}{4 [\alpha(b) - \alpha(a)]}.$$

The remainder of the proof of part 1 is identical to the second half of the proof of Theorem 6.3 with Δx_i, $b - a$, $U(P, f)$, and $L(P, f)$ replaced by $\Delta \alpha_i$, $\alpha(b) - \alpha(a)$, $U(P, f, \alpha)$, and $L(P, f, \alpha)$, respectively.

2. Assume that f is continuous on $[a, b]$ and let $\varepsilon > 0$. By Theorem 10.4, f is in $\mathcal{R}(\alpha)$ on $[a, b]$. Since $\int_a^b f \, d\alpha$ and each $S(P, f, \alpha)$ are in the closed interval $[L(P, f, \alpha), U(P, f, \alpha)]$ for all P in $\mathcal{P}$, to show (4) it suffices to find a $\delta > 0$ such that if P is in $\mathcal{P}$ with $\|P\| < \delta$, then $U(P, f, \alpha) - L(P, f, \alpha) < \varepsilon$. But this is precisely what is done in the proof of Theorem 10.4.

3. Assume that f is in $\mathcal{R}(\alpha)$ and α is continuous on $[a, b]$. Let $B > 0$ with $|f(x)| \leq B$ for all x in $[a, b]$, and let $\varepsilon > 0$. Since $\int_a^b f \, d\alpha = \overline{\int_a^b} f \, d\alpha$, there is a $P' = \{x_i\}_{i=0}^n$ in $\mathcal{P}$ with

$$U(P', f, \alpha) < \int_a^b f \, d\alpha + \frac{\varepsilon}{2}.$$

Since α is uniformly continuous on $[a, b]$, there is a $\delta_1 > 0$ such that if x and y are in $[a, b]$ with $|x - y| < \delta_1$, then $|\alpha(x) - \alpha(y)| < \varepsilon/(4nB)$. Hence, if $\|P\| < \delta_1$, then each $\Delta \alpha_i < \varepsilon/(4nB)$. With the usual modifications as in part 1 above, the remainder of the proof is identical to the first part of the proof of Theorem 6.3. ∎

Example 10.6 The purpose of this example is to show that the continuity assumptions made in Theorem 10.10 are necessary. Let

$$f(x) = \begin{cases} 0 & \text{if } 0 \leq x \leq 1 \\ 1 & \text{if } 1 < x \leq 2 \end{cases}$$

and

$$\alpha(x) = \begin{cases} 0 & \text{if } 0 \leq x < 1 \\ 1 & \text{if } 1 \leq x \leq 2. \end{cases}$$

Since f is continuous from the left at 1, Example 10.2 implies that f is in $\mathcal{R}(\alpha)$ on $[0, 2]$ and $\int_0^2 f \, d\alpha = f(1)(1 - 0) = 0$.

We now show that the limit of $S(P, f, \alpha)$ does not exist. Let $P = \{x_i\}_{i=0}^n$ be in $\mathcal{P}[0, 2]$ and let t_i be in $[x_{i-1}, x_i]$ for $i = 1, 2, \ldots, n$. If 1 is in P, say $1 = x_j$, then $S(P, f, \alpha) = \sum_{i=1}^n f(t_i) \Delta \alpha_i = f(t_j) = 0$ since $x_{j-1} \leq t_j \leq 1$. But if 1 is not in P, say $x_{j-1} < 1 < x_j$, then

$$S(P, f, \alpha) = f(t_j) = \begin{cases} 0 & \text{if } t_j \leq 1 \\ 1 & \text{if } t_j > 1. \end{cases}$$

Since Definition 10.5 must hold for all possible t_i's, $\lim_{\|P\| \to 0} S(P, f, \alpha)$ does not exist. Thus, taking refinements and taking the limit as $\|P\| \to 0$ are not equivalent for Riemann-Stieltjes integrals.

Further Properties of the Integral

In Example 10.5, the integration by parts formula allowed us to reduce a Riemann-Stieltjes integral to a Riemann integral. The following theorem permits us to make the same type of reduction in a manner the reader may have already been wondering about, namely, when can we replace $d\alpha(x)$ in $\int_a^b f(x)\,d\alpha(x)$ by $\alpha'(x)dx$, where α' is the derivative of α?

Theorem 10.11 If f is Riemann integrable on $[a, b]$ and α' is continuous on $[a, b]$, then f is in $\mathcal{R}(\alpha)$ on $[a, b]$ and

$$\int_a^b f(x)\,d\alpha(x) = \int_a^b f(x)\alpha'(x)dx.$$

Proof   Since α' is continuous on $[a, b]$, α' is in $\mathcal{R}[a, b]$ and so $f\alpha'$ is in $\mathcal{R}[a, b]$ by Corollary 6.1 (or Corollary 10.1). Also,

$$\lim_{\|P\|\to 0} S(P, f\alpha') = \int_a^b f\alpha'$$

by Theorem 6.3 (or part 3 of Theorem 10.10).

Let $\varepsilon > 0$. Choose $\delta_1 > 0$ such that if P is in $\mathcal{P}[a, b]$ with $\|P\| < \delta_1$, then

$$\left| S(P, f\alpha') - \int_a^b f\alpha' \right| < \frac{\varepsilon}{2}. \tag{5}$$

[The idea is to show that $S(P, f, \alpha)$ is "close to" $S(P, f\alpha')$ for appropriate P.]

Let $P = \{x_i\}_{i=0}^n$ be in $\mathcal{P}[a, b]$ and let t_i be in $[x_{i-1}, x_i]$ for $i = 1, 2, \ldots, n$. Then

$$S(P, f\alpha') = \sum_{i=1}^n f(t_i)\alpha'(t_i)\Delta x_i$$

and

$$S(P, f, \alpha) = \sum_{i=1}^n f(t_i)\Delta\alpha_i.$$

By the Mean Value Theorem (Theorem 5.5) applied to α on $[x_{i-1}, x_i]$,

$$\Delta\alpha_i = \alpha'(c_i)\Delta x_i,$$

where c_i is in (x_{i-1}, x_i) for each $i = 1, 2, \ldots, n$. Therefore,

$$S(P, f, \alpha) - S(P, f\alpha') = \sum_{i=1}^n f(t_i)\left[\alpha'(c_i) - \alpha'(t_i)\right]\Delta x_i. \tag{6}$$

Let $M > 0$ with $|f(x)| \le M$ for all x in $[a, b]$. Since α' is uniformly continuous on $[a, b]$, there is a $\delta_2 > 0$ such that if x and y are in $[a, b]$ with $|x - y| < \delta_2$, then $\left|\alpha'(x) - \alpha'(y)\right| < \varepsilon/[2M(b - a)]$. If $P = \{x_i\}_{i=0}^n$ is in $\mathcal{P}[a, b]$ with $\|P\| < \delta_2$, then (6) implies that

$$\left| S(P, f, \alpha) - S(P, f\alpha') \right| < M\frac{\varepsilon}{2M(b - a)}\sum_{i=1}^n \Delta x_i = \frac{\varepsilon}{2}.$$

Combining this last inequality with (5) gives us that

$$\left| S(P, f, \alpha) - \int_a^b f\alpha' \right| < \varepsilon$$

whenever $\|P\| < \delta = \min\{\delta_1, \delta_2\}$. Therefore,

$$\lim_{\|P\|\to 0} S(P, f, \alpha) = \int_a^b f\alpha',$$

and this completes the proof by appealing to part 1 of Theorem 10.10. ∎

A theorem similar to Theorem 10.11 is given in Exercise 5.

Remark   Theorem 10.11 may fail if we assume only that α' is Riemann integrable on $[a, b]$. If $\alpha(x) = [x]$ on $[0, 1]$, then $\alpha'(x) = 0$ for $0 \le x < 1$ and $\alpha'(1)$ does not exist. By Theorem 6.5, defining $\alpha'(1)$ arbitrarily, α' is in $\mathcal{R}[0, 1]$. However, for any continuous function f on $[0, 1]$, $\int_0^1 f \, d\alpha = f(1)$ but $\int_0^1 f\alpha' = 0$.

Remark   One way to generalize what we have been doing would be to allow our integrator α to be an arbitrary function g defined on $[a, b]$, and to define $\int_a^b f \, dg$ as the limit of Riemann-Stieltjes sums, $S(P, f, g)$, as the mesh of P tends to zero, whenever this limit exists. Our proof of Theorem 10.11 still holds in this generalized setting.

Example 10.7
$$\int_0^{\pi/2} x^2 \, d(\sin x) = \int_0^{\pi/2} x^2 \cos x \, dx \qquad \text{(Theorem 10.11)}$$

$$= \frac{\pi^2}{4} - 2 \int_0^{\pi/2} x \sin x \, dx \qquad \text{(integration by parts)}$$

$$= \frac{\pi^2}{4} - 2 \qquad \text{(integration by parts).}$$

Example 10.8   By Theorem 10.11,
$$\int_0^1 x^2 \, d(x^2) = 2 \int_0^1 x^3 \, dx = \left. \frac{x^4}{2} \right|_0^1 = \frac{1}{2}.$$
Using integration by parts,
$$\int_0^1 x^2 \, d(x^2) = 1 - \int_0^1 x^2 \, d(x^2),$$
and so $2 \int_0^1 x^2 \, d(x^2) = 1$ or $\int_0^1 x^2 \, d(x^2) = \dfrac{1}{2}$.

We end this section with a proposition that shows how to substitute or change variables in a Riemann-Stieltjes integral.

Proposition 10.2   Suppose that f and β are continuous on $[a, b]$, β is strictly increasing on $[a, b]$, and α is the inverse function of β. Then
$$\int_{\beta(a)}^{\beta(b)} f(\alpha(y)) \, d\alpha(y) = \int_a^b f(x) \, dx. \qquad (7)$$
[Formally, this is obtained by letting $x = \alpha(y)$.]

Proof   By Exercise 8, α is strictly increasing and continuous on $[\beta(a), \beta(b)]$. Hence, both integrals in (7) exist and (4) of Theorem 10.10 holds.

Let $P = \{x_i\}_{i=0}^n$ be in $\mathcal{P}[a, b]$ and set $y_i = \beta(x_i)$ for $i = 0, 1, 2, \ldots, n$. Then $Q = \{y_i\}_{i=0}^n$ is in $\mathcal{P}[\beta(a), \beta(b)]$. Also,
$$\sum_{i=1}^n f(\alpha(y_i)) \left[\alpha(y_i) - \alpha(y_{i-1})\right] = \sum_{i=1}^n f(x_i) \left[x_i - x_{i-1}\right]. \qquad (8)$$
Since β is uniformly continuous on $[a, b]$, $\|P\| \to 0$ implies $\|Q\| \to 0$.

Therefore, as $\|P\| \to 0$, the two members of (8) tend to the corresponding members of (7).   ■

Example 10.9   We reconsider $\int_0^1 y^2 \, d(y^2)$ of Example 10.8. Letting $f(x) = x$, $\beta(x) = \sqrt{x}$, $\alpha(y) = y^2$ on $[0, 1]$ (or simply $x = y^2$), Proposition 10.2 implies that

$$\int_0^1 y^2 \, d(y^2) = \int_0^1 x \, dx = \left. \frac{x^2}{2} \right|_0^1 = \frac{1}{2}.$$

≡ Exercises ≡

1. Use Theorem 10.11 to calculate $\int_0^n [x] \, d(x^2)$ where n is in $\mathbb{N}$. Compare with Exercise 8 in Section 10.2.

2. Evaluate the following.
    (a) $\int_0^{\pi/2} x \, d(\sin x)$ (b) $\int_0^{\pi/2} \sin x \, d(\sin x)$

3. Evaluate the following.
    (a) $\int_0^1 x^3 \, d(x^2)$ (b) $\int_0^1 x^3 \, d(x^3)$

4. Evaluate the following.
    (a) $\int_{-\pi/4}^{\pi/4} \cos^2 x \, d(\tan x)$ (b) $\int_{-\pi/4}^{\pi/4} \cos^2 x \, d[x]$

5. Show that if f is in $\mathcal{R}(\alpha)$ on $[a, b]$ and α' is continuous on $[a, b]$, then $f\alpha'$ is Riemann integrable on $[a, b]$ and

$$\int_a^b f(x) d\alpha(x) = \int_a^b f(x)\alpha'(x) \, dx.$$

6. The purpose of this exercise is to give a proof of the integration by parts formula under the generalization indicated in the Remark preceding Example 10.7. Suppose that f is in $\mathcal{R}(\alpha)$ on $[a, b]$ and that (4) of Theorem 10.10 holds. Use part 1 of Theorem 10.10 to show that α is in $\mathcal{R}(f)$ and

$$\int_a^b f \, d\alpha = f(b)\alpha(b) - f(a)\alpha(a) - \int_a^b \alpha \, df.$$

7. Let

$$f(x) = \begin{cases} 0 & \text{if} \quad -1 \le x < 0 \\ 1 & \text{if} \quad 0 \le x \le 1 \end{cases}$$

and

$$\alpha(x) = \begin{cases} 0 & \text{if} \quad -1 \le x \le 0 \\ 1 & \text{if} \quad 0 < x \le 1. \end{cases}$$

Show that f is in $\mathcal{R}(\alpha)$ on $[-1, 1]$ but $\lim\limits_{\|P\| \to 0} S(P, f, \alpha)$ does not exist.

8. Let β be a strictly increasing and continuous function on $[a, b]$. Show that $\alpha = \beta^{-1}$ is strictly increasing and continuous on $[\beta(a), \beta(b)]$. [*Hint:* See Corollary 4.3.]

10.4 Functions of Bounded Variation

In this section we obtain the standard results concerning functions of bounded variation that are needed (in the next section) to extend our theory of Riemann-Stieltjes integration to integrators of bounded variation. Our main results are Corollaries 10.2 and 10.4.

Basic Results and Examples

For a function $f : [a, b] \to \mathbb{R}$ where $a < b$ and a partition $P = \{x_i\}_{i=0}^n$ of $[a, b]$, we let $\Delta f_i = f(x_i) - f(x_{i-1})$ for $i = 1, 2, \ldots, n$.

Definition 10.6 A function $f : [a, b] \to \mathbb{R}$ is of *bounded variation* on $[a, b]$ if there is an $M > 0$ such that $\sum_{i=1}^n |\Delta f_i| \leq M$ for all partitions $P = \{x_i\}_{i=0}^n$ of $[a, b]$. The *total variation* of a function f on $[a, b]$, denoted by $V_f(a, b)$, is given by

$$V_f(a, b) = \sup \left\{ \sum_{i=1}^n |\Delta f_i| : P = \{x_i\}_{i=0}^n \in \mathcal{P}[a, b] \right\}.$$

We also define f to be of bounded variation on $[a, a]$ with $V_f(a, a) = 0$.

Note that f is of bounded variation on $[a, b]$ if and only if $V_f(a, b)$ is finite. When no confusion can arise, we will use V_f for $V_f(a, b)$.

Example 10.10 If f is monotone on $[a, b]$, then f is of bounded variation on $[a, b]$ and $V_f(a, b) = |f(b) - f(a)|$. To see this, let $P = \{x_i\}_{i=0}^n$ be in $\mathcal{P}[a, b]$. If f is monotone increasing, then

$$\sum_{i=1}^n |\Delta f_i| = \sum_{i=1}^n \Delta f_i = \sum_{i=1}^n \left[f(x_i) - f(x_{i-1}) \right] = f(b) - f(a);$$

while if f is monotone decreasing, then

$$\sum_{i=1}^n |\Delta f_i| = -\sum_{i=1}^n \Delta f_i = f(a) - f(b).$$

Proposition 10.3 If f is of bounded variation on $[a, b]$, then f is bounded on $[a, b]$.

Proof Suppose that $a < x < b$ and consider the partition $P = \{a, x, b\}$. Then

$$|f(x)| - |f(a)| \leq |f(x) - f(a)|$$
$$\leq |f(x) - f(a)| + |f(b) - f(x)|$$
$$\leq V_f(a, b),$$

and so $|f(x)| \leq |f(a)| + V_f(a, b)$ for all x in (a, b). Letting $P = \{a, b\}$ gives $|f(b)| - |f(a)| \leq V_f(a, b)$ and so $|f(x)| \leq |f(a)| + V_f(a, b)$ for all x in $[a, b]$. ∎

Example 10.11   Our purpose here is to provide an example of a continuous bounded function that is not of bounded variation. Let

$$f(x) = \begin{cases} x \sin \dfrac{\pi}{x} & \text{if } 0 < x \le 1 \\ 0 & \text{if } x = 0. \end{cases}$$

Then f is continuous and hence bounded on $[0, 1]$. For n in $\mathbb{N}, n > 2$, consider the partition

$$P = \left\{ 0, \frac{2}{2n}, \frac{2}{2n-1}, \cdots, \frac{2}{6}, \frac{2}{5}, \frac{2}{4}, \frac{2}{3}, 1 \right\}.$$

Then

$$\left| f(1) - f\left(\frac{2}{3}\right) \right| + \left| f\left(\frac{2}{3}\right) - f\left(\frac{2}{4}\right) \right| + \cdots +$$

$$\left| f\left(\frac{2}{2n-1}\right) - f\left(\frac{2}{2n}\right) \right| + \left| f\left(\frac{2}{2n}\right) - f(0) \right|$$

$$= \frac{2}{3} + \frac{2}{3} + \frac{2}{5} + \frac{2}{5} + \cdots + \frac{2}{2n-1} + \frac{2}{2n-1}$$

$$> \frac{1}{2} + \frac{1}{3} + \frac{1}{4} + \frac{1}{5} + \cdots + \frac{1}{2n-2} + \frac{1}{2n-1}.$$

Since $\overset{\infty}{\underset{k=2}{\Sigma}} 1/k$ diverges, the sum above can be made arbitrarily large, and so f is not of bounded variation on $[0, 1]$.

By Examples 10.10 and 10.11, continuity is neither necessary nor sufficient for a function to be of bounded variation. Our next result gives a sufficient condition for a continuous function to be of bounded variation.

Theorem 10.12   If f is continuous on $[a, b]$ and f' exists and is bounded on (a, b), then f is of bounded variation on $[a, b]$. (Hence, polynomials, $\sin x$, and $\cos x$ are all of bounded variation on $[a, b]$.)

Proof   Let $B > 0$ with $\left| f'(x) \right| \le B$ for all x in (a, b). Let $P = \{x_i\}_{i=0}^{n}$ be in $\mathcal{P}[a, b]$. By the Mean Value Theorem there is a c_i in (x_{i-1}, x_i) such that

$$\Delta f_i = f(x_i) - f(x_{i-1}) = f'(c_i)(x_i - x_{i-1})$$

for $i = 1, 2, \ldots, n$. Hence,

$$\sum_{i=1}^{n} |\Delta f_i| = \sum_{i=1}^{n} \left| f'(c_i) \right| (x_i - x_{i-1}) \le B(b - a),$$

and so f is of bounded variation on $[a, b]$.   ■

It should be observed that the function in Example 10.11 does not have a bounded derivative. By considering $f(x) = \sqrt{x}$ on $[0, 1]$, it is clear that a function of bounded variation need not have a bounded derivative.

Example 10.12   Let

$$f(x) = \begin{cases} x^2 \sin \dfrac{\pi}{x} & \text{if } 0 < x \le 1 \\ 0 & \text{if } x = 0. \end{cases}$$

Then $f'(x) = 2x \sin(\pi/x) - \pi \cos(\pi/x)$ if $x \neq 0$, and $f'(0) = 0$. Hence, $|f'(x)| \leq 2 + \pi$ for all x in $[0, 1]$, and so f is of bounded variation on $[0, 1]$. Thus, a function may oscillate infinitely many times and still be of bounded variation.

Although the sum or product of two monotone functions need not be monotone, our next result shows that the set of functions of bounded variation is closed under the operations of addition, subtraction, and multiplication.

Theorem 10.13 If f and g are of bounded variation on $[a, b]$, then $f \pm g$ and fg are of bounded variation on $[a, b]$.

Proof   Let $P = \{x_i\}_{i=0}^n$ be in $\mathcal{P}[a, b]$. Then

$$\Delta(f + g)_i = (f + g)(x_i) - (f + g)(x_{i-1}) = \Delta f_i + \Delta g_i,$$

and so

$$\sum_{i=1}^n |\Delta(f + g)_i| \leq \sum_{i=1}^n |\Delta f_i| + \sum_{i=1}^n |\Delta g_i|$$

$$\leq V_f + V_g.$$

Hence, $V_{f+g} \leq V_f + V_g$, and so $f + g$ is of bounded variation on $[a, b]$.

By Proposition 10.3 there exist A and B such that $|f(x)| \leq A$ and $|g(x)| \leq B$ for all x in $[a, b]$. For P as above,

$$|\Delta(fg)_i| = |f(x_i)g(x_i) - f(x_{i-1})g(x_{i-1})|$$

$$\leq |f(x_i)g(x_i) - f(x_i)g(x_{i-1})| + |f(x_i)g(x_{i-1}) - f(x_{i-1})g(x_{i-1})|$$

$$\leq A|\Delta g_i| + B|\Delta f_i|,$$

and so

$$\sum_{i=1}^n |\Delta(fg)_i| \leq A \sum_{i=1}^n |\Delta g_i| + B \sum_{i=1}^n |\Delta f_i|$$

$$\leq AV_g + BV_f.$$

Therefore, fg is of bounded variation on $[a, b]$ and $V_{fg} \leq AV_g + BV_f$.

Since a constant function is of bounded variation, $-g = (-1)g$ is of bounded variation, and so $f - g = f + (-g)$ is of bounded variation.   ∎

In Exercise 4 we will consider the quotient of two functions of bounded variation, which may or may not be of bounded variation. We next derive an additive property for V_f.

Proposition 10.4 If f is of bounded variation on $[a, b]$ and c is in $[a, b]$, then f is of bounded variation on $[a, c]$ and on $[c, b]$ and

$$V_f(a, b) = V_f(a, c) + V_f(c, b).$$

Proof   The result is clear if $c = a$ or b. Let $a < c < b$. If $P_1 = \{x_i\}_{i=0}^n$ is in $\mathcal{P}[a, c]$ and $P_2 = \{y_i\}_{i=0}^m$ is in $\mathcal{P}[c, b]$, then $P = P_1 \cup P_2$ is in $\mathcal{P}[a, b]$ and so

$$\sum_{i=1}^n |f(x_i) - f(x_{i-1})| + \sum_{i=1}^m |f(y_i) - f(y_{i-1})| \leq V_f(a, b). \tag{9}$$

Since each sum in (9) is bounded by $V_f(a, b)$ for all partitions of $[a, c]$ and $[c, b]$, f is of bounded variation on $[a, c]$ and on $[c, b]$, and

$$V_f(a, c) + V_f(c, b) \le V_f(a, b).$$

To obtain the reverse inequality, let $\varepsilon > 0$. By the definition of $V_f(a, b)$, there is a partition $P = \{x_i\}_{i=0}^n$ of $[a, b]$ such that

$$V_f(a, b) - \varepsilon < \sum_{i=1}^n |f(x_i) - f(x_{i-1})|. \tag{10}$$

We can assume that $c = x_j$ for some j in $\{1, 2, \ldots, n-1\}$, because otherwise we could add c to P and (10) would still hold for $P \cup \{c\}$. [If c were in (x_{k-1}, x_k), then

$$|f(x_k) - f(x_{k-1})| \le |f(x_k) - f(c)| + |f(c) - f(x_{k-1})|.]$$

Since

$$\sum_{i=1}^n |f(x_i) - f(x_{i-1})| = \sum_{i=1}^j |f(x_i) - f(x_{i-1})| + \sum_{i=j+1}^n |f(x_i) - f(x_{i-1})|$$

$$\le V_f(a, c) + V_f(c, b),$$

we have

$$V_f(a, b) - \varepsilon < V_f(a, c) + V_f(c, b).$$

Since ε is arbitrary, $V_f(a, b) \le V_f(a, c) + V_f(c, b)$.   ■

Decomposition

> **Definition 10.7** Let f be of bounded variation on $[a, b]$. The *total variation function* of f, denoted by v, is defined by
>
> $$v(x) = V_f(a, x)$$
>
> for x in $[a, b]$.

Example 10.13   Define $f : [0, 2] \to \mathbb{R}$ by $f(x) = x - [x]$, where $[x]$ denotes the largest integer less than or equal to x (Figure 10.2).

For x in $[0, 1)$, $v(x) = f(x) - f(0) = x - [x] = x$ since f is monotone increasing on $[0, 1)$. To find $v(1) = V_f(0, 1)$, let $P = \{x_i\}_{i=0}^n$ be in $\mathcal{P}[0, 1]$. Then

$$\sum_{i=1}^n |\Delta f_i| = f(x_{n-1}) - f(0) + |f(1) - f(x_{n-1})|$$

$$= 2f(x_{n-1})$$

$$= 2x_{n-1},$$

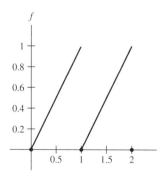

Figure 10.2

and so $v(1) = \sup\{\sum_{i=1}^n |\Delta f_i| : P \in \mathcal{P}[0, 1]\} = 2(1) = 2$.

For $1 < x < 2$, by Proposition 10.4,

$$v(x) = V_f(0, 1) + V_f(1, x)$$
$$= 2 + f(x) - f(1)$$
$$= 2 + x - [x]$$
$$= 2 + x - 1$$
$$= 1 + x$$

since f is monotone increasing on $[1, 2)$.

For $P = \{x_i\}_{i=0}^n$ in $\mathcal{P}[1, 2]$,

$$\sum_{i=1}^n |\Delta f_i| = f(x_{n-1}) - f(1) + |f(2) - f(x_{n-1})|$$
$$= 2f(x_{n-1})$$
$$= 2(x_{n-1} - [x_{n-1}])$$
$$= 2(x_{n-1} - 1),$$

and so $V_f(1, 2) = 2(2-1) = 2$. Hence $v(2) = V_f(0, 1) + V_f(1, 2) = 4$. Thus

$$v(x) = \begin{cases} x & \text{if } 0 \le x < 1 \\ x+1 & \text{if } 1 \le x < 2 \\ 4 & \text{if } x = 2. \end{cases}$$

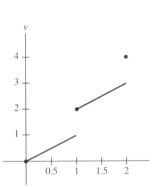

Figure 10.3

In Example 10.13, note that v is monotone increasing and that v is continuous wherever f is continuous. Our next two theorems show that this is always the case.

Theorem 10.14 If f is of bounded variation on $[a, b]$, then both v and $v - f$ are monotone increasing on $[a, b]$.

Proof For $a \le x < y \le b$, Proposition 10.4 implies that $v(y) = v(x) + V_f(x, y)$ or, equivalently, $v(y) - v(x) = V_f(x, y) \ge 0$. Hence, $v(x) \le v(y)$ and so v is monotone increasing on $[a, b]$. Also,

$$(v - f)(y) - (v - f)(x) = [v(y) - v(x)] - [f(y) - f(x)]$$
$$= V_f(x, y) - [f(y) - f(x)].$$

From the definition of $V_f(x, y)$, using $P = \{x, y\}$ as a partition of $[x, y]$, $|f(y) - f(x)| \le V_f(x, y)$, and so $(v - f)(y) - (v - f)(x) \ge 0$. Thus, $v - f$ is monotone increasing on $[a, b]$. ∎

Corollary 10.2 A function f is of bounded variation on $[a, b]$ if and only if f can be expressed as the difference of two monotone increasing functions on $[a, b]$.

Proof If f is of bounded variation on $[a, b]$, then $f = v - (v - f)$. The converse follows from Example 10.10 and Theorem 10.13. ∎

Remark The decomposition of a function of bounded variation as the difference of two monotone increasing functions is not unique. If g is monotone increasing on $[a, b]$, then $v + g$ and $v - f + g$ are monotone increasing on $[a, b]$ and $f = (v + g) - (v - f + g)$.

Corollary 10.3   If f is of bounded variation on $[a, b]$, then $f(x+)$ exists for $a \leq x < b$, $f(x-)$ exists for $a < x \leq b$, and the set of discontinuities of f is countable.

Proof   This follows from Corollary 10.2 and the corresponding results for monotone functions (Theorems 4.8 and 4.9). $\blacksquare$

Theorem 10.15   Let f be of bounded variation on $[a, b]$ and let c be in $[a, b]$. Then f is continuous at c if and only if v is continuous at c.

Proof   First suppose that $a \leq c < b$ and that f is continuous at c. We will show that v is continuous from the right at c. Let $\varepsilon > 0$. Then there exists a $\delta_1 > 0$ such that if x is in $[a, b]$ with $|x - c| < \delta_1$, then $|f(x) - f(c)| < \varepsilon/2$. By definition of $V_f(c, b)$, there is a $P = \{x_i\}_{i=0}^{n}$ in $\mathcal{P}[c, b]$ such that

$$V_f(c, b) - \frac{\varepsilon}{2} < \sum_{i=1}^{n} |f(x_i) - f(x_{i-1})|. \tag{11}$$

Let $\delta = \min\{\delta_1, x_1 - c\}$, and let $x > c$ with $x - c < \delta$. From (11) we have

$$V_f(c, b) - \frac{\varepsilon}{2} \leq |f(x) - f(c)| + |f(x_1) - f(x)| + \sum_{i=2}^{n} |f(x_i) - f(x_{i-1})|$$

$$< \frac{\varepsilon}{2} + V_f(x, b)$$

since $\{x\} \cup \{x_i\}_{i=1}^{n}$ is in $\mathcal{P}[x, b]$. By Proposition 10.4,

$$v(x) - v(c) = V_f(a, x) - V_f(a, c)$$
$$= V_f(c, x)$$
$$= V_f(c, b) - V_f(x, b)$$
$$< \varepsilon.$$

Hence, $0 \leq v(x) - v(c) < \varepsilon$ and so v is continuous from the right at c. A similar argument (see Exercise 6) shows that v is continuous from the left at c whenever f is continuous at c and $a < c \leq b$. Hence, v is continuous at c if f is continuous at c.

Now suppose that v is continuous at c and let $\varepsilon > 0$. Then there is a $\delta > 0$ such that if x is in $[a, b]$ with $|x - c| < \delta$, then $|v(x) - v(c)| < \varepsilon$. Let x be in $[a, b]$. If $x > c$ with $|x - c| < \delta$, then

$$|f(x) - f(c)| \leq V_f(c, x) = v(x) - v(c) < \varepsilon$$

since $\{c, x\}$ is in $\mathcal{P}[c, x]$; while if $x < c$ with $|x - c| < \delta$, then

$$|f(c) - f(x)| \leq V_f(x, c) = v(c) - v(x) < \varepsilon$$

since $\{x, c\}$ is in $\mathcal{P}[x, c]$. Therefore, f is continuous at c. $\blacksquare$

Corollary 10.4   Let f be continuous on $[a, b]$. Then f is of bounded variation on $[a, b]$ if and only if f can be expressed as the difference of two continuous monotone increasing functions.

Proof   The decomposition $f = v - (v - f)$ expresses f appropriately by Theorems 10.14 and 10.15. $\blacksquare$

In summary, we have shown that a function of bounded variation can be written as the difference of two monotone increasing functions, and that a continuous function of bounded variation can be written as the difference of two continuous monotone increasing functions. In either case, this decomposition is not unique. These results will allow us to extend our theory of Riemann-Stieltjes integration to integrators of bounded variation in the next section.

Exercises

1. (a) Show that $V_f(a, b) = 0$ if and only if f is a constant function on $[a, b]$.
   (b) Show that $|f|$ is of bounded variation on $[a, b]$ if f is of bounded variation on $[a, b]$.
   (c) Give an example showing that $|f|$ being of bounded variation need not imply that f is of bounded variation.

2. Show that
$$f(x) = \begin{cases} x \cos\left(\dfrac{\pi}{2x}\right) & \text{if } 0 < x \leq 1 \\ 0 & \text{if } x = 0 \end{cases}$$
is continuous but not of bounded variation on $[0, 1]$. [*Hint:* Let
$$P = \left\{ 0, \frac{1}{2n}, \frac{1}{2n-1}, \ldots, \frac{1}{3}, \frac{1}{2}, 1 \right\}.]$$

3. Show that
$$f(x) = \begin{cases} x^2 \cos\left(\dfrac{\pi}{2x}\right) & \text{if } 0 < x \leq 1 \\ 0 & \text{if } x = 0 \end{cases}$$
is of bounded variation on $[0, 1]$.

4. (a) Give an example of a function f of bounded variation such that $1/f$ is not of bounded variation.
   (b) Let f be of bounded variation on $[a, b]$ and assume that f is bounded away from 0 (that is, assume that there is an $\varepsilon > 0$ such that $|f(x)| \geq \varepsilon$ for all x in $[a, b]$). Show that $1/f$ is of bounded variation on $[a, b]$ and that $V_{1/f}(a, b) \leq (1/\varepsilon^2) V_f(a, b)$.

5. Find the total variation function v if
   (a) $f(x) = \sin x$ on $[0, 2\pi]$;
   (b) $f(x) = \begin{cases} x & \text{if } 0 \leq x < 1 \\ 2 & \text{if } 1 \leq x < 2 \\ 3 - x & \text{if } 2 \leq x \leq 3. \end{cases}$

6. Complete the proof of Theorem 10.15 by showing that if f is continuous at c where $a < c \leq b$, then v is continuous from the left at c.

7. Show that if f is of bounded variation on $[a, b]$, then f is Riemann integrable on $[a, b]$. Show by example that the converse is false.

8. Let f be of bounded variation on $[a, b]$ and suppose that the function $g = f$ except at a finite number of points in $[a, b]$. Show that g is of bounded variation on $[a, b]$. [*Hint:* First consider $g = f$ except at exactly one point in $[a, b]$.]

9. Let f be Riemann integrable on $[a, b]$ and let $F(x) = \int_a^x f$ for x in $[a, b]$. Show that F is of bounded variation on $[a, b]$ and

$$V_F(a, x) \le \int_a^x |f|$$

for all x in $[a, b]$.

10. A function f on $[a, b]$ satisfies a *uniform Lipschitz condition* of order $\alpha > 0$ on $[a, b]$ if there is an $M > 0$ such that $|f(x) - f(y)| \le M |x - y|^\alpha$ for all x and y in $[a, b]$.

    (a) If f is such a function, show that $\alpha > 1$ implies that f is a constant function, and $\alpha = 1$ implies that f is of bounded variation on $[a, b]$.

    (b) Give an example of a function of bounded variation that satisfies no uniform Lipschitz condition on $[a, b]$.

10.5 Integrators of Bounded Variation

In this section, f is a bounded function on $[a, b]$. Suppose that β, γ, β_1, and γ_1 are monotone increasing functions on $[a, b]$ with $\beta - \gamma = \beta_1 - \gamma_1$, and suppose that f is Riemann-Stieltjes integrable with respect to β, γ, β_1, and γ_1 on $[a, b]$. By part 2 of Proposition 10.1,

$$\int_a^b f \, d\beta + \int_a^b f \, d\gamma_1 = \int_a^b f \, d\beta_1 + \int_a^b f \, d\gamma$$

and so

$$\int_a^b f \, d\beta - \int_a^b f \, d\gamma = \int_a^b f \, d\beta_1 - \int_a^b f \, d\gamma_1. \tag{12}$$

Definition 10.8   Let α be of bounded variation on $[a, b]$, and suppose that $\alpha = \beta - \gamma$, where β and γ are monotone increasing functions on $[a, b]$ with f Riemann-Stieltjes integrable with respect to both β and γ on $[a, b]$. Then the *Riemann-Stieltjes integral of f with respect to α* on $[a, b]$ is defined by

$$\int_a^b f \, d\alpha = \int_a^b f \, d\beta - \int_a^b f \, d\gamma.$$

In this situation we say that f is *Riemann-Stieltjes integrable with respect to α* on $[a, b]$, and we denote this by: f is in $R(\alpha)$ on $[a, b]$.

Remark By Corollary 10.2, a function α of bounded variation can be expressed as the difference of two monotone increasing functions. By (12), $\int_a^b f \, d\alpha$ is independent of the decomposition of α that we choose—provided, of course, that f is Riemann-Stieltjes integrable with respect to the monotone increasing integrators. Also note that Definition 10.8 allows us to extend part 2 of Proposition 10.1 to include the case where $c < 0$; in particular, if β is monotone increasing on $[a, b]$, then $\int_a^b f \, d(-\beta) = -\int_a^b f \, d\beta$.

Theorem 10.16 If either (1) f is continuous and α is of bounded variation on $[a, b]$, or (2) f and α are both of bounded variation and α is continuous on $[a, b]$, then f is in $\mathcal{R}(\alpha)$ on $[a, b]$.

Proof For part 1, we appeal to Corollary 10.2 and Theorem 10.4. For part 2, we only split α into the difference of two continuous monotone increasing functions (Corollary 10.4), we split f into the difference of two monotone increasing functions, and we appeal to Theorem 10.8 and part 1 of Proposition 10.1. ∎

Example 10.14 Letting $[x]$ denote the greatest integer function,

$$\int_0^2 (x^2 - x) \, d(x - [x]) = \int_0^2 (x^2 - x) \, dx - \int_0^2 (x^2 - x) \, d[x]$$

$$= \int_0^2 x^2 \, dx - \int_0^2 x \, dx - \int_0^2 x^2 \, d[x] + \int_0^2 x \, d[x]$$

$$= \frac{8}{3} - 2 - (1 + 4) + (1 + 2)$$

$$= -\frac{4}{3}.$$

Remark Note that when either part 1 or part 2 of Theorem 10.16 is satisfied, then (4) of Theorem 10.10 holds. That is,

$$\int_a^b f \, d\alpha = \lim_{\|P\| \to 0} S(P, f, \alpha),$$

where $S(P, f, \alpha)$ is a Riemann-Stieltjes sum for f with respect to the bounded variation integrator α. (Observe that $\Delta\alpha_i = \Delta\beta_i - \Delta\gamma_i$ when $\alpha = \beta - \gamma$.)

Remark When part 1 or part 2 of Theorem 10.16 holds, parts 1, 2, and 4 of Proposition 10.1 hold. However, inequalities such as part 3 of Proposition 10.1 or part 2 of Corollary 10.1 may not hold when the integrator is not monotone increasing. For example,

$$\int_0^1 1 \, d(-[x]) = -\int_0^1 1 \, d[x] = -1 > -2 = \int_0^1 2 \, d(-[x])$$

and

$$\left| \int_0^1 1 \, d(-[x]) \right| = 1 > -1 = \int_0^1 |1| \, d(-[x]).$$

Theorem 10.17 Suppose that f and α satisfy part 1 or part 2 of Theorem 10.16, and let v be the total variation function of α on $[a, b]$. Then

$$\left| \int_a^b f \, d\alpha \right| \le \int_a^b |f| \, dv \le M \, V_\alpha(a, b),$$

where $|f(x)| \le M$ for all x in $[a, b]$.

Proof If part 1 of Theorem 10.16 holds, then $|f|$ is continuous on $[a, b]$. If part 2 of Theorem 10.16 holds, then v is continuous (Theorem 10.15) and $|f|$ is of bounded variation (Exercise 1 in Section 10.4) on $[a, b]$. Since v is monotone increasing (Theorem 10.14), $\int_a^b |f| \, dv$ exists in either case.

Let $P = \{x_i\}_{i=0}^n$ be in $\mathcal{P}[a, b]$, and let t_i be in $[x_{i-1}, x_i]$ for $i = 1, 2, \dots, n$. Then

$$\left| \sum_{i=1}^n f(t_i) \Delta \alpha_i \right| \le \sum_{i=1}^n |f(t_i)| \, |\Delta \alpha_i|$$

$$\le \sum_{i=1}^n |f(t_i)| \, V_\alpha(x_{i-1}, x_i)$$

$$= \sum_{i=1}^n |f(t_i)| \left[V_\alpha(a, x_i) - V_\alpha(a, x_{i-1}) \right] \qquad \text{(by Proposition 10.4)}$$

$$= \sum_{i=1}^n |f(t_i)| \, \Delta v_i. \qquad (13)$$

As $\|P\| \to 0$, the left side of (13) tends to $\left| \int_a^b f \, d\alpha \right|$ and the last sum in (13) tends to $\int_a^b |f| \, dv$. Hence,

$$\left| \int_a^b f \, d\alpha \right| \le \int_a^b |f| \, dv.$$

Also,

$$\int_a^b |f| \, dv \le M \int_a^b dv = M[v(b) - v(a)]$$
$$= M \, v(b) = M \, V_\alpha(a, b)$$

since v is monotone increasing.   ∎

Remark By Exercise 6 in Section 10.3, the integration by parts formula

$$\int_a^b f \, d\alpha = f(b)\alpha(b) - f(a)\alpha(a) - \int_a^b \alpha \, df$$

holds whenever f and α are of bounded variation on $[a, b]$ and one of them is continuous on $[a, b]$. Also, by a completely similar proof, Theorem 10.11 holds whenever part 1 or part 2 of Theorem 10.16 is satisfied.

Example 10.15 We reconsider Example 10.14 using integration by parts and Theorem 10.11.

$$\int_0^2 (x^2 - x)\, d(x - [x]) = (x^2 - x)(x - [x])\big|_0^2 - \int_0^2 (x - [x])\, d(x^2 - x)$$

$$= 0 - \int_0^2 (x - [x])(2x - 1)\, dx$$

$$= \int_0^2 (x - 2x^2)\, dx - \int_0^2 (1 - 2x)[x]\, dx$$

$$= -\frac{10}{3} - \int_1^2 (1 - 2x)\, dx$$

$$= -\frac{10}{3} + 2$$

$$= -\frac{4}{3}.$$

Remark In Exercise 7 we consider the analogue of Theorem 10.6 when α is of bounded variation. We point out that the proof of part 2 of Theorem 10.6 relies on the First Mean Value Theorem (Theorem 10.5), which depends on the monotonicity of α. For example, if $f(x) = x$ and $\alpha(x) = x^2$ on $[-1, 1]$, then

$$\int_{-1}^1 f\, d\alpha = \int_{-1}^1 x(2x)\, dx = \frac{4}{3}$$

but $f(c)[\alpha(1) - \alpha(-1)] = 0$ for all c in $[-1, 1]$. Thus, the First Mean Value Theorem does not hold for integrators of bounded variation.

Exercises

1. Find

   (a) $\int_0^\pi x \, d(\sin x)$,

   (b) $\int_{-1}^2 x \, d\,|x|$.

2. Recall Exercise 5 in Section 10.4. Verify that the first inequality in Theorem 10.17 holds for $\int_0^\pi x \, d(\sin x)$.

3. Find $\int_0^{2\pi} (x - [x]) \, d(\cos x)$.

4. This is usually called the *Second Mean Value Theorem* (compare Exercise 6 in Section 10.2). Let f be monotone on $[a, b]$, and let α be continuous and of bounded variation on $[a, b]$. Show that there is a c in $[a, b]$ such that

   $$\int_a^b f \, d\alpha = f(a) \, [\alpha(c) - \alpha(a)] + f(b) \, [\alpha(b) - \alpha(c)].$$

   [*Hint:* First integrate by parts.]

5. Let α_n be of bounded variation on $[a, b]$ for each n in $\mathbb{N}$. Suppose that α is a function on $[a, b]$ such that $V_{\alpha - \alpha_n}(a, b) \to 0$. If f is continuous on $[a, b]$, show that

   $$\lim_{n \to \infty} \int_a^b f \, d\alpha_n = \int_a^b f \, d\alpha.$$

   [*Hint:* Theorem 10.17.]

6. Let f be continuous and of bounded variation on $[0, 2\pi]$. Show that

   $$\left| \int_0^{2\pi} f(x) \cos nx \, dx \right| \le \frac{1}{n} V_f(0, 2\pi).$$

   A similar result holds for $\int_0^{2\pi} f(x) \sin nx \, dx$ when $f(0) = f(2\pi)$. [*Hint:* See the Remark following the proof of Theorem 10.17, and use Theorem 10.17.]

7. Let f and α satisfy part 1 or part 2 of Theorem 10.16 and define F on $[a, b]$ by

   $$F(x) = \int_a^x f \, d\alpha.$$

   Prove that F is of bounded variation on $[a, b]$ and that every point of continuity of α is a point of continuity of F.

8. Let f and g be continuous on $[a, b]$, and let α be of bounded variation on $[a, b]$. Define $\beta : [a, b] \to \mathbb{R}$ by

   $$\beta(x) = \int_a^x f \, d\alpha.$$

   Prove that g is in $\mathcal{R}(\beta)$ on $[a, b]$ and that

   $$\int_a^b g \, d\beta = \int_a^b gf \, d\alpha.$$

The Topology of $\mathbb{R}$

In general, a topological space *is a pair* $(X, \mathcal{T})$ *where* X *is a nonempty set and* $\mathcal{T}$ *is a collection of subsets of* X *satisfying the following properties: (1)* $\emptyset \in \mathcal{T}$ *and* $X \in \mathcal{T}$*; (2) if* $U_\alpha \in \mathcal{T}$ *for each* α *in an index set* I*, then* $\bigcup_{\alpha \in I} U_\alpha \in \mathcal{T}$ *(that is,* $\mathcal{T}$ *is closed under arbitrary unions); (3) for* n *in* $\mathbb{N}$*, if* $U_1, U_2, \ldots, U_n$ *are each in* $\mathcal{T}$*, then* $\bigcap_{i=1}^{n} U_i \in \mathcal{T}$ *(that is,* $\mathcal{T}$ *is closed under finite intersections). The members of* $\mathcal{T}$ *are called* open sets in X, *and* $\mathcal{T}$ *is called a* topology on X.

There are many topologies that can be imposed on $\mathbb{R}$*. One possible point of confusion is that in some of these topologies an open interval may not be an open set. However, in the usual topology on* $\mathbb{R}$*, which we define in Section 11.1, open intervals will be open sets and closed intervals will be closed sets (also to be defined in Section 11.1).*

In Section 11.2 we generalize our previous definition of neighborhood, and we consider accumulation points (intuitively, points close to other points in a set). Sections 11.1 and 11.2 contain the basic concepts to be used in the rest of the chapter. Perhaps the most important sets in analysis are compact sets, which are defined in Section 11.3. Our main result in Section 11.3 is the Heine-Borel Theorem. We characterize the connected subsets of $\mathbb{R}$ *in Section 11.4, and in Section 11.5 we relate continuity to the concepts previously discussed in this chapter. Many of the results about continuous functions on a closed interval hold for continuous functions on a compact set.*

Throughout this chapter we refer to earlier material in the text that is now implied by our more general results. For example, the Intermediate Value Theorem is an easy consequence of how continuous functions behave on a connected set.

11.1 Open and Closed Sets

Open Sets

Definition 11.1 Let $U \subset \mathbb{R}$. Then U is *open in* $\mathbb{R}$ if for each x in U there is an $\varepsilon > 0$ such that the interval $(x - \varepsilon, x + \varepsilon) \subset U$.

Terminology If U is open in $\mathbb{R}$, we will also refer to U as an *open set in* $\mathbb{R}$, as an *open subset of* $\mathbb{R}$, or simply as *open*.

Example 11.1

1. $\emptyset$ and $\mathbb{R}$ are both open in $\mathbb{R}$.

2. Every "open" interval is open in $\mathbb{R}$.

3. Every "open" ray is open in $\mathbb{R}$; that is, (a, ∞) and $(-\infty, a)$ are open in $\mathbb{R}$ for each a in $\mathbb{R}$.

4. $(0, 1) \cup (2, 3)$ is open in $\mathbb{R}$.

5. $[0, 1)$ is not open in $\mathbb{R}$ because of the point 0.

6. $(0, 1]$ is not open in $\mathbb{R}$ because of the point 1.

7. $(0, 1) \cup \{2\}$ is not open in $\mathbb{R}$ because of the point 2.

8. $\emptyset$ is the only countable set that is open in $\mathbb{R}$ since an interval contains an uncountable number of points.

The next result shows that the open sets just defined form a topology on $\mathbb{R}$.

Proposition 11.1   The arbitrary union and the finite intersection of open sets in $\mathbb{R}$ are open in $\mathbb{R}$.

Proof   Let U_α be an open set in $\mathbb{R}$ for each α in an index set I, and let $x \in \bigcup_{\alpha \in I} U_\alpha$. Then there exists an α_0 in I such that $x \in U_{\alpha_0}$. By Definition 11.1, there is an $\varepsilon > 0$ such that $(x - \varepsilon, x + \varepsilon) \subset U_{\alpha_0}$. Hence, $(x - \varepsilon, x + \varepsilon) \subset \bigcup_{\alpha \in I} U_\alpha$, and so $\bigcup_{\alpha \in I} U_\alpha$ is open in $\mathbb{R}$.

Let n be in $\mathbb{N}$, let $U_1, U_2, \ldots, U_n$ each be open in $\mathbb{R}$, and let $x \in \bigcap_{i=1}^{n} U_i$. For each $i = 1, 2, \ldots, n$, since U_i is open in $\mathbb{R}$, there is an $\varepsilon_i > 0$ such that $(x - \varepsilon_i, x + \varepsilon_i) \subset U_i$. Letting $\varepsilon = \min\{\varepsilon_1, \varepsilon_2, \ldots, \varepsilon_n\}$, we have that $\varepsilon > 0$ and $(x - \varepsilon, x + \varepsilon) \subset \bigcap_{i=1}^{n} U_i$. Thus, $\bigcap_{i=1}^{n} U_i$ is open in $\mathbb{R}$.   ∎

Example 11.2   For each n in $\mathbb{N}$, $(-1/n, 1/n)$ is open in $\mathbb{R}$ but $\bigcap_{n=1}^{\infty} (-1/n, 1/n) = \{0\}$ is not open in $\mathbb{R}$. Thus, the arbitrary intersection of open sets need not be open. Also, neither $(-1, 0]$ nor $[0, 1)$ is open in $\mathbb{R}$, but $(-1, 0] \cup [0, 1) = (-1, 1)$ is open in $\mathbb{R}$.

Remember that $\mathbb{Z}$ denotes the integers. Using Proposition 11.1, $\mathbb{R} \setminus \mathbb{Z} = \bigcup_{n \in \mathbb{Z}} (n, n+1)$ is open in $\mathbb{R}$. The following theorem shows that this decomposition of the open set $\mathbb{R} \setminus \mathbb{Z}$ is typical of the open sets in $\mathbb{R}$.

Theorem 11.1   A nonempty subset of $\mathbb{R}$ is open in $\mathbb{R}$ if and only if it is the union of countably many pairwise disjoint open intervals and/or open rays.

Proof   Since open intervals and open rays are open sets in $\mathbb{R}$, any union of these is open by Proposition 11.1.

Let U be a nonempty open subset of $\mathbb{R}$ and let $x \in U$. Since U is open, x is contained in an open interval that is contained in U. Hence, both $A_x = \{a \in \mathbb{R} : (a, x] \subset U\}$ and $B_x = \{b \in \mathbb{R} : [x, b) \subset U\}$ are nonempty. Let $a_x = \inf A_x$ and $b_x = \sup B_x$. (Recall from Section 2.2 that a_x is a real number if A_x is bounded below, and $a_x = -\infty$ if A_x is not bounded below. Similarly, b_x is either a real number or $+\infty$.) Neither a_x nor b_x is in U. (If a_x were in U, there would be an $\varepsilon > 0$ such that $(a_x - \varepsilon, a_x + \varepsilon) \subset U$; hence, $(a_x - \varepsilon, x] \subset U$, which contradicts the fact that a_x is a lower bound of A_x. Similarly, b_x in U would contradict the fact that b_x is an upper bound of B_x.) Therefore, $U = \underset{x \in U}{\cup} (a_x, b_x)$, where each (a_x, b_x) is either an open interval or an open ray.

Let x and y be two distinct points of U. We now show that (a_x, b_x) and (a_y, b_y) are either disjoint or identical. Suppose that z is in $(a_x, b_x) \cap (a_y, b_y)$. If $a_x < a_y$, then $a_x < a_y < z < b_x$; and so $a_y \in (a_x, b_x) \subset U$, which contradicts the fact that a_y is not in U. If $a_x > a_y$, then $a_y < a_x < z < b_y$, implying that a_x is in U, which again is a contradiction. Therefore, $a_x = a_y$. Similarly, $b_x = b_y$.

We now have that $\mathcal{C} = \{(a_x, b_x) : x \in U\}$ is a collection of pairwise disjoint open intervals and/or open rays and that $U = \cup \mathcal{C}$. That $\mathcal{C}$ is countable now follows as in Exercise 11 in Section 2.4. For each C in $\mathcal{C}$, since $\mathbb{Q}$ is dense in $\mathbb{R}$, choose one rational number q_C in C. Since $\mathcal{C}$ is a pairwise disjoint collection, the q_C's are distinct. Therefore, $\mathcal{C}$ is equivalent to a subset of $\mathbb{Q}$, and hence $\mathcal{C}$ is countable. ∎

By appealing to the fact (see Section 1.2) that the empty set is the union of an empty collection of open intervals, one could eliminate the word "nonempty" from Theorem 11.1.

Closed Sets

We caution the reader beforehand that "open" and "closed" are not antonyms. A set may be neither open nor closed; a set may be both open and closed.

Definition 11.2 Let $F \subset \mathbb{R}$. Then F is *closed in* $\mathbb{R}$ if $\mathbb{R} \setminus F$ is open in $\mathbb{R}$.

Terminology If F is closed in $\mathbb{R}$, we will also refer to F as a *closed set in* $\mathbb{R}$, as a *closed subset of* $\mathbb{R}$, or simply as *closed*.

Example 11.3

1. $\emptyset$ and $\mathbb{R}$ are both closed in $\mathbb{R}$. (That these are the only sets that are both open and closed in $\mathbb{R}$ will be seen in Section 11.4.)

2. $\{a\}$ is closed in $\mathbb{R}$ for each a in $\mathbb{R}$ since $\mathbb{R} \setminus \{a\} = (-\infty, a) \cup (a, \infty)$ is open in $\mathbb{R}$.

3. Every "closed" interval is closed in $\mathbb{R}$ since $\mathbb{R} \setminus [a, b] = (-\infty, a) \cup (b, \infty)$ is open in $\mathbb{R}$.

4. Every "closed" ray $[a, \infty)$ or $(-\infty, a]$ is closed in $\mathbb{R}$.

5. $[0, 1] \cup \{2\}$ is closed in $\mathbb{R}$.

6. $[0, 1)$ is neither open nor closed in $\mathbb{R}$. Why?

7. $\mathbb{N}$ and $\mathbb{Z}$ are closed in $\mathbb{R}$. For example, $\mathbb{R} \setminus \mathbb{N} = (-\infty, 1) \cup \overset{\infty}{\underset{n=1}{\cup}} (n, n+1)$ is open in $\mathbb{R}$.

The next result is the analogue of Proposition 11.1 for closed sets.

Proposition 11.2 The arbitrary intersection and the finite union of closed sets in $\mathbb{R}$ are closed in $\mathbb{R}$.

Proof Let F_α be closed in $\mathbb{R}$ for each α in a nonempty index set I. By DeMorgan's Laws (Proposition 1.8),

$$\mathbb{R} \setminus \bigcap_{\alpha \in I} F_\alpha = \bigcup_{\alpha \in I} (\mathbb{R} \setminus F_\alpha).$$

Since each $\mathbb{R} \setminus F_\alpha$ is open in $\mathbb{R}$, $\bigcup_{\alpha \in I} (\mathbb{R} \setminus F_\alpha)$ is open in $\mathbb{R}$ by Proposition 11.1. Thus, $\mathbb{R} \setminus \bigcap_{\alpha \in I} F_\alpha$ is open in $\mathbb{R}$, and so $\bigcap_{\alpha \in I} F_\alpha$ is closed in $\mathbb{R}$ by Definition 11.2.

For n in $\mathbb{N}$, let $F_1, F_2, \ldots, F_n$ each be closed in $\mathbb{R}$. Since $\mathbb{R} \setminus \bigcup_{i=1}^{n} F_i = \bigcap_{i=1}^{n} (\mathbb{R} \setminus F_i)$ by DeMorgan's Laws and $\bigcap_{i=1}^{n} (\mathbb{R} \setminus F_i)$ is open in $\mathbb{R}$ by Proposition 11.1, $\bigcup_{i=1}^{n} F_i$ is closed in $\mathbb{R}$ by Definition 11.2. ∎

Example 11.4

1. For each n in $\mathbb{N}$, $[1/n, 1]$ is closed in $\mathbb{R}$ but $\bigcup_{n=1}^{\infty} [1/n, 1] = (0, 1]$ is not closed in $\mathbb{R}$. Thus, the arbitrary union of closed sets need not be closed.

2. Every finite subset of $\mathbb{R}$ is closed in $\mathbb{R}$ since $\{x_1, x_2, \ldots, x_n\} = \bigcup_{i=1}^{n} \{x_i\}$ and each $\{x_i\}$ is closed.

Exercise 4 provides an example of a countable subset of $\mathbb{R}$ that is not closed in $\mathbb{R}$. Thus, the word "finite" cannot be replaced by the word "countable" in part 2 of Example 11.4.

The reader may be anticipating a characterization of closed sets similar to the characterization of open sets given in Theorem 11.1. No such characterization exists, as the following example indicates.

Example 11.5 (Cantor set) To construct the Cantor set, we successively remove from the closed interval $[0, 1]$ a countable number of pairwise disjoint open intervals. First we remove the open middle third $(\frac{1}{3}, \frac{2}{3})$ from $[0, 1]$ to obtain

$$F_1 = [0, 1] \setminus \left(\frac{1}{3}, \frac{2}{3}\right) = \left[0, \frac{1}{3}\right] \cup \left[\frac{2}{3}, 1\right].$$

We next remove (see Figure 11.1) the open middle thirds from each of the two disjoint closed intervals in F_1; that is, we remove $(\frac{1}{9}, \frac{2}{9})$ from $[0, \frac{1}{3}]$ and we remove $(\frac{7}{9}, \frac{8}{9})$ from $[\frac{2}{3}, 1]$ to obtain

$$F_2 = \left[0, \frac{1}{9}\right] \cup \left[\frac{2}{9}, \frac{1}{3}\right] \cup \left[\frac{2}{3}, \frac{7}{9}\right] \cup \left[\frac{8}{9}, 1\right].$$

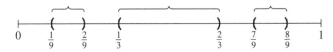

$$
\begin{array}{ccccccccc}
0 & & \frac{1}{9} & & \frac{2}{9} & \frac{1}{3} & & \frac{2}{3} & \frac{7}{9} & \frac{8}{9} & 1
\end{array}
$$

Figure 11.1

For the next step of the construction, we remove (see Figure 11.2) the open middle thirds from each of the 2^2 disjoint closed intervals in F_2 to obtain

$$F_3 = \left[0, \frac{1}{27}\right] \cup \left[\frac{2}{27}, \frac{1}{9}\right] \cup \left[\frac{2}{9}, \frac{7}{27}\right] \cup \left[\frac{8}{27}, \frac{1}{3}\right] \cup$$

$$\left[\frac{2}{3}, \frac{19}{27}\right] \cup \left[\frac{20}{27}, \frac{7}{9}\right] \cup \left[\frac{8}{9}, \frac{25}{27}\right] \cup \left[\frac{26}{27}, 1\right].$$

If the nth step of the construction has been completed, then F_n is the union of 2^n disjoint closed intervals each of the form $[k/3^n, (k+1)/3^n]$ and each of length $1/3^n$. To perform the $(n+1)$st step of the construction, we remove the open middle thirds from each of the 2^n disjoint closed intervals in F_n to obtain F_{n+1} as the union of 2^{n+1} disjoint closed intervals each of length $1/3^{n+1}$.

Figure 11.2

The Cantor set $\mathfrak{C}$ is defined by

$$\mathfrak{C} = \bigcap_{n=1}^{\infty} F_n.$$

$\mathfrak{C}$ is closed in $\mathbb{R}$ by Proposition 11.2, and $\mathfrak{C}$ consists of those points in $[0, 1]$ that are left after removing all of the open middle thirds $(\frac{1}{3}, \frac{2}{3})$, $(\frac{1}{9}, \frac{2}{9})$, etc. What points are left? From the construction, $\mathfrak{C}$ contains all the endpoints of the closed intervals that make up each F_n; that is, $\mathfrak{C}$ contains $0, 1, \frac{1}{3}, \frac{2}{3}, \frac{1}{9}, \frac{2}{9}, \frac{7}{9}, \frac{8}{9}, \frac{1}{27}, \ldots,$ which, of course, forms a countable set. We now show that $\mathfrak{C}$ is uncountable.

Let A be the set of all $0 - 1$ valued sequences; that is, the members of A are sequences $(x_n)_{n \in \mathbb{N}}$ where each x_n is either 0 or 1. Then A is uncountable by Exercise 13 in Section 2.4. Define $f : A \to \mathfrak{C}$ by

$$f\left((x_n)_{n \in \mathbb{N}}\right) = \sum_{n=1}^{\infty} \frac{2x_n}{3^n} \tag{1}$$

for each $(x_n)_{n \in \mathbb{N}}$ in A. The right side of (1) is a ternary (base 3) expansion of a point in $[0, 1]$, where each $2x_n$ is either 0 or 2. By Exercise 5, $\mathfrak{C}$ consists of all such ternary expansions. Since duplicate ternary expansions occur only when one of the expansions contains a 1, f is one-to-one, and so $\mathfrak{C}$ is uncountable by part 2 of Proposition 2.12. Also, f is onto $\mathfrak{C}$ because if $x = \sum_{n=1}^{\infty} x_n/3^n$ is in $\mathfrak{C}$ with $x_n = 0$ or 2 for each n, then $f((\frac{1}{2}x_n)_{n \in \mathbb{N}}) = x$. In Exercise 7 we ask the reader to show that $\mathfrak{C}$ is uncountable by another argument.

In this section we make one last observation about $\mathfrak{C}$. The sum of the lengths of all the removed open middle thirds is

$$\frac{1}{3} + \frac{2}{3^2} + \frac{2^2}{3^3} + \cdots + \frac{2^n}{3^{n+1}} + \cdots = 1.$$

Thus, $\mathfrak{C}$ is a closed subset of $[0, 1]$, $\mathfrak{C}$ has the same cardinality as $[0, 1]$, and the open intervals making up the complement of $\mathfrak{C}$ in $[0, 1]$ have a total length of 1.

☰ Exercises ☰

1. Let U be open in $\mathbb{R}$ and let F be closed in $\mathbb{R}$. Show that $U \setminus F$ is open in $\mathbb{R}$ and $F \setminus U$ is closed in $\mathbb{R}$.

2. For $A \subset \mathbb{R}$, the *interior of A*, denoted by A°, is defined by
   $$A^\circ = \bigcup \{U \subset A : U \text{ is open in } \mathbb{R}\}.$$

   (a) Show that A° is open in $\mathbb{R}$, A° is the largest (by containment) open set contained in A, and
   $$A \text{ is open in } \mathbb{R} \Longleftrightarrow A = A^\circ.$$

   (b) The points in A° are called the *interior points of A*. Show that x is an interior point of A if and only if there exists an open set U in $\mathbb{R}$ such that $x \in U \subset A$.

   (c) What are $\mathbb{Q}^\circ$, $(\mathbb{R} \setminus \mathbb{Q})^\circ$, and $[0, 1]^\circ$?

3. Let A and B be subsets of $\mathbb{R}$. In the notation of Exercise 2, show that

   (a) $(A^\circ)^\circ = A^\circ$,

   (b) $(A \cap B)^\circ = A^\circ \cap B^\circ$, and

   (c) $A^\circ \cup B^\circ \subset (A \cup B)^\circ$, and give an example showing that the inclusion may be proper.

4. Show that $\{1/n : n \in \mathbb{N}\}$ is neither open nor closed in $\mathbb{R}$.

5. Let x be in $[0, 1]$. Then $x = \sum_{n=1}^{\infty} x_n/3^n = x_1/3 + x_2/3^2 + x_3/3^3 + \cdots$, where each x_n is 0, 1, or 2, is a ternary (base 3) expansion of x. This expansion is unique except when x has a certain form. For example,
   $$\frac{1}{3} = \frac{0}{3} + \frac{2}{3^2} + \frac{2}{3^3} + \cdots + \frac{2}{3^n} + \cdots$$
   has two ternary expansions. Also,
   $$\frac{8}{9} = \frac{2}{3} + \frac{2}{3^2} = \frac{2}{3} + \frac{1}{3^2} + \frac{2}{3^3} + \frac{2}{3^4} + \cdots + \frac{2}{3^n} + \cdots$$
   has two ternary expansions. If an x in $[0, 1]$ has two ternary expansions, let us agree to use the one whose x_n's are 0 and 2. Show that the Cantor set
   $$\mathfrak{C} = \left\{ x = \sum_{n=1}^{\infty} \frac{x_n}{3^n} : x_n \in \{0, 2\} \text{ for each } n \text{ in } \mathbb{N} \right\}.$$

   [*Hint:* If x_1 is 0 or 2, then x is not in $(\frac{1}{3}, \frac{2}{3})$; if x_2 is 0 or 2, then x is not in $(\frac{1}{9}, \frac{2}{9})$ or $(\frac{7}{9}, \frac{8}{9})$.]

6. Show that $\frac{1}{4}$ is in $\mathfrak{C}$ and that $\frac{1}{4}$ is not an endpoint of any closed interval contained in any F_n.

7. Each x in $[0, 1]$ has a binary (base 2) expansion of the form
   $$x = \sum_{n=1}^{\infty} \frac{x_n}{2^n} = \frac{x_1}{2} + \frac{x_2}{2^2} + \frac{x_3}{2^3} + \cdots,$$
   where each x_n is 0 or 1. Of course, some x's have two binary expansions;

for example,

$$\frac{1}{2} = \frac{0}{2} + \frac{1}{2^2} + \frac{1}{2^3} + \frac{1}{2^4} + \cdots.$$

In these cases, let us agree to use the expansion that terminates. Show that

$$f\left(\sum_{n=1}^{\infty} \frac{x_n}{2^n}\right) = \sum_{n=1}^{\infty} \frac{2x_n}{3^n}$$

is a one-to-one map of $[0, 1]$ into $\mathfrak{C}$.

11.2 Neighborhoods and Accumulation Points

We first generalize the definition of neighborhood given earlier in the text. We then characterize closed sets in terms of sequences and accumulation points. Finally, we obtain the smallest closed set that contains a given set.

Neighborhoods

In Definition 3.3, we defined a neighborhood of a real number x to be any open interval centered at x. Given $\varepsilon > 0$, the open interval $(x - \varepsilon, x + \varepsilon)$ was our generic neighborhood of x. Many analysts prefer a neighborhood of x to be any open set that contains x; however, topologists usually prefer a more general definition of neighborhood, which we now state.

> **Definition 11.3**   Let $G \subset \mathbb{R}$ and let x be in G. Then G is a *neighborhood of* x if there is an open set U in $\mathbb{R}$ such that $x \in U \subset G$.

Thus, G is a neighborhood of x if and only if G contains an open set that contains x. Combining Definitions 11.1 and 11.3, we have that G is a neighborhood of x if and only if there is an $\varepsilon > 0$ such that $(x - \varepsilon, x + \varepsilon) \subset G$. The difference now is that neighborhoods need not be open intervals or even open sets. Some people call $(x - \varepsilon, x + \varepsilon)$ an *ε-neighborhood* of x or a *basic neighborhood* of x.

Example 11.6

1. $[0, 1]$ is a neighborhood of each point in $(0, 1)$, but $[0, 1]$ is not a neighborhood of 0 or 1.

2. $(0, 1) \cup \{2\}$ is a neighborhood of each point in $(0, 1)$ but is not a neighborhood of 2.

3. $\mathbb{Z}$ is not a neighborhood of any of its points. Neither is $\mathbb{N}$ or $\mathbb{Q}$. This is because none of these sets contains an open interval.

4. The Cantor set is not a neighborhood of any of its points. See Exercise 8.

5. If G is a neighborhood of x, then any set containing G is also a neighborhood of x.

Proposition 11.3   A set is open in $\mathbb{R}$ if and only if it is a neighborhood of each of its points.

Proof   If G is open in $\mathbb{R}$ and x is in G, taking $U = G$ in Definition 11.3 shows that G is a neighborhood of x.

Now suppose that G is a neighborhood of each of its points. For each x in G, by Definition 11.3, there is an open set U_x in $\mathbb{R}$ such that $x \in U_x \subset G$. By Proposition 11.1, $G = \underset{x \in G}{\cup} U_x$ is open in $\mathbb{R}$.   ∎

From Exercise 2 in Section 11.1, it now follows that a set in $\mathbb{R}$ is a neighborhood of precisely those points that lie in its interior.

The reader is probably wondering how this new definition of neighborhood, Definition 11.3, affects earlier results in the text that involved the old definition of neighborhood, Definition 3.3. It does not; the earlier results are still valid under Definition 11.3. This follows since now every neighborhood of x must contain an old type neighborhood of x, $(x - \varepsilon, x + \varepsilon)$, for some $\varepsilon > 0$. For example, we restate Definition 3.4 in our new setting.

Definition 11.4   The sequence $(x_n)_{n \in \mathbb{N}}$ in $\mathbb{R}$ converges to the real number x if for every neighborhood G of x, the sequence $(x_n)_{n \in \mathbb{N}}$ is eventually in G.

Since the sequence $(x_n)_{n \in \mathbb{N}}$ is eventually in every neighborhood G of x if and only if $(x_n)_{n \in \mathbb{N}}$ is eventually in every basic neighborhood $(x - \varepsilon, x + \varepsilon)$ of x, Definitions 3.4 and 11.4 are equivalent.

Accumulation Points

Definition 3.10 of an accumulation point carries over verbatim to our new setting, as do all of the examples and results on pages 56–57. This is because everything on pages 56–57 was phrased in terms of neighborhoods. These results and definitions, which we now restate, are also summarized at the beginning of Section 4.3.

Accumulation Point Results   For $A \subset \mathbb{R}$ and x in $\mathbb{R}$, x is an accumulation point of A

if and only if every neighborhood of x contains a point of A different from x [that is, $(G \setminus \{x\}) \cap A \neq \emptyset$ for all neighborhoods G of x]

if and only if every neighborhood of x contains infinitely many points of A

if and only if there exists a sequence of distinct points in A converging to x.

Although we cannot characterize closed sets as we did open sets in Theorem 11.1, we can characterize closed sets in terms of sequences and accumulation points.

Theorem 11.2   Let $F \subset \mathbb{R}$. The following are equivalent.

1. F is closed in $\mathbb{R}$.

2. If $(x_n)_{n \in \mathbb{N}}$ is a sequence in F and x is in $\mathbb{R}$ with $(x_n)_{n \in \mathbb{N}}$ converging to x, then x is in F. (The reader should compare this with Theorem 3.3.)

3. F contains all of its accumulation points.

Proof   Part 1 implies part 2. Let F be closed in $\mathbb{R}$, and let $(x_n)_{n \in \mathbb{N}}$ be a sequence in F that converges to a point x in $\mathbb{R}$. Suppose that x is not in F. Then $\mathbb{R} \setminus F$ is an open neighborhood of x. Since $x_n \to x$, the sequence $(x_n)_{n \in \mathbb{N}}$ is eventually in $\mathbb{R} \setminus F$ (Definition 11.4), which contradicts the fact that each x_n is in F. Therefore, x is in F.

Part 2 implies part 3. If x is an accumulation point of F, then by the accumulation point results above, there is a sequence (of distinct points) in F that converges to x. By part 2, x is in F.

Part 3 implies part 1. To show that F is closed, we will show that $\mathbb{R} \setminus F$ is open. Let y be in $\mathbb{R} \setminus F$. By part 3, y is not an accumulation point of F. By the accumulation point results above, there is a neighborhood G of y such that $G \cap F = \emptyset$. Therefore, $y \in G \subset \mathbb{R} \setminus F$, and so $\mathbb{R} \setminus F$ is a neighborhood of y. Since y is an arbitrary element of $\mathbb{R} \setminus F$, $\mathbb{R} \setminus F$ is a neighborhood of each of its points. By Proposition 11.3, $\mathbb{R} \setminus F$ is open in $\mathbb{R}$. ∎

Example 11.7

1. $\{1/n : n \in \mathbb{N}\}$ is not closed in $\mathbb{R}$ since $\{1/n : n \in \mathbb{N}\}$ does not contain its accumulation point 0.

2. $\mathbb{Q}$ is not closed in $\mathbb{R}$ since every real number is an accumulation point of $\mathbb{Q}$.

3. Since $\mathbb{N}$ and $\mathbb{Z}$ have no accumulation points, both $\mathbb{N}$ and $\mathbb{Z}$ are closed in $\mathbb{R}$.

We next characterize, in Theorem 11.3 below, the smallest (by containment) closed set that contains a given set. First, we need the following.

Notation   For $A \subset \mathbb{R}$, let A' denote the set of accumulation points of A.

Proposition 11.4 A' is closed in $\mathbb{R}$ for each $A \subset \mathbb{R}$.

Proof   Let $A \subset \mathbb{R}$ and let x be an accumulation point of A'. To show that A' is closed in $\mathbb{R}$, it suffices by Theorem 11.2 to show that x is in A'. That is, we must show that x is an accumulation point of A.

Let G be a neighborhood of x. By Definition 11.3, there is an open set U in $\mathbb{R}$ such that $x \in U \subset G$. Since x is an accumulation point of A' and U is a neighborhood of x,

$$(U \setminus \{x\}) \cap A' \neq \emptyset.$$

Let $y \in (U \setminus \{x\}) \cap A'$. Since y is in A', y is an accumulation point of A. Since $y \in U$ and U is an open neighborhood of y, $U \cap A$ is infinite. Since $U \subset G$, $G \cap A$ is infinite. Therefore, x is in A'. ∎

Theorem 11.3 Let $A \subset \mathbb{R}$. Then $\overline{A} = A \cup A'$ is closed in $\mathbb{R}$. Moreover, if F is closed in $\mathbb{R}$ and $A \subset F$, then $\overline{A} \subset F$. ($\overline{A}$ is called the *closure of A* in $\mathbb{R}$. $\overline{A}$ is the smallest closed set in $\mathbb{R}$ containing A.)

Proof   Let x be an accumulation point of $\overline{A}$. By Exercise 2, either x is an accumulation point of A or x is an accumulation point of A'. If x is an accumulation point of A, then x is in A'. If x is an accumulation point of A', then x is in A' by Proposition 11.4. Therefore, x is in $\overline{A}$, and so $\overline{A}$ is closed in $\mathbb{R}$ by Theorem 11.2.

Let F be closed in $\mathbb{R}$ with $A \subset F$. If x is an accumulation point of A, then, since $A \subset F$, x is an accumulation point of F. That is, $A' \subset F'$. Since F is closed in $\mathbb{R}$, $F' \subset F$ by Theorem 11.2. Therefore, $A' \subset F$ and so $\overline{A} = A \cup A' \subset F$.   ∎

Example 11.8

1. $\overline{[0, 1]} = \overline{(0, 1)} = [0, 1]$.
2. $\overline{\mathbb{Q}} = \mathbb{R}$ and $\overline{\mathbb{R} \setminus \mathbb{Q}} = \mathbb{R}$.
3. $\overline{\{1/n : n \in \mathbb{N}\}} = \{1/n : n \in \mathbb{N}\} \cup \{0\}$.

At this point the reader may be wondering when a set equals its set of accumulation points. That is, when does $A = A'$? By Proposition 11.4, such a set must be closed.

Definition 11.5 Let $A \subset \mathbb{R}$. Then A is *perfect* if $A = A'$.

Example 11.9

1. If $-\infty < a < b < \infty$, then the closed interval $[a, b]$ is perfect. Also, $[a, \infty)$ and $(-\infty, b]$ are perfect.

2. $\mathbb{N}$, $\mathbb{Z}$, and $\mathbb{Q}$ are not perfect.

3. $A = [0, 1] \cup \{2\}$ is not perfect because 2 is an isolated point of A and not an accumulation point of A. (An isolated point was defined in Definition 3.10.)

4. The Cantor set is perfect. That is, every point in the Cantor set $\mathfrak{C}$ is an accumulation point of $\mathfrak{C}$. See Exercise 9.

≡ Exercises ≡

1. Show that if G_1 and G_2 are two neighborhoods of the real number x, then $G_1 \cap G_2$ is a neighborhood of x.

2. Let A and B be subsets of $\mathbb{R}$ and let x be an accumulation point of $A \cup B$. Show that either x is an accumulation point of A or x is an accumulation point of B. [*Hint:* Prove the contrapositive statement.]

3. For $A \subset \mathbb{R}$, show that

$$\overline{A} = \bigcap \{F : A \subset F \text{ and } F \text{ is closed in } \mathbb{R}\}.$$

4. For subsets A and B of $\mathbb{R}$, show that

(a) $\overline{\overline{A}} = \overline{A}$,

(b) $\overline{(A \cup B)} = \overline{A} \cup \overline{B}$, and

(c) $\overline{(A \cap B)} \subset \overline{A} \cap \overline{B}$, and give an example showing that the inclusion may be proper.

5. Let F be a nonempty closed set in $\mathbb{R}$.

(a) If F is bounded above, show that sup F is in F. [*Hint:* If sup F is not in F, use Proposition 2.5 to show that sup F is an accumulation point of F.]

(b) If F is bounded below, show that inf F is in F.

6. Let $A \subset \mathbb{R}$. In Definition 2.6, we defined A to be *dense* in $\mathbb{R}$ if between every two real numbers there exists a point of A. Show that the following are equivalent:

(a) A is dense in $\mathbb{R}$;

(b) for every nonempty open set U in $\mathbb{R}$, $U \cap A \neq \emptyset$;

(c) $\overline{A} = \mathbb{R}$.

7. Show that if U_1 and U_2 are both open and dense in $\mathbb{R}$, then $U_1 \cap U_2$ is dense in $\mathbb{R}$. [*Hint:* Use Exercise 6(b).]

8. A subset A of $\mathbb{R}$ is called *nowhere dense* if the interior (recall Exercise 2 in Section 11.1) of its closure is empty—that is, if $\overline{A}^\circ = \emptyset$. For example, $\mathbb{N}$, $\mathbb{Z}$, and $\{1/n : n \in \mathbb{N}\}$ are nowhere dense; however, $\mathbb{Q}$ is not nowhere dense since $\overline{\mathbb{Q}}^\circ = \mathbb{R}$. Show that the Cantor set is nowhere dense.

9. Show that the Cantor set is perfect. [*Hint:* In the notation of Example 11.5, given x in $\mathfrak{C}$, x is in some closed interval in F_n for each n in $\mathbb{N}$. Given $\varepsilon > 0$, recall the length of the closed intervals in F_n to choose n appropriately. Then the endpoints of this closed interval in F_n will both be in $(x - \varepsilon, x + \varepsilon)$.]

11.3 Compact Sets

The reader can recall that in calculus there were many more theorems about continuous functions on a closed interval than on an open interval. In this section we show that a closed interval has a much stronger property than an open interval—namely, that of compactness. The main result of this section is the Heine-Borel Theorem, Theorem 11.4.

Definition 11.6 Let A be a subset of $\mathbb{R}$ and let $\mathcal{G}$ be a collection of subsets of $\mathbb{R}$. Then $\mathcal{G}$ is a *cover* of A (or $\mathcal{G}$ *covers* A) if $A \subset \cup \mathcal{G}$. If $\mathcal{G}$ is a cover of A and if each member of $\mathcal{G}$ is open, then $\mathcal{G}$ is an *open cover* of A. If a subcollection of $\mathcal{G}$ covers A, then this subcollection is called a *subcover* of A.

We are primarily interested in open covers.

Example 11.10   Let $\mathcal{G} = \{(-n, n) : n \in \mathbb{N}\}$. Then $\mathcal{G}$ is an open cover of $\mathbb{R}$ with no finite subcover of $\mathbb{R}$; that is, no finite number of members of $\mathcal{G}$ has union $\mathbb{R}$ (since a finite number of members of $\mathcal{G}$ will only cover out to the maximum n). Also, $\mathcal{G}$ is an open cover of $[0, 1]$, and $(-2, 2)$ is a finite subcover of $[0, 1]$.

Example 11.11   Let $\mathcal{G} = \{(0, 1/n) : n \in \mathbb{N}\}$. Then $\mathcal{G}$ is an open cover of $(0, 1)$ with a finite subcover—namely, $(0, 1)$.

Example 11.12   Let $\mathcal{G} = \{(1/n, 1) : n = 2, 3, 4, \ldots\}$. Then $\mathcal{G}$ is an open cover of $(0, 1)$ with no finite subcover of $(0, 1)$.

Example 11.13   Let a and b be real numbers with $a < b$. Choose n_0 in $\mathbb{N}$ such that $b - (1/n_0) > a$. Let $\mathcal{G} = \{(a - 1, b - (1/n)) : n \in \mathbb{N}, n \geq n_0\}$. Then $\mathcal{G}$ is an open cover of $[a, b)$ with no finite subcover of $[a, b)$.

In Proposition 11.5 below, we establish the fact that every open cover has a countable subcover. For the purpose of this section, let $\mathcal{B}$ be the collection of all open intervals in $\mathbb{R}$ with rational endpoints. By Exercise 12 in Section 2.4, $\mathcal{B}$ is countable.

Lemma 11.1   For each open set U in $\mathbb{R}$ and for each x in U, there is a member B of $\mathcal{B}$ with $x \in B \subset U$.

Proof   Given x in the open set U, let (a, b) be an open interval in $\mathbb{R}$ such that $x \in (a, b) \subset U$. Since $\mathbb{Q}$ is dense in $\mathbb{R}$, there exist q_1 and q_2 in $\mathbb{Q}$ such that $a < q_1 < x < q_2 < b$. Therefore, $(q_1, q_2) \in \mathcal{B}$ and $x \in (q_1, q_2) \subset U$.   ∎

Proposition 11.5   (Lindelöf property of $\mathbb{R}$) Let A be a subset of $\mathbb{R}$ and let $\mathcal{G}$ be an open cover of A. Then $\mathcal{G}$ has a countable subcover of A; that is, a countable subcollection of $\mathcal{G}$ covers A.

Proof   For each x in A, choose a U in $\mathcal{G}$ with x in U. Then, by Lemma 11.1, choose a B in $\mathcal{B}$ with $x \in B \subset U$. Let $\mathcal{B}_1$ be the collection of all B's chosen this way. Then $\mathcal{B}_1$ is a cover of A and since $\mathcal{B}_1 \subset \mathcal{B}$, $\mathcal{B}_1$ is countable. For each B in $\mathcal{B}_1$, choose one U in $\mathcal{G}$ with $B \subset U$. The collection of all U's chosen this way is countable, is a subcollection of $\mathcal{G}$, and covers A.   ∎

The obvious question is when does an open cover have a finite subcover?

Compactness

Definition 11.7   A subset A of $\mathbb{R}$ is *compact* if every open cover of A has a finite subcover of A.

Example 11.14   Any finite subset of $\mathbb{R}$ is compact (just choose one open set in the cover corresponding to each point in the finite set). $\mathbb{N}$ is not compact because $\{(n - \frac{1}{2}, n + \frac{1}{2}) : n \in \mathbb{N}\}$ is an open cover of $\mathbb{N}$ with no finite subcover. Also, by Example 11.12, $(0, 1)$ is not compact.

The following proposition gives necessary conditions for a subset of $\mathbb{R}$ to be compact.

Proposition 11.6   If A is a compact subset of $\mathbb{R}$, then A is closed and bounded.

Proof   To show that A is closed, we show that $\mathbb{R} \setminus A$ is open. Let $x \in \mathbb{R} \setminus A$. For each a in A, by Lemma 3.1, there exist open sets U_a and V_a with

$$a \in U_a, x \in V_a, \text{ and } U_a \cap V_a = \emptyset. \tag{2}$$

The collection $\{U_a : a \in A\}$ is an open cover of A. Since A is compact, some finite subcollection covers A; say $\{U_{a_i}\}_{i=1}^n$ covers A. Let

$$U = \bigcup_{i=1}^{n} U_{a_i} \text{ and } V = \bigcap_{i=1}^{n} V_{a_i},$$

where V_{a_i} was chosen as in (2). Then U and V are both open in $\mathbb{R}$, $A \subset U$, $x \in V$, and $U \cap V = \emptyset$. Therefore, $x \in V \subset \mathbb{R} \setminus A$ and so $\mathbb{R} \setminus A$ is a neighborhood of x. Thus, $\mathbb{R} \setminus A$ is a neighborhood of each of its points, and so $\mathbb{R} \setminus A$ is open by Proposition 11.3.

To show that A is bounded, consider the open cover $\{(-n, n) : n \in \mathbb{N}\}$ of A. Since A is compact, a finite number of these intervals cover A; say $\{(-n_i, n_i)\}_{i=1}^m$ covers A. Letting $M = \max\{n_i : i = 1, 2, \ldots, m\}$, we have that $A \subset (-M, M)$, and so A is bounded.   ∎

The strength of the following theorem is that it gives us a converse to Proposition 11.6. Although this theorem is true in $\mathbb{R}^n$, it is not true in general topological spaces.

Theorem 11.4   (Heine-Borel Theorem) Let $A \subset \mathbb{R}$. Then A is compact if and only if A is closed and bounded.

Proof   Let A be a closed and bounded subset of $\mathbb{R}$, and let $\mathcal{G}$ be an open cover of A. By Proposition 11.5, we can assume that $\mathcal{G}$ is countably infinite; if $\mathcal{G}$ were finite, $\mathcal{G}$ would be a finite subcover of A. We write $\mathcal{G} = \{G_n\}_{n=1}^{\infty}$, where each G_n is open and $A \subset \bigcup_{n=1}^{\infty} G_n$. We want to show that $\mathcal{G}$ has a finite subcover of A.

Suppose that no finite subcollection of $\mathcal{G}$ covers A. We construct a sequence in A as follows:

choose $x_1 \in A \setminus G_1$, since G_1 cannot cover A;

choose $x_2 \in A \setminus (G_1 \cup G_2)$, since $\{G_1, G_2\}$ cannot cover A;

in general, for each n in $\mathbb{N}$, choose $x_n \in A \setminus \bigcup_{i=1}^{n} G_i$, since $\{G_i\}_{i=1}^n$ cannot cover A.

Then $(x_n)_{n \in \mathbb{N}}$ is a sequence in A. Since A is bounded, the sequence $(x_n)_{n \in \mathbb{N}}$ is bounded. By the Bolzano-Weierstrass Theorem for sequences (Theorem 3.10) and Theorem 11.2 (since A is closed), there exist a subsequence $\left(x_{n_k}\right)_{k=1}^{\infty}$ of $(x_n)_{n \in \mathbb{N}}$ and an x in A with $x_{n_k} \underset{k}{\to} x$.

Since $\mathcal{G}$ covers A, x is in G_m for some m in $\mathbb{N}$. Since G_m is a neighborhood of x, $\left(x_{n_k}\right)_{k=1}^{\infty}$ is eventually in G_m. For $n_k > m$, this contradicts the way x_{n_k} was chosen. Therefore, $\mathcal{G}$ has a finite subcollection that covers A.   ∎

A different proof that closed and bounded in $\mathbb{R}$ implies compact is asked for in Exercise 6. In preparation for that exercise, we have the next proposition. Note that our proof does not use Theorem 11.4.

Proposition 11.7 Let $B \subset A \subset \mathbb{R}$, where B is closed in $\mathbb{R}$ and A is compact. Then B is compact. (That is, a closed subset of a compact set is compact.)

Proof Let $\mathcal{G}$ be an open cover of B. Then the collection $\mathcal{G} \cup \{\mathbb{R} \setminus B\}$ is an open cover of A. Since A is compact, some finite subcollection of $\mathcal{G} \cup \{\mathbb{R} \setminus B\}$ covers A, and hence B. If $\mathbb{R} \setminus B$ is a member of this finite subcollection, by removing it we have a finite subcollection of $\mathcal{G}$ that covers B. ∎

Since closed intervals are compact, our next theorem implies the Nested Intervals Theorem (Theorem 3.8).

Theorem 11.5 Let $\{A_\alpha\}_{\alpha \in I}$ be a collection of compact sets in $\mathbb{R}$ such that the intersection of every finite subcollection of $\{A_\alpha\}_{\alpha \in I}$ is nonempty. Then $\underset{\alpha \in I}{\cap} A_\alpha$ is nonempty.

Proof Suppose that $\underset{\alpha \in I}{\cap} A_\alpha = \emptyset$. By DeMorgan's Laws (Proposition 1.8),

$$\mathbb{R} = \mathbb{R} \setminus \emptyset = \mathbb{R} \setminus \bigcap_{\alpha \in I} A_\alpha = \bigcup_{\alpha \in I} (\mathbb{R} \setminus A_\alpha).$$

Fix α_0 in I. Then $\{\mathbb{R} \setminus A_\alpha\}_{\alpha \in I}$ is an open cover of $\mathbb{R}$, and hence an open cover of A_{α_0}. Since A_{α_0} is compact, $\{\mathbb{R} \setminus A_{\alpha_i}\}_{i=1}^{n}$ covers A_{α_0} for some finite number of indices $\alpha_1, \alpha_2, \ldots, \alpha_n$. Thus,

$$A_{\alpha_0} \subset \bigcup_{i=1}^{n} (\mathbb{R} \setminus A_{\alpha_i}) = \mathbb{R} \setminus \bigcap_{i=1}^{n} A_{\alpha_i}$$

and therefore

$$A_{\alpha_0} \cap \left(\bigcap_{i=1}^{n} A_{\alpha_i} \right) = \emptyset,$$

which is a contradiction to the hypothesis that every finite subcollection of $\{A_\alpha\}_{\alpha \in I}$ has nonempty intersection. Hence, $\underset{\alpha \in I}{\cap} A_\alpha \neq \emptyset$. ∎

Relative Topology

At this point in our development, $\mathbb{R}$ is the only topological space we have considered. We now use the topology on $\mathbb{R}$ to induce a natural topology on a subset of $\mathbb{R}$, thus making a subset of $\mathbb{R}$ into a topological space. This will be very useful in the next section. Recall from the beginning of this chapter that a topology on a set consists of a collection of subsets of the set that satisfy certain properties.

Definition 11.8 Let X be a nonempty subset of $\mathbb{R}$. The *relative topology on X* is

$$\{U \cap X : U \text{ is open in } \mathbb{R}\}.$$

Terminology   If U is open in $\mathbb{R}$, then $U \cap X$ is *open relative to X* or $U \cap X$ is an *open set in X* or, simply, $U \cap X$ is *open in X*. Thus, a set $S \subset X$ is open in X if and only if $S = U \cap X$ for some set U that is open in $\mathbb{R}$.

Note that $\emptyset = \emptyset \cap X$ and $X = \mathbb{R} \cap X$ imply that both $\emptyset$ and X are open in X. Verification that the relative topology on X is closed under arbitrary unions and finite intersections follows from the distributive properties of sets (Exercise 10 in Section 1.2) and is asked for in Exercise 8.

Example 11.15

1. Let U be an open set in $\mathbb{R}$ such that $U \subset (0, 1)$. Then $U = U \cap [0, 1]$ is also open in $[0, 1]$.
2. $[0, \frac{1}{2}) = (-\frac{1}{2}, \frac{1}{2}) \cap [0, 1]$, and so $[0, \frac{1}{2})$ is open in $[0, 1]$ even though $[0, \frac{1}{2})$ is not open in $\mathbb{R}$.
3. $(\frac{3}{4}, 1] = (\frac{3}{4}, 2) \cap [0, 1]$ is open in $[0, 1]$ but not in $\mathbb{R}$.
4. $[e, \pi] \cap \mathbb{Q} = (e, \pi) \cap \mathbb{Q}$ is open in $\mathbb{Q}$.

The point we are making is that if $S \subset X \subset \mathbb{R}$, then S may be open in X without being open in $\mathbb{R}$. Thus, the property of being open depends on the space in which S is embedded. The same is true for the property of being closed (see Exercise 9). However, compactness behaves differently; the next proposition shows that it makes sense to consider compact spaces. To formulate the next proposition, we will temporarily say that A is compact relative to X whenever a cover of A by open sets in X has a finite subcover of A.

Proposition 11.8   Let $A \subset X \subset \mathbb{R}$. Then A is compact relative to $\mathbb{R}$ if and only if A is compact relative to X.

Proof   Suppose that A is compact relative to $\mathbb{R}$, and let $\{S_\alpha\}_{\alpha \in I}$ be a cover of A by open sets in X. By Definition 11.8, each $S_\alpha = U_\alpha \cap X$, where each U_α is an open set in $\mathbb{R}$. Thus, $\{U_\alpha\}_{\alpha \in I}$ is a cover of A by open sets in $\mathbb{R}$. Since A is compact relative to $\mathbb{R}$, there exist a finite number of indices $\alpha_1, \alpha_2, \ldots, \alpha_n$ in I such that $\{U_{\alpha_i}\}_{i=1}^n$ covers A. Since $A \subset X$, $\{S_{\alpha_i}\}_{i=1}^n$ covers A, and so A is compact relative to X.

Conversely, suppose that A is compact relative to X, and let $\{U_\alpha\}_{\alpha \in I}$ be a cover of A by open sets in $\mathbb{R}$. Then $\{U_\alpha \cap X\}_{\alpha \in I}$ is a cover of A by open sets in X. Since A is compact relative to X, a finite number of $\{U_\alpha \cap X\}_{\alpha \in I}$ cover A, and so the corresponding finite number of $\{U_\alpha\}_{\alpha \in I}$ cover A. Thus, A is compact relative to $\mathbb{R}$.   ∎

≡ Exercises ≡

1. Find an open cover for each of the following sets with no finite subcover, where a and b are in $\mathbb{R}$ with $a < b$.

   (a) $(0, 1]$ (b) $[a, \infty)$ (c) $(-\infty, a)$

   (d) $(-\infty, a]$ (e) (a, b) (f) $\mathbb{Z}$

2. (a) Show that the arbitrary intersection of compact sets is compact.

(b) Show that the finite union of compact sets is compact.

(c) Give an example showing that the arbitrary union of compact sets need not be compact.

3. Let $(x_n)_{n \in \mathbb{N}}$ be a sequence in $\mathbb{R}$ converging to the real number x. Show that $\{x_n : n \in \mathbb{N}\} \cup \{x\}$ is compact.

4. If A is a nonempty compact set in $\mathbb{R}$, show that sup A and inf A are both in A. [*Hint:* Exercise 5 in Section 11.2.]

5. Let $A \subset \mathbb{R}$. Show that the following are equivalent.

(a) A is compact.

(b) Every sequence in A has a subsequence that converges to a point of A.

(c) Every infinite subset of A has an accumulation point in A.

[*Hint:* For (c) or (b) implying (a), do closed and bounded separately.]

6. This exercise provides another proof that a closed and bounded subset of $\mathbb{R}$ is compact. By Proposition 11.7, it suffices to show that the closed interval $[a, b]$ is compact. Let $\mathcal{G}$ be an open cover of $[a, b]$ and let

$$E = \{x \in [a, b] : [a, x] \text{ can be covered by a finite subcollection of } \mathcal{G}\}.$$

Show that sup $E = b$ is in E.

7. (a) Show that the Cantor set is compact. (It is also perfect by Exercise 9 in Section 11.2.)

(b) Give an example of a compact subset of $\mathbb{R}$ that is not perfect.

8. For $X \subset \mathbb{R}$, show that the relative topology on X is closed under arbitrary unions and finite intersections.

9. Let $T \subset X \subset \mathbb{R}$. Similar to Definition 11.2, we define T to be *closed in X* if and only if $X \setminus T$ is open in X.

(a) Show that T is closed in X if and only if $T = F \cap X$ for some closed set F in $\mathbb{R}$.

(b) Show that $(0, \frac{1}{2}]$ is closed in $(0, 1)$ but not in $\mathbb{R}$.

10. In the relative topology on $\mathbb{Z}$, show that every subset of $\mathbb{Z}$ is both open and closed in $\mathbb{Z}$. Such a topological space is called *discrete*. [*Hint:* Show that $\{x\}$ is open in $\mathbb{Z}$ for each x in $\mathbb{Z}$.]

11.4 Connected Sets

Intuitively, a connected set is one that consists of a single piece. Our goal in this section is to characterize all the connected subsets of $\mathbb{R}$, which we accomplish in Corollary 11.1.

Definition 11.9 A subset X of $\mathbb{R}$ is *disconnected* if X can be represented as the union of two nonempty disjoint subsets of X, both of which are open in X. X is *connected* if X is not disconnected.

Thus, X is disconnected if there exist nonempty disjoint subsets S and T of X such that S and T are both open in X and $X = S \cup T$. This representation splits X into two pieces.

Example 11.16

1. The empty set and $\{x\}$ are connected for each x in $\mathbb{R}$.

2. $[0, 1] \cup [2, 3]$ is disconnected.

3. $(0, 1) \cup \{2\}$ is disconnected.

4. $\mathbb{N}$, $\mathbb{Z}$, and $\mathbb{Q}$ are disconnected. To show that $\mathbb{Q}$ is disconnected, let $S = \mathbb{Q} \cap (-\infty, \pi)$ and $T = \mathbb{Q} \cap (\pi, \infty)$. Then S and T are both open in $\mathbb{Q}$, S and T are both nonempty, $S \cap T = \emptyset$, and $\mathbb{Q} = S \cup T$.

Note that in part 4 of Example 11.16, $S = \mathbb{Q} \cap (-\infty, \pi) = \mathbb{Q} \cap (-\infty, \pi]$. Thus, S is also closed in $\mathbb{Q}$. Since a similar result holds for T, we are led to the following proposition.

Proposition 11.9   For $X \subset \mathbb{R}$, the following are equivalent.

1. X is disconnected.

2. X can be represented as the union of two nonempty disjoint subsets of X, both of which are closed in X.

3. X and $\emptyset$ are not the only subsets of X that are both open and closed in X.

Proof   Part 1 implies part 2. Let $X = S \cup T$, where S and T are nonempty disjoint subsets of X, both of which are open in X. Since $S = X \setminus T$ and $T = X \setminus S$, both S and T are closed in X.

Part 2 implies part 3. If S and T are nonempty disjoint closed sets in X with $X = S \cup T$, then $\emptyset \neq S \neq X$, and S is also open in X since $S = X \setminus T$.

Part 3 implies part 1. Let S be an open and closed set in X such that $\emptyset \neq S \neq X$. Since S is closed in X, $X \setminus S$ is open in X. Thus $X = S \cup (X \setminus S)$ is a representation of X as the union of two nonempty disjoint subsets of X, both of which are open in X.   ■

It follows from the proposition above that a subset X of $\mathbb{R}$ is connected if and only if X and $\emptyset$ are the only subsets of X that are both open and closed in X.

Example 11.17 $\mathbb{R}$ is connected. Let F be a closed subset of $\mathbb{R}$ such that $\emptyset \neq F \neq \mathbb{R}$. We will show that F cannot be open in $\mathbb{R}$. Let x be in F, let y be in $\mathbb{R} \setminus F$, and assume that $x < y$. Letting $z = \sup\{t \in F : t < y\}$, we have $x \leq z \leq y$. By Exercise 5 in Section 11.2, z is in F. Since y is not in F, $z < y$. By the definition of supremum, $(z, y) \cap F = \emptyset$. Therefore, no open interval containing z is contained in F. That is, F is not a neighborhood of z, and so F is not open in $\mathbb{R}$.

If $y < x$, a similar argument using $\inf\{t \in F : t > y\}$ shows that F cannot be open in $\mathbb{R}$.

To accomplish our goal of characterizing all the connected subsets of $\mathbb{R}$, we will use the following proposition. This proposition also shows that connectedness, like compactness, does not depend on the space in which the set is embedded. Our development follows that of Rudin.

Proposition 11.10 Let $X \subset \mathbb{R}$. Then X is disconnected if and only if there exist disjoint subsets U and V of $\mathbb{R}$ such that U and V are both open in $\mathbb{R}$, $X \cap U \neq \emptyset$, $X \cap V \neq \emptyset$, and $X \subset U \cup V$.

Proof   Suppose that such U and V exist. Letting $S = X \cap U$ and $T = X \cap V$, we have that $X = S \cup T$, where S and T are nonempty disjoint subsets of X both of which are open in X. Thus, X is disconnected by Definition 11.9.

Conversely, suppose that X is disconnected, and let $X = S \cup T$ where S and T are nonempty disjoint subsets of X, both of which are open in X. We need to construct U and V. (The problem is getting $U \cap V = \emptyset$.)

Since S is open in X, for each s in S there is a $\delta_s > 0$ such that if x is in X and $|x - s| < \delta_s$, then x is in S. [By the definition of the relative topology, $S = A \cap X$ for some open set A in $\mathbb{R}$. Given s in S, there is an open interval (a, b) such that $s \in (a, b) \subset A$. Then $(a, b) \cap X \subset S$.] Similarly, for each t in T there is a $\delta_t > 0$ such that if x is in X and $|x - t| < \delta_t$, then x is in T. Hence, if s is in S and t is in T, $|s - t| \geq \delta_s$ and $|s - t| \geq \delta_t$, and so

$$|s - t| \geq \frac{1}{2}(\delta_s + \delta_t). \tag{3}$$

For each s in S and t in T, let $U_s = \{y \in \mathbb{R} : |y - s| < \frac{1}{2}\delta_s\}$ and let $V_t = \{y \in \mathbb{R} : |y - t| < \frac{1}{2}\delta_t\}$. Then $U = \bigcup_{s \in S} U_s$ and $V = \bigcup_{t \in T} V_t$ are both open in $\mathbb{R}$, $S \subset U$, $T \subset V$, and so $X = S \cup T \subset U \cup V$.

It remains to show that $U \cap V = \emptyset$. If y is in $U \cap V$, then y is in $U_s \cap V_t$ for some s in S and some t in T. But then

$$|s - t| \leq |s - y| + |y - t| < \frac{1}{2}(\delta_s + \delta_t),$$

which contradicts (3).   ■

Theorem 11.6 Let $X \subset \mathbb{R}$. Then X is connected if and only if whenever x and y are in X with $x < z < y$, then z is in X.

Proof   First suppose that x and y are in X with $x < z < y$, but z is not in X. Then $X \cap (-\infty, z) = X \cap (-\infty, z]$ is a nonempty proper subset of X that is both open and closed in X. By Proposition 11.9, X is disconnected.

Now suppose that X is disconnected. By Proposition 11.10, there are disjoint subsets U and V of $\mathbb{R}$ such that U and V are both open in $\mathbb{R}$, $X \cap U \neq \emptyset$, $X \cap V \neq \emptyset$, and $X \subset U \cup V$. Let x be in $X \cap U$ and $y \in X \cap V$. By altering our notation if necessary, or by making two cases as in Example 11.17, we can assume that $x < y$. Let $z = \sup(U \cap [x, y])$.

Since U contains an open interval centered at x, $x < z$. Since V contains an open interval centered at y and $U \cap V = \emptyset$, $z < y$.

If z were in U, then U would contain an open interval centered at z, implying that z is not an upper bound of $U \cap [x, y]$. Therefore, z is not in U. If z were in V, then V would contain an open interval centered at z, implying, since $U \cap V = \emptyset$, that z is not the least upper bound of $U \cap [x, y]$. Therefore, z is not in V.

Since $X \subset U \cup V$ and z is not in $U \cup V$, z is not in X.   ■

Corollary 11.1   Let a and b be in $\mathbb{R}$ with $a < b$. A subset X of $\mathbb{R}$ is connected if and only if X is one of the following sets:

$$\emptyset, \{a\}, (a, b), (a, b], [a, b), [a, b],$$
$$(-\infty, b), (-\infty, b], (a, \infty), [a, \infty), \mathbb{R}.$$

Proof   This characterization of the connected subsets of $\mathbb{R}$ follows directly from Theorem 11.6.   ∎

Allowing $\{a\}$ to be the closed interval $[a, a]$, Corollary 11.1 can be restated as: a nonempty subset of $\mathbb{R}$ is connected if and only if it is either an interval or a ray in $\mathbb{R}$. No such characterization of connected sets exists in general topological spaces, not even in the plane. We will use this characterization in the next section.

Exercises

1. Provide the details in Example 11.17 for the case in which $y < x$.

2. (a) Show that the intersection of connected subsets of $\mathbb{R}$ is connected. (This is not true in general topological spaces, not even in the plane.)
   (b) Show that the union of two connected subsets of $\mathbb{R}$ need not be connected.
   (c) Suppose that A_α is a connected subset of $\mathbb{R}$ for each α in an index set I. If $\bigcap_{\alpha \in I} A_\alpha \neq \emptyset$, show that $\bigcup_{\alpha \in I} A_\alpha$ is connected. [In $\mathbb{R}$, a proof can be given using Theorem 11.6. However, this result is true in general topological spaces, and a harder proof can be given by assuming that $\bigcup_{\alpha \in I} A_\alpha$ is disconnected and using Proposition 11.10.]

3. Suppose that A is a connected subset of $\mathbb{R}$ and $A \subset B \subset \overline{A}$. Show that B is connected. In particular, $\overline{A}$ is connected. (This is also true in general topological spaces.)

4. Let $X \subset \mathbb{R}$. A *component* of X is a maximal connected subset of X; that is, $K \subset X$ is a component of X if K is connected, and whenever $K \subsetneq A \subset X$, then A is disconnected. For example, $[0, 1]$ and $\{2\}$ are the two components of $[0, 1] \cup \{2\}$; a connected set has only one component—namely, itself.

   (a) What are the components of $\mathbb{N}$ and $\mathbb{Z}$?
   (b) Let x be in X. Show that

   $$K_x = \bigcup \{A \subset X : x \in A \text{ and } A \text{ is connected}\}$$

   is a component of X. [*Hint:* See Exercise 2(c).]
   (c) Show that a component of X is closed in X. [*Hint:* See Exercise 3.]
   (d) Give an example showing that a component of X need not be open in X. [*Hint:* Consider K_0 in $\{1/n : n \in \mathbb{N}\} \cup \{0\}$.]

5. Let $X \subset \mathbb{R}$. Then X is *totally disconnected* if every component of X consists of a single point. For example, $\mathbb{N}$ and $\mathbb{Z}$ are totally disconnected. Oddly, a one-point set is both connected and totally disconnected.

   (a) Show that $\mathbb{Q}$ is totally disconnected. Also note that these components are not open in $\mathbb{Q}$.
   (b) Show that the Cantor set is totally disconnected. [*Hint:* See Exercise 8 in Section 11.2.]

6. Let $X \subset \mathbb{R}$. Show that X is totally disconnected if and only if whenever $x \neq y$ in X there exist nonempty disjoint subsets S and T of X such that S and T are both open in X, $X = S \cup T$, $x \in S$, and $y \in T$.

11.5 Continuous Functions

In this section we relate continuity to the concepts previously discussed in this chapter—namely, open, closed, compact, and connected sets. Our main results are Theorems 11.7, 11.8, 11.9, and 11.10. We end this section with a characterization of closed and bounded intervals in terms of the fixed points of continuous functions. We first restate Definition 4.1.

Definition 11.10   Let $X \subset \mathbb{R}$, let $f : X \to \mathbb{R}$ be a function, and let c be in X. Then f is *continuous at* c if for every neighborhood V of $f(c)$ there is a neighborhood U of c such that if x is in $U \cap X$, then $f(x)$ is in V.

Since every neighborhood (Definition 11.3) in this chapter contains a basic neighborhood (Definition 3.3), Definitions 4.1 and 11.10 are equivalent. In fact, since every neighborhood contains an open neighborhood, a function $f : X \to \mathbb{R}$ is continuous at a point c in X if and only if for every open neighborhood V of $f(c)$ there is an open neighborhood U of c such that $f(U \cap X) \subset V$.

Definition 4.2 remains the same: f is *continuous on* X if and only if f is continuous at each point of X; and Theorem 4.1 characterizing continuity at a point in terms of sequential convergence is still valid. Note that in the proof of Theorem 4.1 we used a neighborhood-type argument, which carries over verbatim to our present setting.

Continuity and Open and Closed Sets

We next characterize continuity on a set in terms of open and closed sets.

Theorem 11.7   For a function $f : X \to \mathbb{R}$, the following are equivalent.

1. f is continuous on X.

2. $f^{-1}(V)$ is open in X for each open set V in $\mathbb{R}$ (the inverse image of an open set is open).

3. $f^{-1}(F)$ is closed in X for each closed set F in $\mathbb{R}$ (the inverse image of a closed set is closed).

Proof   Part 1 implies part 2. Let V be open in $\mathbb{R}$ and let $c \in f^{-1}(V)$. Then V is an open neighborhood of $f(c)$. Since f is continuous at c by part 1, there is an open neighborhood U_c of c such that $f(U_c \cap X) \subset V$. Since $f^{-1}(V) = \bigcup\limits_{c \in f^{-1}(V)} (U_c \cap X)$ and since each $U_c \cap X$ is open in X by Definition 11.8, $f^{-1}(V)$ is open in X.

Part 2 implies part 1. Let c be in X and let V be a neighborhood of $f(c)$. By Definition 11.3 there is an open set G in $\mathbb{R}$ such that $f(c) \in G \subset V$. Since $f^{-1}(G)$ is open in X by part 2, there is an open set U in $\mathbb{R}$ such that $f^{-1}(G) = U \cap X$. Hence, U is a neighborhood of c and $f(U \cap X) = f(f^{-1}(G)) \subset G \subset V$. That is, f is continuous at c. Since c is an arbitrary element of X, f is continuous on X.

The equivalence of part 3 with parts 1 and 2 is left as Exercise 1.   ■

Example 11.18   Let $X = (0, 1) \cup (1, 2)$ and define $f : X \to \mathbb{R}$ by

$$f(x) = \begin{cases} 3 & \text{if } 0 < x < 1 \\ 7 & \text{if } 1 < x < 2. \end{cases}$$

Let V be open in $\mathbb{R}$. If $V \cap \{3, 7\} = \emptyset$, then $f^{-1}(V) = \emptyset$ is open in X; if $V \cap \{3, 7\} = \{3, 7\}$, then $f^{-1}(V) = X$ is open in X; if $V \cap \{3, 7\} = \{3\}$, then $f^{-1}(V) = (0, 1)$ is open in X; if $V \cap \{3, 7\} = \{7\}$, then $f^{-1}(V) = (1, 2)$ is open in X. Therefore, f is continuous on X. Also note that $\emptyset$, X, $(0, 1)$, and $(1, 2)$ are closed in X.

In the example above, $(0, 1)$ is open in X but $f((0, 1)) = \{3\}$ is not open in $\mathbb{R}$; thus, the direct image of an open set under a continuous function need not be open. The reader should be able to find an example showing that the direct image of a closed set under a continuous function need not be closed. Also, in Example 11.18 each point of X is an accumulation point of X, but the range of f has no accumulation points. Hence, continuous functions need not map accumulation points onto accumulation points (a simpler example is a constant function on $\mathbb{R}$). For the following positive result, recall from Theorem 11.3 that for $A \subset \mathbb{R}$, A' is the set of accumulation points of A and $\overline{A} = A \cup A'$ is the smallest closed set in $\mathbb{R}$ that contains A.

Proposition 11.11   Let $X \subset \mathbb{R}$ and $f : X \to \mathbb{R}$. Then f is continuous on X if and only if for all $A \subset X$, $f(\overline{A} \cap X) \subset \overline{f(A)}$.

Proof   Suppose that f is continuous on X, $A \subset X$, and $y \in f(\overline{A} \cap X)$. Then $y = f(x)$ for some x in $\overline{A} \cap X$. If $x \in A$, then $y = f(x) \in f(A) \subset \overline{f(A)}$. If $x \in A'$, then there is a sequence $(x_n)_{n \in \mathbb{N}}$ in A such that $x_n \to x$. Since f is continuous at x, $f(x_n) \to f(x)$. By Theorem 11.2, $y = f(x) \in \overline{f(A)}$.

Conversely, suppose that f is not continuous on X. By Theorem 11.7, there is a closed set F in $\mathbb{R}$ such that $f^{-1}(F)$ is not closed in X. Let $A = f^{-1}(F)$. Since $A \subset X$, A is not closed in $\mathbb{R}$ by Exercise 9(a) in Section 11.3. By Theorem 11.2, A has an accumulation point in $\mathbb{R} \setminus A$. By letting $B = A' \cap (\mathbb{R} \setminus A)$, we have that $B \neq \emptyset$ and $\overline{A} = A \cup A' = A \cup B$. If $B \cap X = \emptyset$, then $A = \overline{A} \cap X$, implying that A is a closed set in X, which is a contradiction. Therefore, $B \cap X \neq \emptyset$.

Let $x \in B \cap X$. Then $x \in \overline{A} \cap X$, and x is not in A. We claim that $f(x)$ is not in $\overline{f(A)}$. Since $\overline{f(A)} = \overline{f(f^{-1}(F))} \subset \overline{F} = F$, if $f(x) \in \overline{f(A)} \subset F$, then $x \in f^{-1}(F) = A$, which is a contradiction. Thus, the claim holds.   ∎

Example 11.19 Recall Exercise 6 in Section 4.2. Let f be a continuous function from $\mathbb{R}$ into $\mathbb{R}$ with $f(x) = c$ for all x in $\mathbb{Q}$. By Proposition 11.11, $f(\mathbb{R}) = f(\overline{\mathbb{Q}}) \subset \overline{f(\mathbb{Q})} = \overline{\{c\}} = \{c\}$; that is, $f(x) = c$ for all x in $\mathbb{R}$.

Continuity and Compactness

We first show that the continuous image of a compact set is compact.

Theorem 11.8 Let X be a compact subset of $\mathbb{R}$ and let $f : X \to \mathbb{R}$ be a continuous function on X. Then $f(X)$ is compact.

Proof Let $\{U_\alpha\}_{\alpha \in I}$ be a cover of $f(X)$ by open sets in $\mathbb{R}$. By Theorem 11.7, $\{f^{-1}(U_\alpha)\}_{\alpha \in I}$ is a cover of X by open sets in X. Since X is compact, there is a finite number of indices $\alpha_1, \alpha_2, \ldots, \alpha_n$ such that $\{f^{-1}(U_{\alpha_i})\}_{i=1}^{n}$ covers X. Hence, $f(X) \subset \bigcup_{i=1}^{n} U_{\alpha_i}$, and so $f(X)$ is compact.   ∎

In $\mathbb{R}$, compact is equivalent to closed and bounded by the Heine-Borel Theorem (Theorem 11.4). Given the hypothesis of Theorem 11.8, $f(X)$ is a closed and bounded subset of $\mathbb{R}$. Thus, f is bounded since $f(X)$ is bounded (Definition 2.8); and for X nonempty, f assumes its bounds since $f(X)$ is closed (by Exercise 5 in Section 11.2 or Exercise 4 in Section 11.3). Hence, Proposition 4.8 and Theorem 4.2 are special cases of Theorem 11.8. We now show that Theorem 4.4 is a special case of the following more general result.

Theorem 11.9 If f is a continuous real valued function on the compact subset X of $\mathbb{R}$, then f is uniformly continuous on X.

Proof Let $\varepsilon > 0$. Recall from Definition 4.6 that we want a $\delta > 0$ (δ depending only on ε) such that if x and y are in X with $|x - y| < \delta$, then $|f(x) - f(y)| < \varepsilon$. For each x in X, since f is continuous at x, there exists a $\delta_x > 0$ such that if y is in X and $|y - x| < \delta_x$, then $|f(y) - f(x)| < \varepsilon/2$. For each x in X, let U_x be the open interval centered at x of radius $\frac{1}{2}\delta_x$; that is,

$$U_x = \left\{ a \in \mathbb{R} : |a - x| < \frac{1}{2}\delta_x \right\}.$$

Then $\{U_x\}_{x \in X}$ is a cover of X by open sets in $\mathbb{R}$. Since X is compact, there is a finite set of points $x_1, x_2, \ldots, x_n$ in X such that $X \subset \bigcup_{i=1}^{n} U_{x_i}$. Let

$$\delta = \frac{1}{2} \min\{\delta_{x_1}, \delta_{x_2}, \ldots, \delta_{x_n}\}.$$

Then $\delta > 0$. Let x and y be in X with $|x - y| < \delta$. Then x is in U_{x_i} for some i in $\{1, 2, \ldots, n\}$. Since

$$|y - x_i| \leq |y - x| + |x - x_i|$$
$$< \delta + \frac{1}{2}\delta_{x_i}$$
$$\leq \frac{1}{2}\delta_{x_i} + \frac{1}{2}\delta_{x_i}$$
$$= \delta_{x_i},$$

$$|f(x) - f(y)| \leq |f(x) - f(x_i)| + |f(x_i) - f(y)|$$
$$< \frac{\varepsilon}{2} + \frac{\varepsilon}{2}$$
$$= \varepsilon$$

by the way δ_{x_i} was chosen. Hence, f is uniformly continuous on X. ■

Remark   Theorem 6.4 can now be strengthened as follows. Let f be in $\mathcal{R}[a, b]$ with the range of f contained in the compact subset X of $\mathbb{R}$. If $\varphi : X \to \mathbb{R}$ is continuous on X, then $\varphi \circ f$ is in $\mathcal{R}[a, b]$. The proof is almost identical to the proof of Theorem 6.4; the uniform continuity of φ now comes from Theorem 11.9 instead of Theorem 4.4. With this new version of Theorem 6.4, Exercise 1(c) in Section 6.3 can also be strengthened as follows. Let f be in $\mathcal{R}[a, b]$. If there is a $\delta > 0$ such that $|f(x)| \geq \delta$ for all x in $[a, b]$, then $1/f$ is in $\mathcal{R}[a, b]$. The argument here is the same as the argument in Exercise 1(c) in Section 6.3; the same $\varphi(x) = 1/x$ works.

Continuity and Connectedness

We first show that the continuous image of a connected set is connected.

Theorem 11.10   Let X be a connected subset of $\mathbb{R}$ and let $f : X \to \mathbb{R}$ be a continuous function on X. Then $f(X)$ is connected.

Proof   Suppose that $f(X)$ is not connected. By Proposition 11.10 there exist disjoint subsets U and V of $\mathbb{R}$ such that U and V are both open in $\mathbb{R}$, $U \cap f(X) \neq \emptyset$, $V \cap f(X) \neq \emptyset$, and $f(X) \subset U \cup V$. Since f is continuous on X, $f^{-1}(U)$ and $f^{-1}(V)$ are both open in X by Theorem 11.7. Also, $f^{-1}(U)$ and $f^{-1}(V)$ are disjoint (since U and V are disjoint); $f^{-1}(U)$ and $f^{-1}(V)$ are both nonempty [since both U and V meet $f(X)$]; and $X = f^{-1}(U) \cup f^{-1}(V)$ [since $f(X) \subset U \cup V$]. Therefore, X is not connected. ■

We now observe that Theorem 11.10 implies the Intermediate Value Theorem (Theorem 4.3). Let f be a continuous function from $[a, b]$ into $\mathbb{R}$ with $f(a) < k < f(b)$. Since $[a, b]$ is connected by Corollary 11.1, $f([a, b])$ is connected by Theorem 11.10; hence, k is in $f([a, b])$ by Theorem 11.6. Our next result implies Exercise 7 in Section 4.4 as a special case.

Proposition 11.12 A nonempty subset X of $\mathbb{R}$ is connected if and only if every continuous function from X into $\mathbb{Z}$ (the integers) is constant.

Proof Suppose that X is connected and let f be a continuous function from X into $\mathbb{Z}$. By Theorem 11.10, $f(X)$ is a connected subset of $\mathbb{Z}$, and hence of $\mathbb{R}$. If $f(X)$ contained two distinct integers, then $f(X)$ would contain every real number between these two integers by Theorem 11.6, contradicting the fact that $f(X) \subset \mathbb{Z}$. Therefore, $f(X)$ consists of a single value.

Conversely, suppose that X is not connected. By Proposition 11.9 there is a subset A of X such that A is both open and closed in X and $\emptyset \neq A \neq X$. Define $f : X \to \mathbb{Z}$ by

$$f(x) = \begin{cases} 1 & \text{if} \quad x \in A \\ 0 & \text{if} \quad x \in X \setminus A. \end{cases}$$

Then f is continuous on X (mimic the argument in Example 11.18), and f is nonconstant. ∎

The reader should compare the following corollary with Exercise 2(c) in Section 11.4. Recall that $f|_A$ denotes the restriction of a function f to a set A.

Corollary 11.2 Suppose that A_α is a connected subset of $\mathbb{R}$ for each α in an index set I. If $\bigcap_{\alpha \in I} A_\alpha \neq \emptyset$, then $\bigcup_{\alpha \in I} A_\alpha$ is connected.

Proof Let $f : \bigcup_{\alpha \in I} A_\alpha \to \mathbb{Z}$ be a continuous function on $\bigcup_{\alpha \in I} A_\alpha$. By Proposition 11.12, $f|_{A_\alpha}$ is constant for each α in I. Since $\bigcap_{\alpha \in I} A_\alpha \neq \emptyset$, this constant must be the same for all α in I. Thus, f is a constant function, and so $\bigcup_{\alpha \in I} A_\alpha$ is connected by Proposition 11.12. ∎

Continuity and Fixed Points

We end this book with an interesting characterization of closed and bounded intervals in Theorem 11.11 below. Although we would like to give credit to someone for these ideas, we have no reference; this material is taken from a talk given by one of the authors. Recall that a fixed point of a function f is a point c such that $f(c) = c$.

Lemma 11.2 Let $A \subset \mathbb{R}$. If every continuous function from A into A has a fixed point, then A is connected.

Proof Suppose that A is not connected. By Theorem 11.6 there exist a and b in A and a c in $\mathbb{R} \setminus A$ such that $a < c < b$. Define $f : A \to A$ by

$$f(x) = \begin{cases} b & \text{if} \quad x < c \\ a & \text{if} \quad x > c. \end{cases}$$

Then f is a continuous function from A into A with no fixed point. ∎

Theorem 11.11 Let $A \subset \mathbb{R}$ and suppose that A has more than one point. Then A is a closed and bounded interval if and only if every continuous function from A into A has a fixed point.

Proof Let $a < b$. The proof that every continuous function from $[a, b]$ into $[a, b]$ has a fixed point is identical to the proof of Proposition 4.9 (just replace 0 by a and 1 by b). We ask for these details in Exercise 6.

Conversely, suppose that every continuous function from A into A has a fixed point. By Lemma 11.2, A is connected. By Corollary 11.1, A is one of the following: $[a, b], (a, b), (a, b], [a, b), (-\infty, b), (-\infty, b], (a, \infty), [a, \infty)$, or $\mathbb{R}$ where $a < b$. To show that $A = [a, b]$, we eliminate the rest by constructing a continuous function on each of the remaining sets into itself with no fixed point.

First observe that $f(x) = x + 1$ is such a continuous function for the sets $\mathbb{R}, [a, \infty)$, and (a, ∞); while $g(x) = x - 1$ is such a continuous function for the sets $(-\infty, b)$ and $(-\infty, b]$. The other three cases take a little more thought: $f(x) = (a + x)/2$ takes (a, b) or $(a, b]$ onto $(a, (a + b)/2)$ or $(a, (a + b)/2]$, respectively, with no fixed point in the interval, and $g(x) = (x + b)/2$ takes $[a, b)$ onto $[(a + b)/2, b)$ with no fixed point in $[a, b)$. ∎

Exercises

In Exercises 1 to 6, $X \subset \mathbb{R}$.

1. Complete the proof of Theorem 11.7 by showing that parts 2 and 3 are equivalent. [*Hint:* $f^{-1}(\mathbb{R} \setminus F) = X \setminus f^{-1}(F)$.]

2. Let $f : X \to \mathbb{R}$ be continuous on X, and let a be in $\mathbb{R}$. Show that $\{x \in X : f(x) < a\}$ and $\{x \in X : f(x) > a\}$ are open in X and that $\{x \in X : f(x) \le a\}, \{x \in X : f(x) \ge a\}$, and $\{x \in X : f(x) = a\}$ are closed in X.

3. Let f and g be continuous functions from X into $\mathbb{R}$. Then show that $\{x \in X : f(x) = g(x)\}$ is closed in X. Conclude that if $X = \mathbb{R}$ and $f = g$ on a dense subset of $\mathbb{R}$, then $f = g$ on $\mathbb{R}$. [*Hint:* See Exercise 6 in Section 11.2.]

4. Let f be a continuous function from $\mathbb{R}$ onto $\mathbb{R}$. Show that if A is dense in $\mathbb{R}$, then $f(A)$ is dense in $\mathbb{R}$. [*Hint:* Use Proposition 11.11.]

5. Use Propositions 11.11 and 11.12 to show that if A is a connected subset of $\mathbb{R}$, then $\overline{A}$ is also connected. Then compare with Exercise 3 in Section 11.4.

6. Show that a continuous function from $[a, b]$ into $[a, b]$ has a fixed point.

In Exercises 7 to 13, X and Y are subsets of $\mathbb{R}$; and we consider X and Y as topological spaces, each endowed with its own relative topology induced by $\mathbb{R}$ (see Definition 11.8 and Exercise 9 in Section 11.3).

7. Let f be a function from X into Y. Show that the following are equivalent.
    (a) f is continuous on X.
    (b) $f^{-1}(S)$ is open in X for each open set S in Y.
    (c) $f^{-1}(T)$ is closed in X for each closed set T in Y.

8. A function $f : X \to Y$ is called an *open map* if $f(U)$ is open in Y for each open set U in X and is called a *closed map* if $f(F)$ is closed in Y for each closed set F in X. Suppose that f is a one-to-one function from X onto Y. Show that the following are equivalent.

   (a) $f^{-1} : Y \to X$ is continuous on Y.

   (b) f is an open map.

   (c) f is a closed map.

9. Let f be a continuous function from X into Y. If X is compact, show that f is a closed map. [*Hint:* Use Proposition 11.7, Theorem 11.8, and then Proposition 11.6.]

10. A *homeomorphism* from X onto Y is a continuous one-to-one function f from X onto Y such that f^{-1} is continuous on Y. Show that if f is a continuous one-to-one function from X onto Y and X is compact, then f is a homeomorphism from X onto Y.

11. Two topological spaces are called *homeomorphic* if there exists a homeomorphism from one onto the other. In Example 1.11 we actually constructed a homeomorphism from $\mathbb{R}$ onto the open interval $(0, 1)$. A topologist considers homeomorphic spaces to be identical because they have the same topological properties; that is, they are topologically indistinguishable. For example, a topologist considers $\mathbb{R}$ and $(0, 1)$ to be the same topological space.

   Show that $[0, 1]$ and $(0, 1)$ are not homeomorphic.

12. Let X be connected. A point c in X is called a *cut point* of X if $X \setminus \{c\}$ is disconnected. Show that if f is a homeomorphism from X onto Y, then c in X is a cut point of X if and only if $f(c)$ is a cut point of Y.

13. Show that neither $[0, 1)$ nor $(0, 1]$ is homeomorphic to $(0, 1)$. Are $[0, 1)$ and $(0, 1]$ homeomorphic?

Bibliography

Apostol, T. M., *Mathematical Analysis*, Addison-Wesley, Reading, MA, 1964.

Bartle, R. G. and Sherbert, D. R., *Introduction to Real Analysis*, Second Edition, John Wiley and Sons, New York, 1992.

Fridy, J. A., *Introductory Analysis*, Harcourt Brace Jovanovich, San Diego, 1987.

Gelbaum, B. R. and Olmsted, J. M., *Counterexamples in Analysis*, Holden-Day, San Francisco, 1964.

Hewitt, E. and Stromberg, K., *Real and Abstract Analysis*, Springer-Verlag, New York, 1965.

Kirkwood, J. R., *An Introduction to Analysis*, Second Edition, PWS Publishing Company, Boston, 1995.

McArthur, W. G., *An Introduction to the Art of Mathematical Proof*, Shippensburg Collegiate Press, Shippensburg, PA, 1969.

Rudin, W., *Principles of Mathematical Analysis*, Second Edition, McGraw-Hill, New York, 1964.

Simmons, G. F., *Calculus Gems*, McGraw-Hill, New York, 1992.

Spiegel, M. R., *Advanced Calculus*, Schaum's Outline Series, McGraw-Hill, New York, 1963.

Hints and Answers

Section 1.1

3. If $a \neq 0$, then $a^2 + b^2 > 0$.

6. If $a > 0$ and $1/a \leq 0$, then $1 = a \cdot (1/a) \leq a \cdot 0 = 0$.

Section 1.2

1. For the second equality in part 6, let $x \in A \cap (B \cup C)$. Then $x \in A$ and $x \in B \cup C$. So either $x \in A$ and $x \in B$ or $x \in A$ and $x \in C$. Therefore, $x \in (A \cap B) \cup (A \cap C)$ and so $A \cap (B \cup C) \subset (A \cap B) \cup (A \cap C)$. For the other containment, reverse the steps above.

3. Let $x \in X \setminus \bigcap_{\alpha \in I} A_\alpha$. Then $x \in X$ and $x \notin \bigcap_{\alpha \in I} A_\alpha$. So $\exists\, \alpha_0 \in I$ such that $x \notin A_{\alpha_0}$. Then $x \in X \setminus A_{\alpha_0}$ and so $x \in \bigcup_{\alpha \in I}(X \setminus A_\alpha)$. To show the other containment, let $x \in \bigcup_{\alpha \in I}(X \setminus A_\alpha)$. Then $\exists\, \alpha_0 \in I$ with $x \in X \setminus A_{\alpha_0}$. So $x \in X$ and $x \notin A_{\alpha_0}$ and thus $x \in X$ and $x \notin \bigcap_{\alpha \in I} A_\alpha$. Therefore, $x \in X \setminus \bigcap_{\alpha \in I} A_\alpha$.

5. If $x \in B \setminus (B \setminus A)$, then $x \in B$ and $x \notin B \setminus A$. If $x \notin A$, then $x \in B \setminus A$. Therefore $x \in A$. Thus, $B \setminus (B \setminus A) \subset A$.

8. Draw a picture. $\bigcup_{n=1}^{\infty} A_n = (1, \infty)$. $\bigcap_{n=1}^{\infty} A_n = \emptyset$, because if $x \in \bigcap_{n=1}^{\infty} A_n$, then $x > n$ for all n in $\mathbb{N}$.

9. $\bigcup_{n=1}^{\infty} A_n = [0, 1]$. $\bigcap_{n=1}^{\infty} A_n = \{0\}$.

10.
$$x \in X \cap \left(\bigcup_{\alpha \in I} A_\alpha \right) \Leftrightarrow x \in X \text{ and } x \in \bigcup_{\alpha \in I} A_\alpha$$
$$\Leftrightarrow x \in X \text{ and } x \in A_{\alpha_0} \text{ for some } \alpha_0 \in I$$
$$\Leftrightarrow x \in X \cap A_{\alpha_0} \text{ for some } \alpha_0 \in I$$
$$\Leftrightarrow x \in \bigcup_{\alpha \in I}(X \cap A_\alpha)$$

Section 1.3

2. (a) $\{-1\} \cup [0, 1]$ (b) $(-\infty, 1]$ (c) $(-\infty, 0)$

   (d) $\emptyset$ (e) $(-\infty, 0)$ (f) $\{-1\}$

4. One can easily show that $A \subset f^{-1}(f(A))$ always $[x \in A \Rightarrow f(x) \in f(A) \Rightarrow x \in f^{-1}(f(A))]$.

Let $x \in f^{-1}(f(A))$. Then $f(x) \in f(A)$ and so $\exists$ an $a \in A$ such that $f(x) = f(a)$. Since f is 1-1, $x = a \in A$. Thus, $f^{-1}(f(A)) \subset A$ if f is 1-1.

Example. Let $f(x) = x^2$ on $\mathbb{R}$. Let $A = [0, 1]$. Then $f^{-1}(f(A)) = [-1, 1] \supsetneq A$.

7. For x_1 and x_2 in X, if $f(x_1) = f(x_2)$, then $g(f(x_1)) = g(f(x_2))$. Since $g \circ f$ is 1-1, $x_1 = x_2$.

Example. Define f and g on $\mathbb{R}$ by $f(x) = e^x$ and $g(x) = x^2$. Then g is not 1-1 but $g \circ f(x) = e^{2x}$ is 1-1. Note that g is 1-1 on the range of f.

10. $f(x) = (b - a)x + a$ is the equation of the straight line segment joining $(0, a)$ to $(1, b)$. So f is a bijection of $(0, 1)$ onto (a, b).

11. Let $y_1, y_2 \in Y$ with $f^{-1}(y_1) = f^{-1}(y_2)$. By Proposition 1.13, $y_1 = f(f^{-1}(y_1)) = f(f^{-1}(y_2)) = y_2$. So f^{-1} is 1-1.

If $x \in X$, then $f(x) \in Y$. By Proposition 1.13, $x = f^{-1}(f(x))$. So f^{-1} is onto X.

Section 1.4

Below we give the step in going from $p(k)$ is true to $p(k + 1)$ is true.

2. $1^3 + 2^3 + \cdots + k^3 + (k + 1)^3 = \left[\dfrac{k(k + 1)}{2} \right]^2 + (k + 1)^3$

$$= (k + 1)^2 \left[\dfrac{k^2}{4} + (k + 1) \right]$$

$$= (k + 1)^2 \dfrac{(k + 2)^2}{4}$$

$$= \left[\dfrac{(k + 1)(k + 2)}{2} \right]^2$$

4. $2^{k+1} = 2(2^k) < 2(k!) \leq (k + 1)(k!) = (k + 1)!$

5. $7^{k+1} - 3^{k+1} = (7^k \cdot 7 - 7 \cdot 3^k) + (7 \cdot 3^k - 3^k \cdot 3)$

$$= 7(7^k - 3^k) + 4 \cdot 3^k$$

Since 4 divides $7^k - 3^k$ by the induction hypothesis and 4 divides $4 \cdot 3^k$, 4 divides $7^{k+1} - 3^{k+1}$.

9. Let $X = \{x_1, \ldots, x_k, x_{k+1}\}$. Then $\{x_1, \ldots, x_k\}$ has 2^k subsets by the induction hypothesis. Since all other subsets of X are of the form $A \cup \{x_{k+1}\}$ where $A \subset \{x_1, \ldots, x_k\}$, there are 2^k subsets of X containing x_{k+1}. So the number of subsets of X is $2^k + 2^k = 2^{k+1}$.

10. $(1 + x)^{k+1} = (1 + x)^k (1 + x)$

$$\geq (1 + kx)(1 + x)$$

$$= 1 + (k + 1)x + kx^2$$

$$\geq 1 + (k + 1)x$$

Section 2.1

1. If $x + r = s \in \mathbb{Q}$, then $x = s - r \in \mathbb{Q}$ since $\mathbb{Q}$ is a field.

2. $-\sqrt{2} + \sqrt{2} = 0$ and $\sqrt{2} \cdot \sqrt{2} = 2$

6. $\frac{1657}{4950}$

9. $a \le b \Rightarrow a^2 = a \cdot a \le a \cdot b$ (since $a \ge 0$) and $a \le b \Rightarrow a \cdot b \le b \cdot b = b^2$. Transitivity $\Rightarrow a^2 \le b^2$.

Let $a^2 \le b^2$ and assume $a \ne b$. Then $b^2 - a^2 > 0$ or $0 < (b-a)(b+a)$. Since $b + a > 0$, $b - a > 0$ and so $b > a$.

10. From the proof of the triangle inequality, equality holds $\Leftrightarrow |ab| = ab \Leftrightarrow ab \ge 0$.

Section 2.2

1. (a) $\sup = \sqrt{2}$, $\inf = -\sqrt{2}$ (b) $\sup = \sqrt{2}$, $\inf = -\sqrt{2}$

(c) $\sup = \infty$, $\inf = \sqrt{2}$ (d) $\sup = \infty$, $\inf = -\infty$

3. Since $A \ne \emptyset$, $\inf A \le \sup A$. Since $A \subset B$, $\sup B$ is an upper bound of A. Since $\sup A$ is the least upper bound of A, $\sup A \le \sup B$. Similarly, $\inf B$ is a lower bound of A and so $\inf B \le \inf A$.

6. To show $\sup(a + S) = a + \sup S$, let $\alpha = \sup S \in \mathbb{R}$. $\forall s \in S$, $s \le \alpha \Rightarrow a + s \le a + \alpha \Rightarrow a + \alpha$ is an upper bound of $a + S$.

Let γ be a real upper bound of $a + S$. We need to show that $a + \alpha \le \gamma$. $\forall s \in S$, $a + s \le \gamma \Rightarrow s \le \gamma - a \Rightarrow \gamma - a$ is an upper bound of S. Since $\alpha = \sup S$, $\alpha \le \gamma - a$ or $a + \alpha \le \gamma$.

7. Let $\alpha = \sup A$, $\beta = \sup B$, $\eta = \max\{\alpha, \beta\}$. We need to show that $\eta = \sup(A \cup B)$.

If $x \in A$, then $x \le \alpha \le \eta$, and if $x \in B$, then $x \le \beta \le \eta$. So, η is an upper bound of $A \cup B$.

If γ is an upper bound of $A \cup B$, then γ is an upper bound of A and γ is an upper bound of B. So $\gamma \ge \alpha$ and $\gamma \ge \beta$ and thus, $\gamma \ge \eta$. Therefore, $\eta = \sup(A \cup B)$.

Section 2.3

3. (a) $\sup = 1$, $\inf = 0$ (b) $\sup = +\infty$, $\inf = -\infty$

(c) $\sup = +\infty$, $\inf = 0$ (d) $\sup = \sqrt{2}$, $\inf = -\sqrt{2}$

6. Note that $\sup\{f(x)\} + \sup\{g(x)\}$ is an upper bound of $\{f(x) + g(x)\}$. The functions in Example 2.3 will serve as an example.

7. (b) $\forall y \in X$, $g(y)$ is an upper bound of $\{f(x) : x \in X\}$, and so $\sup f(X) \le g(y)$ $\forall y \in X$. Thus $\sup f(X)$ is a lower bound of $\{g(y) : y \in X\}$.

(c) Let $f, g : [0, 1] \to \mathbb{R}$ be defined by $f(x) = x^2$ and $g(x) = x$.

Section 2.4

3. Let f be a bijection of $\{1, \ldots, n\}$ onto A and let g be a bijection of $\{1, \ldots, m\}$ onto B. Define h from $\{1, \ldots, n, n+1, \ldots, n+m\}$ onto $A \cup B$ by

$$h(i) = \begin{cases} f(i) & \text{if } i \in \{1, \ldots, n\} \\ g(i - n) & \text{if } i \in \{n+1, \ldots, n+m\}. \end{cases}$$

Use induction for the second part.

4. Let $(x_n)_{n=1}^{\infty}$ be a sequence of distinct points in $X \setminus \{x\}$. Map $x \to x_1$, $x_n \to x_{n+1} \forall n$, $y \to y$ otherwise.

7. $A \sim$ subset of B

11. Choose a rational in each member of $\mathfrak{A}$ (since $\mathbb{Q}$ is dense in $\mathbb{R}$). Since $\mathfrak{A}$ is pairwise disjoint, these rationals are distinct. Thus, $\mathfrak{A} \sim$ subset of $\mathbb{Q}$.

13. Either assume A is countable and use a Cantor diagonalization argument or note that every x in $[0, 1]$ has a binary representation—that is, a representation as a sequence of 0's and 1's.

Section 3.1

1. (b) Choose $n_0 \in \mathbb{N}$ with $1/n_0 < \varepsilon/6$.
  (c) Choose $n_0 \in \mathbb{N}$ with $1/n_0 < \frac{25}{29}\varepsilon$.
  (d) Choose $n_0 \in \mathbb{N}$ with $1/n_0 < (2/\sqrt{3})\sqrt{\varepsilon}$.

2. Yes.

3. No.

7. Both sequences are unbounded.

9. (a) Note that $0 < e/\pi < 1$.
  (c) Write $n^{1/n} = e^{(1/n) \ln n}$ and proceed as in Example 3.6.
  (e) $$\frac{n^2}{n!} = \frac{n}{(n-1)(n-2)\cdots(3)(2)(1)} < \frac{n}{(n-1)(n-2)} < \frac{n}{n^2 - 3n} = \frac{1}{n-3}$$

Section 3.2

1. (a) 1   (b) Limit does not exist.   (c) $\frac{1}{2}$
  (d) 9   (e) 0   (f) $-\frac{1}{2}$

3. Write $x_n = [(x_n + y_n) + (x_n - y_n)]/2$ and use Theorem 3.2.

6. If $|x_n| \le B \ \forall n \in \mathbb{N}$, then $0 \le |x_n y_n| \le B|y_n| \to 0$.

Section 3.3

1. (c) No. See Section 3.5.

2. (a) e (b) $e^{1/2}$

4. For the one direction, note that $(x_n)_{n \in \mathbb{N}}$ and $(y_n)_{n \in \mathbb{N}}$ are both subsequences of $(z_n)_{n \in \mathbb{N}}$. For the other direction,
$$z_n = \begin{cases} x_{(n+1)/2} & \text{if } n \text{ is odd} \\ y_{n/2} & \text{if } n \text{ is even}. \end{cases}$$

5. (a) Let n_1 be the first positive integer with $x_{n_1} > 1$. Let n_2 be the first positive integer greater than n_1 with $x_{n_2} > 2$. Continue, getting $x_{n_k} > k$ at each stage.

Section 3.4

1. $(x_n)_{n \in \mathbb{N}}$ is bounded above by 2 and strictly increasing, $x_n \to \frac{3}{2}$.

2. $x_n \to 1$

3. $x_n \to 2$

6. $\bigcap_{n=1}^{\infty} (0, 1/n] = \emptyset$ and $\bigcap_{n=1}^{\infty} [1 - (1/n), 1) = \emptyset$

8. (c) Let $\varepsilon > 0$. Choose $n_0 \in \mathbb{N}$ with $b_{n_0} - a_{n_0} < \varepsilon$. Then $0 \le \beta - \alpha \le b_{n_0} - a_{n_0} < \varepsilon$.

Section 3.5

1. $[0, 7]$

3. First use Theorem 3.10. Then use Proposition 3.3.

4. $\mathbb{R}$

5. $\{1/n\}_{n \in \mathbb{N}} \cup \{1 + (1/n)\}_{n \in \mathbb{N}} \cup \{2 + (1/n)\}_{n \in \mathbb{N}}$ has 0, 1, and 2 as accumulation points.

7. Sequences that are eventually the constant x.

Section 3.6

2. (a) Combine Exercise 4 in Section 3.4 and Theorem 3.12.

3. Let $\varepsilon = 1$ in Definition 3.11. Then eventually $|x_n - x_m| < 1$.

4. (c) For $m > n$,

$$|x_m - x_n| \le |x_m - x_{m-1}| + |x_{m-1} - x_{m-2}| + \cdots + |x_{n+1} - x_n|$$

$$= \frac{L}{2^{n-1}} \left[1 + \frac{1}{2} + \left(\frac{1}{2}\right)^2 + \cdots + \left(\frac{1}{2}\right)^{m-n-1} \right]$$

$$= \frac{L}{2^{n-1}} \cdot \frac{1 - \left(\frac{1}{2}\right)^{m-n}}{1 - \frac{1}{2}}.$$

5. (b) In the induction step, use

$$x_{2k+3} = \frac{x_{2k+2} + x_{2k+1}}{2} = \frac{x_{2k+1} + (L/2^{2k}) + x_{2k+1}}{2} \quad (by \; part \; (a))$$

$$= x_{2k+1} + \frac{L}{2^{2k+1}}.$$

(c) $x = \frac{1}{3}a + \frac{2}{3}b$

7. (b) $x_n \to -1 + \sqrt{2}$

Section 3.7

1. Let $\alpha > 0$ and $\beta < 0$.

(c) Choose $n_0 \in \mathbb{N}$ with $n_0 > 6 + \alpha$.

(d) Choose $n_0 \in \mathbb{N}$ with $n_0 > (6 + \alpha)^2$.

(e) Choose $n_0 \in \mathbb{N}$ with $n_0 > 7 + \beta^2$.

3. (a) If $|x_n| \to \infty$ and $\beta < 0$, then $(|x_n|)_{n \in \mathbb{N}}$ is eventually in $(-\beta, \infty)$. Since $x_n < 0 \ \forall n$, $(x_n)_{n \in \mathbb{N}}$ is eventually in $(-\infty, \beta)$.

   (b) Let $(x_n)_{n \in \mathbb{N}} = (1, -2, 3, -4, 5, -6, \ldots)$.

5. Since $|x_n| \to \infty$, we can assume that $x_n \neq 0 \ \forall n \in \mathbb{N}$. Then $1/|x_n| \to 0$. Since $x_n y_n \to L \in \mathbb{R}$, $|y_n| = (1/|x_n|) |x_n y_n| \to 0 \cdot |L| = 0$.

7. (a) Eventually, $x_n/y_n < 1$ or $x_n < y_n$.

   (b) If $y_n \leq B \ \forall n \in \mathbb{N}$, then $0 < x_n = y_n(x_n/y_n) \leq B(x_n/y_n) \to 0$.

9. Since $\lim\limits_{n \to \infty} x_n \neq \infty$ and $x_n > 0 \ \forall n$, $(x_n)_{n \in \mathbb{N}}$ has a subsequence $(y_n)_{n \in \mathbb{N}}$ that is bounded. Apply Theorem 3.10 to $(y_n)_{n \in \mathbb{N}}$.

Section 3.8

1. (a), (c), (d), (e) $\limsup x_n = 1$ and $\liminf x_n = -1$.

   (b) $\limsup x_n = \infty$ and $\liminf x_n = -\infty$.

2. Any sequence with limit ∞

3. If $\beta = \liminf y_n$ is ∞, the result is clear. So assume that $\beta < \infty$. Let $(y_{n_k})_{k=1}^{\infty}$ be a subsequence of $(y_n)_{n \in \mathbb{N}}$ with $y_{n_k} \to \beta$. The corresponding subsequence $(x_{n_k})_{k=1}^{\infty}$ of $(x_n)_{n \in \mathbb{N}}$ has a subsequence with limit in $\mathbb{R}^{\#}$. Since $x_{n_k} \leq y_{n_k} \ \forall k$, this limit must be $\leq \beta$.

Section 4.1

1. Choose $\delta = \varepsilon$.

2. Choose $\delta = \min\{|c|/2, c^2\varepsilon/2\}$.

4. For $f(c) < h$, let $\varepsilon = h - f(c) > 0$. Then $\exists$ a neighborhood U of c such that $x \in U \cap D \Rightarrow f(x) \in (f(c) - \varepsilon, f(c) + \varepsilon)$.

5. Let $\varepsilon = 1$.

6. g is the composition of two continuous functions.

9. To show that fg is continuous at c, note that
$$|f(x)g(x) - f(c)g(c)| \leq |f(x)||g(x) - g(c)| + |g(c)||f(x) - f(c)|$$
and f is bounded on a neighborhood of c by Exercise 5. Now make each summand less than $\varepsilon/2$.

Section 4.2

2. For $c \in \mathbb{Q}$, let $(x_n)_{n \in \mathbb{N}}$ be a sequence in $\mathbb{R} \setminus \mathbb{Q}$ with $x_n \to c$. Then $f(x_n) \to 0 \neq f(c)$. For $c \in \mathbb{R} \setminus \mathbb{Q}$, let $(y_n)_{n \in \mathbb{N}}$ be a sequence in $\mathbb{Q}$ with $y_n \to c$. Then $f(y_n) \to 1 \neq f(c)$.

4. To show that f is continuous at $\frac{1}{2}$, let $\delta = \varepsilon$. To show that f is discontinuous at $c \neq \frac{1}{2}$, proceed as in Exercise 2.

6. For $x \in \mathbb{R} \setminus \mathbb{Q}$, let $(x_n)_{n \in \mathbb{N}}$ be a sequence in $\mathbb{Q}$ with $x_n \to x$.

7. Let $f(x) = 1/x$ and consider $(1/n)_{n=2}^{\infty}$.

Section 4.3

2. For $c \in \mathbb{R}$, let $(x_n)_{n \in \mathbb{N}}$ be a sequence in $\mathbb{Q} \setminus \{c\}$ and $(y_n)_{n \in \mathbb{N}}$ be a sequence in $(\mathbb{R} \setminus \mathbb{Q}) \setminus \{c\}$ with $x_n \to c$ and $y_n \to c$. Then $f(x_n) \to 1$ and $f(y_n) \to -1$. Use Proposition 4.6.

4. If $\lim\limits_{x \to c} f(x) < a$, then $f(x) < a$ for $x \neq c$ in some neighborhood of c.

6. Let $V = (L - 1, L + 1)$, where $L = \lim\limits_{x \to c} f(x)$.

7. Let $V = (L/2, (3/2)L)$, where $L = \lim\limits_{x \to c} f(x)$.

10. (a) $f(0) = f(0) + f(0) \Rightarrow f(0) = 0.$ $f(2) = f(1) + f(1) = 2f(1).$
    $\forall n \in \mathbb{N},\ 0 = f(0) = f(-n + n) = f(-n) + f(n)$, so that $f(-n) = -f(n)$.

    (b) Let $p \in \mathbb{Z}$ and $q \in \mathbb{N}$. Then

$$pf(1) \underset{(a)}{=} f(p) = f\left(q\left(\frac{p}{q}\right)\right) = f\left(\frac{p}{q} + \cdots + \frac{p}{q}\right) = qf\left(\frac{p}{q}\right).$$

Section 4.4

2. Any constant function or $f(x) = 1/(x^2 + 1)$ maps $(-1, 1)$ onto $(\frac{1}{2}, 1]$.

4. $f(0) > 0$, $f(1) < 0$, $f(2) > 0$. Use the Intermediate Value Theorem.

5. Recall Example 4.15 and use the Intermediate Value Theorem.

7. Suppose that f is nonconstant and use the Intermediate Value Theorem.

8. Define $g : [0, \frac{1}{2}] \to \mathbb{R}$ by $g(x) = f(x) - f(x + \frac{1}{2})$. If $g(0) \neq 0$, apply the Intermediate Value Theorem to g.

Section 4.5

1. Let $\varepsilon = 1$ and $\delta > 0$. Let $x > 0$ and let $y = x + (\delta/2)$. Show that $|f(x) - f(y)| \geq (\delta/2)(3x^2)$. Now choose x.

3. $(1/[(\pi/2) + n\pi])_{n \in \mathbb{N}}$ is Cauchy, but its image under f is not Cauchy.

5. On $[1, \infty)$, $\delta = \varepsilon/2$ will work. On $(0, 1]$, consider the Cauchy sequence $(1/n)_{n \in \mathbb{N}}$.

8. Suppose that f is not bounded on D. Then $\exists$ a sequence $(x_n)_{n \in \mathbb{N}}$ in D with $|f(x_n)| > n\ \forall n \in \mathbb{N}$. Use the Bolzano-Weierstrass Theorem to obtain a Cauchy subsequence of $(x_n)_{n \in \mathbb{N}}$.

9. For fg, let $f(x) = g(x) = x$ on $\mathbb{R}$.

11. Let $f(x) = \begin{cases} -1 & \text{if } x < \sqrt{2} \\ 1 & \text{if } x > \sqrt{2}. \end{cases}$

Section 4.6

1. f has a discontinuity of the first kind at each integer. Also, f is monotone increasing.

2. f has a discontinuity of the first kind at each integer. f is not monotone.

4. f has a discontinuity of the second kind at every $c \neq 0$.

9. Let $f(x) = \tan x$ on $(-\pi/2, \pi/2)$. The difference between this and Theorem 4.8 is that $-\pi/2$ and $\pi/2$ are not in the domain of f.

10. If f is discontinuous at $c \in I$, then no number between $f(c-)$ and $f(c+)$ can be in the range of f [except $f(c)$].

Section 5.1

2. For part 2 of Proposition 5.1, note that
$$\frac{f(x)g(x) - f(c)g(c)}{x - c} = f(x) \cdot \frac{g(x) - g(c)}{x - c} + g(c) \cdot \frac{f(x) - f(c)}{x - c}.$$
By Theorem 5.1, f is continuous at c. As $x \to c$, the left side approaches $(fg)'(c)$ while the right side approaches $f(c)g'(c) + g(c)f'(c)$. Letting g be the constant function a, part 3 follows from part 2.

5. $f'(x) = \sin(1/x) - (1/x)\cos(1/x)$ for $x \neq 0$. $f'(0) = \lim\limits_{x \to 0} \sin(1/x)$ does not exist.

6. $f(x) - f(c)$ may be 0 for some x in any neighborhood of c.

8. Theorem 4.2 implies that f has an absolute maximum and minimum on $[a, b]$. If f is not identically zero, at least one of these occurs at an interior point.

9. (a) $x > c \Rightarrow [f(x) - f(c)]/(x - c) \geq 0$ and $x < c \Rightarrow [f(x) - f(c)]/(x - c) \geq 0$. Hence, $f'(c) \geq 0$.

11. No such function exists by Theorem 5.3.

Section 5.2

1. (a) and (b) follow from the Mean Value Theorem, while (c) follows from the intermediate value property of f'.

3. Use the Mean Value Theorem on $[x_1, x_2]$, where x_1 and x_2 are two consecutive roots of f. Example: $f(x) = x^2 + 1$.

5. Use the Mean Value Theorem to show that $(a_n)_{n \in \mathbb{N}}$ is Cauchy.

7. For $a < x < b$, the Mean Value Theorem implies that $[f(x) - f(a)]/(x - a) = f'(c_x)$, where $a < c_x < x$. Let $x \to a$.

8. $f(x) = \sqrt{x}$

11. For $x > 1$, apply the Mean Value Theorem to $f(t) = \ln t$ on $[1, x]$.

Section 5.3

1. $P_7(x) = x - \dfrac{x^3}{3!} + \dfrac{x^5}{5!} - \dfrac{x^7}{7!}$. $R_7(x) = \dfrac{(\sin c)x^8}{8!}$, c between 0 and x.

2. $P_7(x) = 1 - \dfrac{x^2}{2!} + \dfrac{x^4}{4!} - \dfrac{x^6}{6!}$. $R_7(x) = \dfrac{(\cos c)x^8}{8!}$, c between 0 and x.

3. $P_7(x) = x - \dfrac{x^2}{2} + \dfrac{x^3}{3} - \dfrac{x^4}{4} + \dfrac{x^5}{5} - \dfrac{x^6}{6} + \dfrac{x^7}{7}$. $R_7(x) = \dfrac{-x^8}{8(c + 1)^8}$, $0 < c < x$.

4. For $\sin x$ and $\cos x$, $n = 9$. For $\ln(1 + x)$, $n = 10^6$.

Section 5.4

1. $-\infty$
2. 0
3. 1
4. $e^{-1/2}$
5. 2
6. 0. Note that $0 < e/\pi < 1$.
7. 0
8. e^3

Section 6.1

2. $\int_0^1 f = 1$
3. $U(P, f) - L(P, f) \le 2 \|P\| \, \forall P; \int_0^1 f = 1$
4. $U(P, f) \le 6 \|P\|$ whenever $1 \in P; \int_0^2 f = 0$
5. $U(P, f) = 1 \, \forall P$. Let $0 < \varepsilon < 1$ and choose n such that $1/n < \varepsilon/2$. Let
$$P = \left\{0, \frac{1}{n}, \frac{1}{n-1}, \ldots, \frac{1}{3}, \frac{1}{2}, 1\right\} = \{x_i\}_{i=0}^n.$$
$\forall i = 1, \ldots, n-1$, add $y_i < z_i$ between x_i and x_{i+1} with $y_i - x_i < \varepsilon/4n$ and $x_{i+1} - z_i < \varepsilon/4n$. For this refinement, the upper sum minus the lower sum is less than ε. $\int_0^1 f = 1$.
7. $L(P, f) = 0 \, \forall P$. Show that $U(P, f) \ge \frac{1}{4} \, \forall P$ by passing to a refinement containing $\frac{1}{2}$.

Section 6.2

2. $\int_0^1 x \, dx$
3. $\int_0^1 [1/(x - 2)] \, dx$
4. $\int_0^1 [1/(1 + x)] \, dx$
5. $\int_0^1 \sin[(\pi/2)x] dx = (2/\pi) \int_0^{\pi/2} \sin x \, dx$

Section 6.3

1. Use Theorem 6.4.
3. Use Theorem 6.2 with the union of the appropriate partitions of $[a, c]$ and $[c, b]$.
5. Theorem 6.5 implies that $f \in \mathcal{R}[x_{i-1}, x_i] \, \forall i = 1, 2, \ldots, n$. Extend Exercise 3 to n subintervals of $[a, b]$.
8. By Proposition 6.5 and Corollary 6.1, $\left| \int_a^b f - \int_c^b f \right| = \left| \int_a^c f \right| \le \int_a^c |f|$. Now use the fact that f is bounded.
9. Note that each Riemann sum for f on $[0, a]$ has a corresponding Riemann sum for f on $[-a, 0]$.

10. Corollary 6.1 implies that fg, f^2, and g^2 are in $\mathcal{R}[a, b]$. Follow the hint to obtain $Ax^2 + Bx + C \geq 0$, where $A = \int_a^b f^2$, $B = 2\int_a^b fg$, and $C = \int_a^b g^2$.

Section 6.4

1. Since f agrees with a continuous function on $[0, 1]$ except at 0, the result follows from Theorem 6.5.

3. Use Theorem 6.6.

7. Apply Theorem 6.8 to $\int_a^b (f - g) = 0$.

9. Let m and M be the minimum and maximum values of f on $[a, b]$. Then $m\int_a^b g \leq \int_a^b fg \leq M\int_a^b g$. For $\int_a^b g \neq 0$, the Intermediate Value Theorem implies that $f(c) = \int_a^b fg / \int_a^b g$ for some c in $[a, b]$.

10. $c = 0$

Section 6.5

1. $F(x) = \begin{cases} 0 & \text{if} \ \ 0 \leq x \leq 1 \\ x - 1 & \text{if} \ \ 1 \leq x \leq 2 \end{cases}$

2. $F(x) = x$

5. $3x^2 \cos(x^6) - 2x \cos(x^4)$

7. Use Theorem 6.12.

9. Since $\int_0^x f$ is differentiable by Theorem 6.12, f is differentiable. Differentiate $f^2(x) = 2\int_0^x f$ to obtain $f'(x) = 1 \ \forall x > 0$.

Section 6.6

1. For $x \geq 1$, $1/(x\sqrt{x+1}) \leq 1/x^{3/2}$. So $\int_1^\infty [1/(x\sqrt{x+1})]\,dx$ converges by the Comparison test.

For $0 < x \leq 1$, $1/(x\sqrt{x+1}) \geq 1/(x\sqrt{2})$. So $\int_0^1 [1/(x\sqrt{x+1})]\,dx$ diverges by the Comparison test.

3. $|e^{-x}\sin(1/x)| \leq e^{-x} \ \forall x > 0$. Use the Comparison test to obtain that both $\int_1^\infty e^{-x}\sin(1/x)\,dx$ and $\int_0^1 e^{-x}\sin(1/x)\,dx$ converge.

4. Let $y = 1/x$ and use Example 6.17.

6. Use the Limit Comparison test with x^{-p}.

10. (b) Letting $u = x^{p-1}$ and $dv = e^{-x}dx$, integration by parts yields

$$\Gamma(p) = -e^{-x}x^{p-1}\Big|_0^\infty + (p - 1)\int_0^\infty e^{-x}x^{p-2}\,dx.$$

The first term on the right is 0 since $p - 1 > 0$.

11. $\int_0^\infty [1/(x^2 + 1)]\,dx = \lim_{t \to \infty}\left(\arctan x\Big|_0^t\right) = \lim_{t \to \infty} \arctan t = \pi/2$

Section 7.1

3. $\frac{1}{2}$ (this telescopes).

4. For $n \geq n_0$, $\sum\limits_{k=1}^{n} a_k = c + \sum\limits_{k=1}^{n} b_k$ for some constant c.

5. 1

7. Eventually, $a_n < 1$ and so $a_n^2 \leq a_n$.

9. (b) Use $b_n = 1/n^{3/2}$.

Section 7.2

3. The limit in Raabe's test is $\frac{1}{2}$.

4. (a) Let $s_0 = 0$, $s_n = \sum\limits_{k=1}^{n} a_k$ for n in $\mathbb{N}$, and $A = \sum\limits_{n=1}^{\infty} a_n$. Then $r_n = A - s_{n-1}$.

(b) Use Dirichlet's test.

(c) For $n > m$, $\sum\limits_{k=m+1}^{n} a_k/r_k = 1 - (r_{n+1}/r_{m+1})$. For fixed m, there is an $n > m$ such that $r_{n+1}/r_{m+1} < 1/2$ since $r_{n+1} \to 0$.

5. Use Dirichlet's test.

7. $(b_n)_{n\in\mathbb{N}}$ converges to some b in $\mathbb{R}$. Either $(b - b_n)_{n\in\mathbb{N}}$ or $(b_n - b)_{n\in\mathbb{N}}$ is monotone decreasing with limit 0.

9. (a) For $n > m$, $\sum\limits_{k=m+1}^{n} a_k/s_k = 1 - (s_m/s_n)$ and $s_n \to \infty$.

Section 7.3

1. Use Theorem 7.11.

3. (a) Yes

(b) No. Consider $a_n = (-1)^{n+1}$.

(c) Yes. Use the Comparison test.

5. No. See Theorem 7.12.

7. In the notation of the proof of Theorem 7.13, choose $p_1 + \cdots + p_{n_1} > 1$, then $p_1 + \cdots + p_{n_1} - q_1 + p_{n_1+1} + \cdots + p_{n_2} > 2$, then $p_1 + \cdots + p_{n_1} - q_1 + p_{n_1+1} + \cdots + p_{n_2} - q_2 + p_{n_2+1} + \cdots + p_{n_3} > 3$, and so on.

9. Given $x > 0$, $\exists n$ with $a_1 + \cdots + a_n > x$. Then $\{1, 2, \ldots, n\} \subset \{\varphi(1), \ldots, \varphi(m)\}$ for some m.

Section 7.4

3. $\frac{3}{4}$

4. $\frac{4}{9}$

5. (b) Consider the series in Example 7.17.

Section 8.1

1. On $[0, b]$, choose $n_0 \in \mathbb{N}$ such that $1/n_0 < \varepsilon/b$. On $[0, \infty)$, note that $f_n(n) = 1 \ \forall n \in \mathbb{N}$.

3. $\forall n \in \mathbb{N}$ the maximum value of f_n on $[0, 1]$ is $f_n(1/(n+1)) \underset{n}{\to} 1/e$.

4. For Example 8.3, the maximum value of f_n is $f_n(1/\sqrt{2n+1}) \underset{n}{\to} \infty$.

6. $f_n(1/n) = \frac{1}{2} \ \forall n \Rightarrow M_n$ (of Exercise 5) $\geq \frac{1}{2} \ \forall n$

9. (b) Note that $|f_n(x)g_n(x) - F(x)G(x)| \leq |f_n(x)||g_n(x) - G(x)| + |G(x)||f_n(x) - F(x)|$.

Section 8.2

1. $\displaystyle \lim_{n\to\infty} f_n(x) = \begin{cases} 1 & \text{if } x = 0 \\ 0 & \text{if } x > 0. \end{cases}$

3. $\int_0^2 f_n = \frac{1}{2}(2/n)(n) = 1 \ \forall n \in \mathbb{N}$

4. From Example 8.3, $\int_0^1 f_n = n^2/(2n+2) \to \infty$.

7. First use Definition 8.2. Then use Theorem 8.1 to conclude that $F(x_n) \underset{n}{\to} F(x)$.

9. Let $\varepsilon > 0$ and choose $0 < \delta < \frac{1}{2}(b-a)$ by the uniform continuity of F corresponding to $\varepsilon/5$. Consider the partition $\{x_i\}_{i=0}^{k+1}$ of $[a,b]$, where $x_0 = a$, $x_{k+1} = b$, and $x_i = a + i(\delta/2) \ \forall i = 0, 1, 2, \ldots, k$, where k is the smallest positive integer such that $a + (k+1)(\delta/2) \geq b$. Then use the pointwise convergence of $(f_n(x_i))_{n\in\mathbb{N}} \ \forall i = 0, 1, 2, \ldots, k+1$.

Section 8.3

1. (d) $|f_n(x)| \leq \left(\frac{1}{3}\right)^n \ \forall x \in D$

2. $\displaystyle \sum_{n=0}^{\infty} x(1-x)^n = \begin{cases} 0 & \text{if } x = 0 \\ 1 & \text{if } 0 < x \leq 1 \end{cases}$

4. From Lemma 7.1, $\sum_{k=1}^{n} f_k(x)g_k(x) = \sum_{k=1}^{n} s_k(x)[g_k(x) - g_{k+1}(x)] + g_{n+1}(x) \cdot s_n(x)$, where $s_n = \sum_{k=1}^{n} f_k$. For $n > m$, argue that $\left| \sum_{k=m+1}^{n} f_k(x)g_k(x) \right| \leq 2Mg_{m+1}(x)$, where M is the uniform bound on $(s_n)_{n\in\mathbb{N}}$.

6. Use Exercise 4 for uniform convergence. Consider $x = 0$ for nonabsolute convergence.

8. Use the Weierstrass M-test for uniform convergence and Theorem 8.7 for the derivative.

Section 8.4

1. (a) $B_1(x) = B_2(x) = B_3(x) = 1 - x$
   (b) $B_1(x) = x$; $B_2(x) = \frac{1}{2}x + \frac{1}{2}x^2$; $B_3(x) = \frac{1}{3}x + \frac{2}{3}x^2$
   (c) $B_1(x) = 1 + (e-1)x$; $B_2(x) = [1 + (e^{1/2} - 1)x]^2$; $B_3(x) = [1 + (e^{1/3} - 1)x]^3$

3. Let $Q_n(x) = P_n(x) - P_n(0)$, where $(P_n)_{n\in\mathbb{N}}$ is a sequence of polynomials chosen according to Theorem 8.10.

Section 9.1

2. (a) $R = \frac{1}{2}$, absolute convergence on $\left[-\frac{1}{2}, \frac{1}{2}\right]$.
   (b) $R = \infty$, interval of convergence is $(-\infty, \infty)$.
   (c) $R = 1$, interval of convergence is $[3, 5)$, absolute convergence on $(3, 5)$.
   (d) $R = e$, interval of convergence is $(2 - e, 2 + e)$.

4. (b) For $x \neq 0$, the limit inferior is 0 and the limit superior is ∞.

6. After the integration, you should have $\sum\limits_{n=0}^{\infty} x^{n+1}/(n+1) = \sum\limits_{n=1}^{\infty} x^n/n = -\ln|1-x|$ for $|x| < 1$.

7. After the integration, you should have $\sum\limits_{n=0}^{\infty} (-1)^n x^{2n+1}/(2n+1) = \arctan x$ for $|x| < 1$.

8. Let $R_n(x) = a_n x^n + a_{n+1}x^{n+1} + \cdots$ for each x in $[0, 1]$. Then $\sum\limits_{n=0}^{\infty} a_n x^n$ converges uniformly on $[0, 1] \Leftrightarrow \forall \varepsilon > 0, \exists\, n_0$ in $\mathbb{N}$ (n_0 independent of x) such that if $n \geq n_0$, then $|R_n(x)| < \varepsilon\; \forall x$ in $[0, 1]$. Let $R_n = R_n(1)$ and note that $R_n(x) = (R_n - R_{n+1})x^n + (R_{n+1} - R_{n+2})x^{n+1} + (R_{n+2} - R_{n+3})x^{n+2} + \cdots$. For Theorem 9.4, use Theorem 8.5.

Section 9.2

3. (a) $2^x = e^{x\ln 2} = \sum\limits_{n=0}^{\infty} (\ln 2)^n x^n/(n!)$

(b) $e^x = \sum\limits_{n=0}^{\infty} e(x-1)^n/(n!)$

5. (b) $\sin x = \sum\limits_{n=0}^{\infty} \dfrac{(-1)^n}{(2n)!}\left(x - \dfrac{\pi}{2}\right)^{2n}$, the interval of convergence is $(-\infty, \infty)$.

7. $\dfrac{1}{2}\ln\left(\dfrac{1+x}{1-x}\right) = \sum\limits_{n=0}^{\infty} \dfrac{x^{2n+1}}{2n+1} = x + \dfrac{x^3}{3} + \dfrac{x^5}{5} + \dfrac{x^7}{7} + \cdots$, the interval of convergence is $(-1, 1)$.

8. $(3+x)^{2/3} = 3^{2/3}\left(1 + \dfrac{x}{3}\right)^{2/3}$

$$= 3^{2/3}\left[1 + \dfrac{2}{3^2}x - \dfrac{2\cdot 1}{3^4\cdot 2!}x^2 + \dfrac{2\cdot 1\cdot 4}{3^6\cdot 3!}x^3 - \cdots\right]$$

$$= 3^{2/3}\left[1 + \dfrac{2}{3^2}x + \sum\limits_{n=2}^{\infty} \dfrac{(-1)^{n+1}\,2\cdot 1\cdot 4\cdot 7\cdot\,\cdots\,\cdot(3n-5)}{3^{2n}\cdot n!}x^n\right]$$

with absolute convergence on $[-3, 3]$.

10. For $p < 0$, rewrite the binomial series at $x = 1$ and $x = -1$ using $p = -|p|$. For (a), note that the nth term does not have limit 0. For (b), use Raabe's test at $x = -1$ and use the Alternating Series test at $x = 1$.

11. $\arcsin x = x + \sum\limits_{n=1}^{\infty} \dfrac{1\cdot 3\cdot 5\cdot\,\cdots\,\cdot(2n-1)}{2\cdot 4\cdot 6\cdot\,\cdots\,\cdot(2n)}\dfrac{x^{2n+1}}{2n+1}$

$$= x + \dfrac{1}{2}\cdot\dfrac{x^3}{3} + \dfrac{1\cdot 3}{2\cdot 4}\cdot\dfrac{x^5}{5} + \dfrac{1\cdot 3\cdot 5}{2\cdot 4\cdot 6}\cdot\dfrac{x^7}{7} + \cdots$$

with absolute convergence on $[-1, 1]$. (At the endpoint 1, use Raabe's test.)

Section 10.1

3. $U(P, f, \alpha) = 2$ and $L(P, f, \alpha) = 1$ for all partitions P.

4. $\int_0^1 g\, d\alpha = 1$

6. For c in (a, b) with $c = x_j$ in $P = \{x_i\}_{i=0}^n \in \mathcal{P}[a, b]$, $U(P, f, \alpha) = [\alpha(c) - \alpha(x_{j-1})] + [\alpha(x_{j+1}) - \alpha(c)]$. Now use the continuity of α at c to choose an appropriate P.

7. For $f_1 + f_2$, first obtain that

$$L(P, f_1, \alpha) + L(P, f_2, \alpha) \leq L(P, f_1 + f_2, \alpha)$$
$$\leq U(P, f_1 + f_2, \alpha)$$
$$\leq U(P, f_1, \alpha) + U(P, f_2, \alpha)$$

$\forall \ P \in \mathcal{P}[a, b]$. Then use Theorem 10.2.

11. By Corollary 10.1, all of the integrals exist. Expand $\int_a^b (xf + g)^2 \, d\alpha$ to obtain a quadratic in x.

Section 10.2

2. Obtain $m \int_a^b g \, d\alpha \leq \int_a^b fg \, d\alpha \leq M \int_a^b g \, d\alpha$ and use the Intermediate Value Theorem.

3. For c in (a, b) with $c = x_j$ in $P = \{x_i\}_{i=0}^n \in \mathcal{P}[a, b]$,

$$U(P, f, \alpha) - U(P, f, \beta) = (M_j - M_{j+1})[\alpha(c) - \beta(c)].$$

As an example, let $\alpha(x) = 0$ on $[0, 1]$ and $\beta(1) = 1$ ($\beta = \alpha$ otherwise). If f is continuous on $[0, 1]$, then $\int_0^1 f \, d\alpha = 0$ but $\int_0^1 f \, d\beta = f(1)$.

5. Let $a \leq c < b$ and consider $f(x) = \begin{cases} 0 & \text{if} \quad a \leq x \leq c \\ 1 & \text{if} \quad c < x \leq b. \end{cases}$

8. $(n - 1)n(4n + 1)/6$

Section 10.3

2. (a) $(\pi/2) - 1$ \qquad (b) $\frac{1}{2}$

3. (a) $\frac{2}{5}$ \qquad (b) $\frac{1}{2}$

4. (a) $\pi/2$ \qquad (b) 1

6. For $P = \{x_i\}_{i=0}^n \in \mathcal{P}[a, b]$ and $t_i \in [x_{i-1}, x_i]$, $f(b)\alpha(b) - f(a)\alpha(a) - S(P, \alpha, f) = S(P', f, \alpha)$, where $P' = P \cup \{t_i\}_{i=1}^n \in \mathcal{P}[a, b]$.

Section 10.4

1. (c) On $[0, 1]$, try $f(x) = \begin{cases} 1 & \text{if} \quad x \in \mathbb{Q} \\ -1 & \text{if} \quad x \in \mathbb{R} \setminus \mathbb{Q} \end{cases}$ with

$P = \{0, t_n, 1/n, t_{n-1}, 1/(n - 1), \ldots, t_2, \frac{1}{2}, t_1, 1\}$, where the t_i's $\in \mathbb{R} \setminus \mathbb{Q}$.

4. (a) If $f(x) \to 0$ as $x \to c$, then $1/f$ will not be bounded and hence $1/f$ will not be of bounded variation on any interval containing c.

5. (a) $v(x) = \begin{cases} \sin x & \text{if} \quad 0 \leq x \leq \pi/2 \\ 2 - \sin x & \text{if} \quad \pi/2 \leq x \leq 3\pi/2 \\ 4 + \sin x & \text{if} \quad 3\pi/2 \leq x \leq 2\pi \end{cases}$

(b) $v(x) = \begin{cases} x & \text{if} \quad 0 \leq x < 1 \\ 2 & \text{if} \quad 1 \leq x < 2 \\ x + 1 & \text{if} \quad 2 \leq x \leq 3 \end{cases}$

7. Use Corollary 10.2 and Theorem 6.9. See Example 10.11.

8. If $g = f$ except at c in $[a, b]$, then $\sum_{i=1}^{n} |\Delta g_i| \leq V_f(a, b) + 2|g(c)| + 2 \sup_{x \in [a,b]} |f(x)|$.

10. (a) For $\alpha > 1$, show that $f'(x) = 0$ on $[a, b]$. For $\alpha = 1$, $V_f(a, b) \leq M(b - a)$.

(b) Note that if f satisfies a uniform Lipschitz condition on $[a, b]$, then f is continuous on $[a, b]$.

Section 10.5

1. (a) -2 (b) $\frac{5}{2}$

2. $\int_0^\pi |x| \, dv = \pi$

3. $2\pi - 6 + \sum_{i=1}^{6} \cos i$

5. $\left| \int_a^b f \, d\alpha - \int_a^b f \, d\alpha_n \right| \leq M V_{\alpha - \alpha_n}(a, b)$

8. For $P = \{x_i\}_{i=0}^{n} \in \mathcal{P}[a, b]$,

$$\int_a^b gf \, d\alpha - S(P, g, \beta) = \sum_{i=1}^{n} \int_{x_{i-1}}^{x_i} [g(t) - g(t_i)] f(t) \, d\alpha(t).$$

Now use the uniform continuity of g and Theorem 10.17.

Section 11.1

1. $U \setminus F = U \cap (\mathbb{R} \setminus F)$

2. (c) $\mathbb{Q}^\circ = (\mathbb{R} \setminus \mathbb{Q})^\circ = \emptyset$ and $[0, 1]^\circ = (0, 1)$

3. (b) Note that if $C \subset D$, then $C^\circ \subset D^\circ$.

(c) For the example, let $A = \mathbb{Q}$ and $B = \mathbb{R} \setminus \mathbb{Q}$.

6. $\frac{1}{4} = \frac{2}{3^2} + \frac{2}{3^4} + \frac{2}{3^6} + \frac{2}{3^8} + \cdots$. If $\frac{1}{4}$ were an endpoint, then $\frac{1}{4} = \frac{k}{3^n}$ for some positive integers k and n.

Section 11.2

3. $\bigcap \{F : A \subset F \text{ and } F \text{ is closed in } \mathbb{R}\}$ is a closed set that contains A; hence it must contain $\overline{A}$ since $\overline{A}$ is the smallest closed set containing A.

4. (b) Note that if $C \subset D$, then $\overline{C} \subset \overline{D}$.

(c) For the example, let $A = \mathbb{Q}$ and $B = \mathbb{R} \setminus \mathbb{Q}$.

6. (a) $\Rightarrow$ (b). Fix $x \in U$. Then $(x - \varepsilon, x + \varepsilon) \subset U$ for some $\varepsilon > 0$. By (a) there is an a in A with $x - \varepsilon < a < x + \varepsilon$.

(c) $\Rightarrow$ (a). If $x < y$ and $A \cap (x, y) = \emptyset$, then $(x + y)/2$ is not in A and not in A'. Thus $(x + y)/2 \in \mathbb{R} \setminus \overline{A}$.

8. If the open interval $(a, b) \subset \mathfrak{C} = \bigcap\limits_{n=1}^{\infty} F_n$, then $b - a \leq 1/3^n$ for each n in $\mathbb{N}$.

9. Choose n in $\mathbb{N}$ such that $1/3^n < \varepsilon$.

Section 11.3

1. (a) $\left\{(1/n, \frac{3}{2})\right\}_{n=2}^{\infty}$

   (b) $\{(a - 1, n) : n \in \mathbb{N}, n > a\}$

   (d) $\{(-n, a + 1) : n \in \mathbb{N}, -n < a\}$

   (f) $\left\{\left(n - \frac{1}{2}, n + \frac{1}{2}\right) : n \in \mathbb{Z}\right\}$

2. (c) $\mathbb{N} = \bigcup\limits_{n \in \mathbb{N}} \{n\}$

3. Let $\mathcal{G}$ be an open cover of $\{x_n : n \in \mathbb{N}\} \cup \{x\}$. Then some $G_0 \in \mathcal{G}$ contains x and all but a finite number of the x_n's. (This also follows from the Heine-Borel Theorem.)

5. (a) $\Rightarrow$ (b). Use the Bolzano-Weierstrass Theorem.

   (b) $\Rightarrow$ (c). An infinite subset B of A contains a sequence of distinct points, which by (b) has a subsequence that converges to a point y of A. Show that y is an accumulation point of B.

   (c) $\Rightarrow$ (a). To show that A is closed, show that A contains all of its accumulation points. If A is unbounded, then A contains the infinite set $\{x_n : n \in \mathbb{N}\}$ such that $|x_n| > n$ for each n in $\mathbb{N}$.

7. (b) $[0, 1] \cup \{2\}$

9. (a) If $T = F \cap X$, then $X \setminus T = (\mathbb{R} \setminus F) \cap X$.

   (b) $\left(0, \frac{1}{2}\right] = \left[-\frac{1}{2}, \frac{1}{2}\right] \cap (0, 1)$

Section 11.4

2. (a) Use Theorem 11.6.

   (c) If $X = \bigcup\limits_{\alpha \in I} A_\alpha$ is not connected, then by Proposition 11.10, $X \subset U \cup V$, where U and V are disjoint open subsets of $\mathbb{R}$ and both have nonempty intersection with X. Since each A_α is connected, each A_α is entirely contained in one of U or V. Since $\bigcap\limits_{\alpha \in I} A_\alpha \neq \emptyset$, all the A_α's are contained in U or all the A_α's are contained in V.

4. (a) Single point sets

   (c) If K is a component of X, then $\overline{K} \cap X$ is connected (use Proposition 11.10). Hence, $K = \overline{K} \cap X$.

   (d) $K_0 = \{0\}$

5. (a) If A is a connected subset of $\mathbb{Q}$ containing two distinct points x and y with $x < y$, then $[x, y] \subset A$; and so A contains irrational numbers.

   (b) If a connected subset of $\mathfrak{C}$ contained more than one point, then $\mathfrak{C}$ would contain an open interval.

Section 11.5

2. Use Theorem 11.7. For example, $\{x \in X : f(x) < a\} = f^{-1}((-\infty, a))$ and $\{x \in X : f(x) \geq a\} = f^{-1}([a, \infty))$.

3. $\{x \in X : f(x) = g(x)\} = (f - g)^{-1}(\{0\})$

7. (a) $\Rightarrow$ (b). If S is open in Y, then $S = V \cap Y$, where V is open in $\mathbb{R}$. By Theorem 11.7, $f^{-1}(V)$ is open in X. Show that $f^{-1}(S) = f^{-1}(V)$.
   (c) $\Rightarrow$ (a). Given F closed in $\mathbb{R}$, $f^{-1}(F) = f^{-1}(F \cap Y)$ is closed in X by (c).

10. Use Exercises 8 and 9.

11. $[0, 1]$ is compact but $(0, 1)$ is not compact. A homeomorphism must preserve compactness.

13. Every point of $(0, 1)$ is a cut point of $(0, 1)$, whereas 0 is not a cut point of $[0, 1)$. Also, $f(x) = 1 - x$ is a homeomorphism from $[0, 1)$ onto $(0, 1]$.

Index